APPLIED RIVER MORPHOLOGY

DAVE ROSGEN
Wildland Hydrology
Pagosa Springs, Colorado

Illustrations
HILTON LEE SILVEY
Western Hydrology
Lakewood, Colorado

The photograph on the Cover, taken on South Chicago Creek,
Front Range Colorado, by Eric Wunrow, shows an A2, step/pool stream type.
(Note lichen line coincident with the break in slope indicating the stage of bankfull discharge.)

Cover Design by David Grassnick, Colorline, *Lakewood, Colorado.*

Contributions from Catena, A Classification of Natural Rivers, 1994,
    D.L. Rosgen, reprinted with permission from Elsevier Science, *Amsterdam.*

Library of Congress Catalog Card Number: 96-60962

Rosgen, Dave, 1942 –
    Applied River Morphology
    International Standard Book Number: 0-9653289-0-2

    Includes bibliographies and Indexes
    1. Geomorphology;
    2. Hydrology;
    3. Environmental Engineering;
    4. Fisheries Management

Prepress (including page layout, design and cover design):
Colorline, *Lakewood, Colorado*
Printing: Printed Media Companies, *Minneapolis, Minnesota*

Copyright © 1996 by Wildland Hydrology
1481 Stevens Lake Road
Pagosa Springs, Colorado

All Illustrations Copyright © 1996 by Hilton Lee Silvey

No part of this book may be reproduced by any mechanical, photographic,
or electronic process, or in the form of a photographic recording,
nor may it be stored in a retrieval system, transmitted, or otherwise copied
for public or private use, without written permission from the publisher.

Printed in the United States of America

DEDICATION

*This book is dedicated
to my mentor and good friend,
Luna B. Leopold,
whose wise counsel and teaching
have benefitted many,
and whose exemplary contributions
have advanced science and
promoted the river ethic.*

FOREWORD

by
Luna B. Leopold

In modern science there is always some tension between the theoretical and the practical application of basic knowledge. The pressures of development and the desire of the present society for profits at the expense of anything natural, wild, or free has degraded landscapes throughout the world by river engineering, river straightening, construction of levees, dams, and concrete channels. These actions have been the essence of our approach to problems of flood control, navigation, irrigation, hydroelectric development, municipal and industrial water needs, even though at the same time much has been learned about river mechanics, geomorphology, hydraulics, and sedimentation. Despite the new knowledge, the traditional engineering approach to river development has not only dominated valley land management, but has failed to incorporate the practical, physical, aesthetic, and financial advantages of approaching river management as maintenance of natural tendencies in river channel behavior.

Finally there appeared a man on the scene a man far-sighted enough to see how that new knowledge should be put to work in channel restoration and maintenance. This new approach did not arise from pure cerebration, for David Rosgen had already two decades of field measurement of rivers and associated analysis of the collected data. When the opportunity arose to design river restoration programs and carry out plans in the field, this background of field experience was an essential ingredient.

But that was only part of the qualifications that he brought to the task, for he was perfectly at home at the helm of a bulldozer, the levers of a trackhoe, or as the driver of an earth mover.

More than anything, Rosgen was imaginative, creative, and a keen field observer. He knows the theory and has a feel for the ways a river works. In the course of this long association with rivers and creeks, he developed a practical and universally applicable scheme for classifying channels. This scheme involves the main parameters that operate in the processes of river mechanics and maintenance. Because the classification depends on knowledge of processes, it is useful not only to describe channels but also to evaluate how a stream will react to change through time. His approach includes data collection both before and after any restoration, so the initial conditions are described and reactions can be monitored.

As a result he quickly established himself as the premier leader in river restoration, river control planning, and channel maintenance without dependence on steel and concrete. His designs use native materials applied in ways that enhance natural tendencies for the channel to seek quasi-equilibrium between sediment and water, both at low flow and in flood.

This book is a generous and detailed explanation of the classification system and how it might be used to incorporate the observed processes of river mechanics into restoration designs that enhance the beauty and health of channels.

TABLE OF CONTENTS

PREFACE	xv
CHAPTER 1: NEW CHALLENGES	1-1
LEARNING FROM THE PAST	1-1
PREDICTION - A KEY TO PREVENTION	1-2
TRADITIONS DIE HARD	1-3
NATURAL STABILITY CONCEPTS	1-3
EFFORTS TO IMPROVE THE RIVER	1-3
NEW EMPHASIS IN RIVERINE, RIPARIAN AND WATERSHED MANAGEMENT	1-5
CHAPTER 2: FUNDAMENTAL PRINCIPLES OF RIVER SYSTEMS	2-1
THE BANKFULL DISCHARGE	2-2
STREAM CHANNEL DIMENSIONS	2-4
STREAM CHANNEL PATTERNS	2-5
STREAM CHANNEL PROFILE	2-8
APPLYING THE PRINCIPLES	2-9
CHAPTER 3: STREAM CLASSIFICATION	3-1
STREAM CLASSIFICATION REVIEW	3-1
OBJECTIVES OF STREAM CLASSIFICATION	3-3
HIERARCHY OF RIVER MORPHOLOGY	3-3
CONTINUUM OF STREAM MORPHOLOGY	3-4
GENERAL APPLICATIONS OF STREAM CLASSIFICATION AND HIERARCHICAL INVENTORY	3-6
Communication	3-6
General Applications	3-7

TABLE OF CONTENTS

CHAPTER 4: GEOMORPHIC CHARACTERIZATION 4-1

BROAD LEVEL DESCRIPTIONS OF THE MAJOR STREAM TYPES 4-4
Stream Descriptions "A" through "G" 4-4
The "Aa+" Stream Type 4-4
The "A" Stream Type 4-6
The "B" Stream Type 4-6
The "C" Stream Type 4-6
The "D" Stream Type 4-8
The "DA" Stream Type 4-8
The "E" Stream Type 4-9
The "F" Stream Type 4-9
The "G" Stream Type 4-10
Morphological Descriptions 4-10

ASSOCIATIONS BETWEEN LANDFORMS AND STREAM TYPES USED IN LEVEL I CLASSIFICATION 4-10
Valley Morphology Related to Stream Types 4-12
Valley Type I 4-12
Valley Type II 4-12
Valley Type III 4-12
Valley Type IV 4-12
Valley Type V 4-12
Valley Type VI 4-15
Valley Type VII 4-15
Valley Type VIII 4-15
Valley Type IX 4-16
Valley Type X 4-16
Valley Type XI 4-16

DELINEATION METHODS 4-20
Delineation of Valley Types and Landforms 4-20
Plan-view Morphology 4-20
Cross-section Morphology 4-22
Channel Sinuosity 4-23
Channel Slope 4-23
Bed Features 4-23
Examples of Broad Level Delineation 4-24
Discussion 4-29

TABLE OF CONTENTS

CHAPTER 5: LEVEL II: THE MORPHOLOGICAL DESCRIPTION 5-1
   LEVEL II DELINEATIVE CRITERIA 5-2
      Cross-section 5-2
         Entrenchment Ratio 5-2
         Width/Depth Ratio 5-2
         Dominant Channel Materials 5-2
      Longitudinal Profile 5-2
         Slope 5-2
         Bed Features 5-2
      Plan-form (pattern) 5-2
         Sinuosity 5-2
         Meander Width Ratio 5-2

   THE REFERENCE REACH 5-2

   THE CONTINUUM OF CHANNEL FORM 5-7

   THE ROLE OF BANKFULL DISCHARGE 5-7
      Field Determination of Bankfull Stage 5-8
      Calibrating Bankfull Stage to Known Streamflows 5-9
      Drainage Area vs. Bankfull Channel Dimensions by Stream Type 5-14

   FIELD METHODS FOR STREAM TYPE DELINEATION 5-15
      Entrenchment 5-15
         Significance 5-15
         Methods 5-19
      Width/Depth Ratio 5-21
         Significance 5-21
         Methods 5-23
      Sinuosity 5-23
         Significance 5-23
         Methods 5-25
      Channel Materials 5-25
         Significance 5-25
         Methods 5-25
      Slope 5-27
         Significance 5-27
         Methods 5-27

   MORPHOLOGICAL DESCRIPTIONS AND EXAMPLES 5-29
      Remote Sensing Methods for Level II Delineation 5-30

TABLE OF CONTENTS

CHAPTER 5: LEVEL II: THE MORPHOLOGICAL DESCRIPTION (continued)

    MORPHOLOGICAL DESCRIPTION AND EXAMPLES OF STREAM TYPES 5-35
        "A1" **Stream Type** - *Summary, Delineative Criteria and Illustration,*
            Delineative Criteria Data Distribution, Photographs ....5-36 – 5-39
        "A2" **Stream Type** - *Summary, Delineative Criteria and Illustration,*
            Delineative Criteria Data Distribution, Photographs ....5-40 – 5-43
        "A3" **Stream Type** - *Summary, Delineative Criteria and Illustration,*
            Delineative Criteria Data Distribution, Photographs ....5-44 – 5-47
        "A4" **Stream Type** - *Summary, Delineative Criteria and Illustration,*
            Delineative Criteria Data Distribution, Photographs ....5-48 – 5-51
        "A5" **Stream Type** - *Summary, Delineative Criteria and Illustration,*
            Delineative Criteria Data Distribution, Photographs ....5-52 – 5-55
        "A6" **Stream Type** - *Summary, Delineative Criteria and Illustration,*
            Delineative Criteria Data Distribution, Photographs ....5-56 – 5-59

        "B1" **Stream Type** - *Summary, Delineative Criteria and Illustration,*
            Delineative Criteria Data Distribution, Photographs ....5-60 – 5-63
        "B2" **Stream Type** - *Summary, Delineative Criteria and Illustration,*
            Delineative Criteria Data Distribution, Photographs ....5-64 – 5-67
        "B3" **Stream Type** - *Summary, Delineative Criteria and Illustration,*
            Delineative Criteria Data Distribution, Photographs ....5-68 – 5-71
        "B4" **Stream Type** - *Summary, Delineative Criteria and Illustration,*
            Delineative Criteria Data Distribution, Photographs ....5-72 – 5-75
        "B5" **Stream Type** - *Summary, Delineative Criteria and Illustration,*
            Delineative Criteria Data Distribution, Photographs ....5-76 – 5-79
        "B6" **Stream Type** - *Summary, Delineative Criteria and Illustration,*
            Delineative Criteria Data Distribution, Photographs ....5-80 – 5-83

        "C1" **Stream Type** - *Summary, Delineative Criteria and Illustration,*
            Delineative Criteria Data Distribution, Photographs ....5-84 – 5-87
        "C2" **Stream Type** - *Summary, Delineative Criteria and Illustration,*
            Delineative Criteria Data Distribution, Photographs ....5-88 – 5-91
        "C3" **Stream Type** - *Summary, Delineative Criteria and Illustration,*
            Delineative Criteria Data Distribution, Photographs ....5-92 – 5-95
        "C4" **Stream Type** - *Summary, Delineative Criteria and Illustration,*
            Delineative Criteria Data Distribution, Photographs ....5-96 – 5-99
        "C5" **Stream Type** - *Summary, Delineative Criteria and Illustration,*
            Delineative Criteria Data Distribution, Photographs ....5-100 – 5-103
        "C6" **Stream Type** - *Summary, Delineative Criteria and Illustration,*
            Delineative Criteria Data Distribution, Photographs ....5-104 – 5-107

TABLE OF CONTENTS

CHAPTER 5: LEVEL II: THE MORPHOLOGICAL DESCRIPTION (continued)

"D3" **Stream Type** - *Summary, Delineative Criteria and Illustration,
Delineative Criteria Data Distribution, Photographs* ....5-108 – 5-111

"D4" **Stream Type** - *Summary, Delineative Criteria and Illustration,
Delineative Criteria Data Distribution, Photographs* ....5-112 – 5-115

"D5" **Stream Type** - *Summary, Delineative Criteria and Illustration,
Delineative Criteria Data Distribution, Photographs* ....5-116 – 5-119

"D6" **Stream Type** - *Summary, Delineative Criteria and Illustration,
Delineative Criteria Data Distribution, Photographs* ....5-120 – 5-121

"DA" **Stream Type** - *Summary, Delineative Criteria and Illustration,
Delineative Criteria Data Distribution, Photographs* ....5-122 – 5-125

"E3" **Stream Type** - *Summary, Delineative Criteria and Illustration,
Delineative Criteria Data Distribution, Photographs* ....5-126 – 5-129

"E4" **Stream Type** - *Summary, Delineative Criteria and Illustration,
Delineative Criteria Data Distribution, Photographs* ....5-130 – 5-133

"E5" **Stream Type** - *Summary, Delineative Criteria and Illustration,
Delineative Criteria Data Distribution, Photographs* ....5-134 – 5-137

"E6" **Stream Type** - *Summary, Delineative Criteria and Illustration,
Delineative Criteria Data Distribution, Photographs* ....5-138 – 5-141

"F1" **Stream Type** - *Summary, Delineative Criteria and Illustration,
Delineative Criteria Data Distribution, Photographs* ....5-142 – 5-145

"F2" **Stream Type** - *Summary, Delineative Criteria and Illustration,
Delineative Criteria Data Distribution, Photographs* ....5-146 – 5-149

"F3" **Stream Type** - *Summary, Delineative Criteria and Illustration,
Delineative Criteria Data Distribution, Photographs* ....5-150 – 5-153

"F4" **Stream Type** - *Summary, Delineative Criteria and Illustration,
Delineative Criteria Data Distribution, Photographs* ....5-154 – 5-157

"F5" **Stream Type** - *Summary, Delineative Criteria and Illustration,
Delineative Criteria Data Distribution, Photographs* ....5-158 – 5-161

"F6" **Stream Type** - *Summary, Delineative Criteria and Illustration,
Delineative Criteria Data Distribution, Photographs* ....5-162 – 5-165

"G1" **Stream Type** - *Summary, Delineative Criteria and Illustration,
Delineative Criteria Data Distribution, Photographs* ....5-166 – 5-169

"G2" **Stream Type** - *Summary, Delineative Criteria and Illustration,
Delineative Criteria Data Distribution, Photographs* ....5-170 – 5-173

"G3" **Stream Type** - *Summary, Delineative Criteria and Illustration,
Delineative Criteria Data Distribution, Photographs* ....5-174 – 5-177

"G4" **Stream Type** - *Summary, Delineative Criteria and Illustration,
Delineative Criteria Data Distribution, Photographs* ....5-178 – 5-181

"G5" **Stream Type** - *Summary, Delineative Criteria and Illustration,
Delineative Criteria Data Distribution, Photographs* ....5-182 – 5-185

"G6" **Stream Type** - *Summary, Delineative Criteria and Illustration,
Delineative Criteria Data Distribution, Photographs* ....5-186 – 5-189

TABLE OF CONTENTS

CHAPTER 6: LEVEL III: ASSESSMENT OF STREAM CONDITION AND DEPARTURE FROM ITS POTENTIAL 6-1

STREAM CONDITION AND STREAM DEPARTURE ANALYSIS 6-3
Stream Potential 6-3

EVOLUTIONARY STAGES OF CHANNEL ADJUSTMENT 6-7
Adjustment Examples 6-7
Equilibrium Interpretations 6-13

LEVEL III FIELD PARAMETERS: THE CHANNEL INFLUENCE VARIABLES 6-13
Introduction 6-13
Description of Level III Channel Influence Variables 6-14
Riparian Vegetation 6-14
Flow Regime 6-15
Size and Stream Order 6-15
Depositional Patterns (sediment) 6-18
Meander Patterns (channels) 6-19
Debris and Channel Blockages 6-25
Stream Channel Stability 6-26
Sediment Supply 6-28
Bed Stability 6-31
Width/Depth Ratio Shifts 6-31
Streambank Erosion Potential 6-35
Altered Stream Channels 6-47

SUMMARY OF LEVEL III ASSESSMENTS OF STREAM CONDITION 6-47

CHAPTER 7: LEVEL IV: FIELD DATA VERIFICATION 7-1
INTRODUCTION 7-1
Stream Flow Measurements 7-3
Sediment Analyses 7-3
Ratio of Bedload to Total Load 7-6
Bedload Size Distribution at or Near the Bankfull Discharge 7-7
Sediment Rating Curve Relations 7-8
Stream Stability Validation 7-11
Field Methods for Monitoring Channel Stability 7-12
Channel Stability 7-12
####### Vertical or Bed Stability 7-12
####### Lateral Stability 7-13
####### Bed Material Size Distribution 7-13
Methods 7-13
####### Vertical or Bed Stability 7-13
####### Lateral Stability 7-14
####### Bed Material Size Distribution 7-15

SUMMARY 7-16

TABLE OF CONTENTS

CHAPTER 8: APPLICATIONS 8-1
   INTRODUCTION 8-1

   HYDRAULIC GEOMETRY RELATIONS 8-1

   FLOW RESISTANCE 8-3
      Mannings "N" Roughness Coefficient 8-3
      Shear Stress and Velocity Relationships 8-4
      Critical Shear Stress Estimates 8-4
      Sediment Relations 8-6

   GRAZING MANAGEMENT 8-8

   FISH HABITAT 8-15
      Habitat Improvement Structures 8-16
         Rearing Habitat Enhancement 8-16
            Low Stage Check Dam 8-16
            Medium Stage Check Dam 8-16
            Boulder Placement 8-16
            Bank Placed Materials 8-16
            Single Wing Deflectors 8-16
            Double Wing Deflectors 8-17
            Channel Constrictor 8-17
            Bank Cover 8-17
            Floating Log Cover 8-17
            Submerged Shelters 8-17
            Half Log Cover 8-17
            Migration Barrier 8-17
         Spawning Habitat Enhancement 8-17
            V-Shaped Gravel Traps 8-17
            Log Sill Gravel Traps 8-17
            Gravel Placement 8-17
         Channel Stability/Habitat Improvement Structures 8-20
            Vortex Rock Weir 8-20
            Native Material Revetment - Root Wads/Vegetation Transplants 8-20
            "W" Weir 8-20
            Native Material Revetment - Rock and/or Log Spurs 8-20
      Application of Guidelines 8-20
         Fish Habitat Structures - Suitability Guidelines (Tables 8-2a, 8-2b) .... 8-24 – 8-25
         Limitations/Discussions of Fish Habitat Improvement Structures .... 8-26 – 8-30

TABLE OF CONTENTS

CHAPTER 8: APPLICATIONS (continued)

RESTORATION 8-31
River Restoration Principles 8-31
What are the observed problems? 8-31
What caused the problem? 8-31
What stream type should this be? 8-33
What is the probable stable form of the stream type
  under the present hydrology and sediment regime? 8-33
Case Studies 8-34
Weminuche River, Southwestern Colorado 8-35
East Fork of the San Juan River and Blanco Rivers in Southwestern Colorado 8-37

ADDITIONAL RESOURCE MANAGEMENT INTERPRETATIONS 8-38

SUMMARY OF APPLICATIONS OF THE HIERARCHICAL STREAM INVENTORY LEVELS 8-41

APPLICATION EXAMPLES 8-41

BIBLIOGRAPHY I

SUBJECT INDEX XI

PREFACE

At one time in my career, I was asked to review flood damage immediately following a major runoff event and to propose designs for stream channel restoration. A great effort ensued to find and document all of the disturbed stream systems within a large, flood-ravaged watershed. Design concepts and corresponding large budget requests were forthcoming under tight time constraints. Unfortunately, we were given the budget as requested and implemented these designs. Upon returning years later to review this and other "restoration" work imposed on the river, I found that, contrary to the design objectives, most of the structures had made things worse. What went wrong?

It wasn't until years later following another flood event, did the same opportunity present itself. This time however, I spent as much time inventorying stream systems in the same watershed that were not adversely affected by the flood as I did reviewing those that were severely impacted. For the first time, I began to observe river morphology in quantitative detail. I found that the valley morphology was the same for both the un-impacted and impacted reach. The basin relief and depositional history which expressed itself in the channel materials, imposed the same set of processes on both reaches. If the un-impacted reach was stable and could accommodate such a flood in the same watershed, why not make the impacted reach match the morphological characteristics of the stable reach?

The morphological features of the un-impacted reach became the blueprint for natural stability. Instead of installing large check dams, rip-rap and gabion baskets on the bed and banks of rivers using "hard control" methods, the dimensions, patterns and profile of the stable "blueprint" reach was emulated on the unstable reach. Following subsequent floods, applying a "natural stability" approach to channel restoration was found to not only be highly successful, but improved the biological function as well, and at a fraction of the "hard control" costs. I found that we were working with the natural self-stabilizing tendencies of rivers rather than imposing our will to make a stream into something it could never be.

However, the natural stability "blueprint" could only be applied to streams of similar morphological character. When valley types changed along with corresponding changes in materials, gradients, shapes, meander patterns and dimensions, the morphological features of the river also changed.

The initial development of a stream classification system emerged based on morphological features because it became necessary to be able to extrapolate data, predict process response and communicate clearly about rivers for the purposes of restoration, and watershed management planning.

I have focused my efforts throughout 31 years on investigating natural stream and river systems, documenting characteristics of a diverse assemblage of natural rivers. Building on geomorphological principles established by Luna B. Leopold, I began in 1968 collecting an array of quantitative data that included water surface slope, valley morphology, valley types, depositional history, stream widths, depths, channel stability, sediment supply, streamflow, stream order, sinuosity, channel materials, the degree of incision within landforms and calculated hydraulic geometry relationships for the rivers studied.

Based on the resultant extensive database, a consistent pattern of natural river geometry emerged. A prototype classification system was first presented in 1985, serving as an introduction to the classification system prior to the publication in the geomorphological journal, Catena, (Rosgen,1994). The hierarchy of river morphology and assessment presented in this book has evolved from quantitative field observations of hundreds of rivers in all climatic regions of North America. The assessment process was refined through experience in restoring streams and rivers, extensive teaching, and feedback from many hydrologists, geomorphologists, engineers, fisheries biologists, and plant ecologists who have used this classification system.

When you gain insight through observations of the river, you want to help others gain the same insight. This is the only way collective progress may be made to prevent the decline of river quality.

We all tend to learn directly from the past experience of trial and error, but if we do not understand - if we do not take a closer, quantitative look at our rivers over time - we have a tendency to unknowingly repeat the errors of the past. For the good of the river as well as our profession, we have to remove ourselves from the darkness of some of those desk bound offices and spend more time in the field.

This is a book on the applied science. It is intended as a reference document that transfers some of my field observations over the years and the observations of others to those who may find similar uses for such information. This book and the data that went into the analysis and thought process would not be possible without the field assistance, extensive discussions, and the shared river ethic of my fellow river comrades: Dr. Luna B. Leopold, H.Lee Silvey, Dale Pfankuch, Owen Williams, Alice Johns, Bob Kasun, Lela Chavez, Brenda Mitchell, Pete Bengeyfield, Bob Delk, Jim Nankervis, Steve Belz, and Al Issacson. Additional recognition goes to the USDA Forest Service Personnel, Water Division I Project, Colorado, for their dedication and efforts in field data collection.

It is extremely difficult to describe complex subjects, however, my good friend and fellow hydrologist, Hilton Lee Silvey, has skillfully presented these concepts through his illustrations. His work is a tribute not only to his ability, but to his significant contribution to rivers.

I would like to personally thank Lee Silvey, Iris Goodman, Luna Leopold, and Jack Ellis for their editing and review, for without it, this book would have required wheels and an interpreter. The competent typing credits go specifically to Kay McElwain. Graphic presentation preparation was done by Cheri Pomeroy and production management handled by Jennifer Franson both with Colorline *(Lakewood, Colorado)*.

And finally I would like to thank my family who have generously tolerated my absence to the river for so long....it's now time to go fishing!

CHAPTER 1

NEW CHALLENGES

"The difficulty lies not in the new ideas, but in escaping from the old ones."
JOHN MAYNARD KEYNES 1936

LEARNING FROM THE PAST

It has been said that the rivers are the life blood of civilization. At no other time in the history of modern man have the cumulative impacts associated with development along the river had a greater impact on water resource values. As populations continue to enlarge, the number of competing and conflicting uses associated with our rivers will increase. Rivers have been a focus of development over time, and as such, an understanding of the naturally stable character of the river is necessary if maintenance of its function and health is to be secured. This has not always been the case, as "flood control" works which change the dimension, pattern and profile of rivers, such as along the Mississippi River have often resulted in an increased frequency and magnitude of flood impacts. If we are to successfully co-exist with the river, we must better understand the consequences of our actions on river systems and associated values. Rivers are sensitive indicators of environmental stress, as shown when excess sediment supply and channel adjustments occur due to deforestation, overgrazing, sub-division development, and other watershed activities that create their cumulative impact on stream systems.

When the works of man run contrary to the natural, stable tendencies of the river, the river eventually dominates. It appears then to be of monumental importance to understand the interrelated process variables which shape the dimension, pattern and profile of the modern river. Underlying a presumably complex set of channel and watershed variables which follow the laws of physics, is a predictable adjustment process of rivers toward their most probable stable form. Natural rivers, which are self-constructed and self-maintained, constantly seek their own stability (Leopold et al. 1964). Unfortunately, for well intended flood control and drainage purposes, rivers have been straightened, leveed (severed from their floodplain), deepened, over-widened, lined with foreign materials, steepened, diverted, and altered in a manner to decrease their natural function and stability. Such works as these have not only created major continuing maintenance problems and promoted a high risk for failure, but, as a consequence, many of the natural resource values of the river have been lost.

Historically, the river has provided not only a source of water for domestic, municipal, and irrigation uses, but has also served as a major transportation corridor. Throughout recent history there has been a consistent increase in the diversity of water resource values associated with the river, including uses of the streams and rivers for aesthetic, recreational, and wildlife purposes. Recent river and watershed management direction has placed a strong emphasis on such amenity values, which often seem to take higher priority over traditional "consumptive" water resource uses and flood control projects. The real challenge we all face is how to meet the current and future demands for traditional uses and intrinsic values of the river - without impairing its stability and function.

PREDICTION - A KEY TO PREVENTION

Proper management of rivers and their watersheds involves balancing a host of resource uses of the river and the ability to predict the response of the river to imposed change. Reliable predictions necessitate a clear understanding about the functions of the river and the physical variables which influence river behavior. Researchers over many years have been fervently working on dissecting the many complex processes associated with river behavior, and have provided valuable dimensions to the discreet individual physical processes associated with the natural functioning of the river and its watershed. The real challenge is how to arrange the results of research into a form that can be utilized as a common statement about the river.

As we are approaching a new age of specialization, it is important that we accumulate an organized, extensive catalog of information about rivers. A challenge posed here is how do we know, with all the available data from a wide variety of rivers, and knowledge acquired for a given reach of river, what set of treatments are applicable to the stream of concern? Traditionally, we spend more time designing a future project or plan than we do understanding how the river adjusted to previous designs and plan implementations. If the mistakes that have been made in the past continue today, then our river and watershed management is progressing in a "backwards" direction. It is then of critical importance to be able to learn from observations of river response to past treatments and develop a clear understanding of the following:

- What caused the problem?
- How did the river respond?
- What was the consequence of the river response?
- How can we remedy the problem?
- How can we PREVENT the problem from recurring?

Prediction of change is not a simple matter. To predict change before change occurs is a fine tribute to a good field observer. However, a field observer can start to put many "clues" together when studying disturbed reaches. A comparison with an undisturbed reach immediately upstream of the source of impact can provide substantial insight, particularly if the undisturbed reach has the same stream morphology. A "control" reach located in a separate, but nearby sub-watershed can also provide a basis for quantifying changes in a disturbed reach, if the control reach is of the same morphological "character" and stream type. The morphological characteristics that separate one stream type from another are the same characteristics which provide a basis for interpretations of river process. Thus, it becomes apparent that not all streams respond the same to a given set of imposed changes. In other words, a "one size fits all" management scenario is not appropriate for river systems.

Management strategies must be designed to accommodate wide variations in sensitivity exhibited by rivers to land and water-use activities and the subsequent mechanisms of river response to imposed change from development impacts. River function and response differences, as well as similarities, only serve to describe the great variety of river types in various hydro-physiographic provinces which pose a challenge for understanding. Those who are assigned management or design tasks which will influence the river must have this understanding. Those who are new to the task must make a miraculous quick study!

TRADITIONS DIE HARD

There exists an immense diversity of interest groups and disciplines dealing with rivers. Communication within and between these various disciplines about rivers is not only varied but often convoluted. For instance, engineers and fisheries biologists (as well as many other disciplines) may have a rather difficult time talking with one another about a common concern, the river. Mathematical inference about hydraulics sometimes draws little attention from the biologist whose training perhaps avoided such terms, even though the subject is critical to a study of habitat components. Similarly, the engineer who designs a concrete revetment for streambank stability does not always have the same understanding for habitat lost as does the biologist. Future use demands placed on the natural resources of the river will require an integration of the knowledge of many disciplines in order to bring about a common understanding and communication. Traditional methods and philosophy of river "control" are very difficult to change, but even more difficult is the driving influence not of process, but of policy.

The greatest challenge for one who is experienced in understanding the processes and combined complexities of rivers is to effectively communicate with those who are not. Journals are filled with academic and research writings which have often been written for a peer audience rather than provide for general knowledge and communication. In many cases, significant research findings have not been applied by field practitioners due to this problem. The "state of the science" of physical process research may be considered more advanced than the "state of the art" for applications. Given these conditions, it would then seem imperative that management and research further intensify their efforts to put research principles into practice.

NATURAL STABILITY CONCEPTS

Natural stream channel stability is achieved by allowing the river to develop a stable dimension, pattern, and profile such that, over time, channel features are maintained and the stream system neither aggrades nor degrades. For a stream to be stable it must be able to consistently transport its sediment load, both in size and type, associated with local deposition and scour. Channel instability occurs when the scouring process leads to degradation, or excessive sediment deposition results in aggradation. When the stream laterally migrates, but maintains its bankfull width and width/depth ratio, stability is achieved even though the river is considered to be an "active" and "dynamic" system.

The consistency of dimension, pattern, and profile that exists among rivers is more than chance or spurious correlation. Mathematical relations exist illustrating a stratification of river systems by unique morphological forms, that provide meaning to an otherwise random appearing, complex set of interrelated variables. Whenever proper attention to the "rules of the river" is not respected, adverse channel adjustments often result in damage to personal property and loss of life.

EFFORTS TO IMPROVE THE RIVER

Recently there has been a resurgence of river restoration projects and "naturalization" applica-

NEW CHALLENGES

tions in developed countries in an effort to improve, mitigate, or to enhance lost water resource values. Millions of dollars have been spent in recent years on rivers to improve stability, fish habitat, visual values, flood control works, and a multitudinous array of manufactured "improvements". Many of these projects, although well intentioned, have changed the dimension, pattern, and the profile of rivers without first developing a firm understanding of how these morphological variables should be arranged to promote a "natural" or "geomorphic potential state." Questions worthy of consideration during project planning include the following: what are the channel characteristics of the natural stable form? How do these morphological characteristics vary by different rivers and river reaches? Listed are but a few of many important questions that need study and resolution prior to embarking on any river restoration project.

In many instances, flood control works along the river, river management plans, water resource development, and other land-use activities are focused only on a particular reach or just that portion of the river immediately adjacent to the project in question. Prescribed solutions must also be linked with an integration of the existing, upstream watershed condition and related physiographic variables. Proposed restoration efforts are often concentrated towards attacking symptoms rather than applying an appropriate cure to the cause(s) of the problem. Without taking into consideration the valley morphology, the watershed, the valley slope, and upstream and downstream adjustment processes as a result of localized river change, it is very difficult to responsibly prescribe river works. Unfortunately, much of this work has proceeded without this understanding. Rivers do not respect political boundaries, and as such, the management directions of man often are not consistent with the functional adjustments of rivers for their own stability and maintenance.

Rarely, if at all, will one observe that all rivers have or exhibit the same set of morphological characteristics. Yet, we, as land and water resource managers, continue to impose on the river a series of common forms and practices to meet site-specific objectives. An example of this is provided where hundreds of rivers were "enhanced" for fish habitat purposes, by changing the naturally occurring sequence of pools (the flat and deep features of a channel) to a 50/50 sequence of pool spacing with the riffles (the steeper, shallow bed features). Dedicated biologists were doing this work but they sometimes lacked an understanding that rarely does a 50/50 riffle/pool sequence (pool-to-pool spacing) occur in nature, and that riffle/pool and step/pool sequences are directly related to a particular stream type and its corresponding channel slope and width. The shallower steeps (riffles, rapids, glides, runs, cascades, etc.) are constructed of coarser bed material and vary in spacing with the deeper flats or pools containing finer bed material, and are controlled as to frequency of occurrence by the bankfull width, meander geometry, valley slope and morphology, channel materials, channel slope, large organic debris, and channel morphology. These undulating bed features are critical elements in natural energy dissipation, just as is the meander geometry. The nature, type, sequence, and function of riffle/pool bed features are well known among fluvial geomorphologists, and need to be of a common knowledge among engineers, biologists, and others working with the river. The stream classification system described herein that is based on the morphology and associated bed features of various stream "types" helps communicate among those engaged in river work the overall process relationships existing between stream channel geometry and system functioning.

To further emphasize these kinds of site-specific problems, approximately 25 years ago, biologists and foresters in the Pacific Northwest, thinking they would enhance fish habitat, were prescribing removal of the majority of woody debris found in stream types with moderate to steep gradients. Nearly all of these streams were of a type that utilized large organic debris to help establish stable bed forms necessary for energy dissipation. The wholesale removal of large woody debris led to the development of severe channel instability as the normal flow regimes created numerous debris

NEW CHALLENGES

torrent events that scoured the treated channels down to bedrock, effectively destroying the very fisheries habitat that was targeted for "enhancement." Today, along the same reaches, large organic debris is being "re-introduced" into the river with the assistance of chainsaws. The question again is, how many? how high? and in what sequence? Anadromous fish habitat conditions were recently established for "Desired Future Conditions" (DFC) utilizing a study team approach as part of a Federal Guide for Pilot Watershed Analysis, based on outputs from President Clinton's Forest Ecosystem Management Assessment Team (USDA, USDI, 1993). Guidelines to establish DFC's have recently been recommended for Pacific Northwest streams under the same planning effort. What does "Desired Future Condition" really mean in terms of today's stream systems? To many, it is the most natural stable channel form to be achieved as a result of current management efforts. To others, it is the long-term "ecological potential" of the stream. In any objective management assessment effort, both of these definitions should be considered. Many guidelines are currently being proposed to achieve "Desired Future Condition." It is critical that prescribed fisheries habitat improvement guidelines such as: desired width/depth ratios; riffle/pool and step/pool sequences; and debris spacing be established as functions of slope and stream width. These morphological variables are mathematically described and integrated with the inherent capabilities and system response characteristics by specific morphological stream types. The natural variability and diversity that exists within a given stream type as well as between stream types creates a wide range of "acceptable" desired future conditions. Thus, to be successful in implementing any established guidelines, it is important to understand the functional relationships between dimension, pattern, and profile of the natural stable channel form for a wide variety of river types. The challenge is to design and utilize efficient inventory procedures which will identify existing states or conditions of streams versus their geomorphic potential state.

NEW EMPHASIS IN RIVERINE, RIPARIAN AND WATERSHED MANAGEMENT

A number of United States government agencies, including but not limited to, the Environmental Protection Agency, USDA Forest Service, Natural Resources Conservation Service, U.S. Army Corps of Engineers, and the Bureau of Land Management are engaged in a process of developing river design criteria and new management standards and guidelines related to resource values associated with rivers, riparian, and watershed areas. Equally noteworthy is the fact that grazing strategies are now being designed to provide "proper functioning condition" of rivers and riparian areas. Since the mid-to late 1960's a number of cumulative effects models designed to assess the response of watersheds to various land management strategies have been under development in national forests in Idaho, Montana, and Colorado. It was widely recognized even then that watershed changes which significantly altered streamflow magnitude and timing would change channel characteristics and stability indices. A wealth of supporting research data was available at that time from Forest Service research experimental watersheds in Oregon, Washington, Colorado, Idaho, North Carolina, and other locations throughout the United States. More recently, efforts related to prediction of sediment supply changes resulting from introduced sediment sources have been evaluated, with emphasis on road construction and channel source sediment increases due to changes in flow regime. A cumulative watershed effects analysis procedure - "An Approach To Water Resources Evaluation of Non-Point Silvicultural Sources (A Procedural Handbook)", known as WRENSS- was developed to quantitatively assess the potential channel and water quality impacts associated with silvicultural activities (USEPA, 1980). The WRENSS model developed by the USDA Forest Service and EPA emphasized the importance of recognizing inherent differences between river types and the diversity of river response to a variety of land use activities. In an effort to assess the

NEW CHALLENGES

channel condition or "channel stability" in the field, USDA Forest Service hydrologists inventoried thousands of miles of stream using a channel stability rating procedure developed by Pfankuch (1975). The channel condition inventory process was one of the first to provide a consistent, quantitative assessment of conditions one could observe along river reaches.

The public is continually demanding that those who manage the watersheds and the rivers provide for the optimum water resource and related values. It is up to the scientist to assist in providing the conduit for observing, analyzing, documenting, and communicating about river process. Such a challenge is not new, nor is the concept of watershed management; however, problems must be solved utilizing new information, improved techniques and process-based methodologies.

CHAPTER 2
FUNDAMENTAL PRINCIPLES OF RIVER SYSTEMS

*Playfair's Law: "Every river appears to consist of a main trunk, fed from
a variety of branches, each running in a valley proportioned to its size, and all
of them together forming a system of valleys connecting with one another,
and having such a nice adjustment of their declivities that none of them join
the principal valley either on too high or too low a level;
a circumstance which would be infinitely improbable if each of these valleys
were not the work of the stream which flows in it."*
JOHN PLAYFAIR, 1802

River form and fluvial processes evolve simultaneously and operate through mutual adjustments toward self-stabilization (Rosgen, 1994). The physical appearance and operational character of the modern-day river is a product of the adjustment of the river's boundaries to the magnitude of streamflow and erosional debris produced from an attendant watershed. The individual river characteristics are further modified by the influence of channel materials, basin relief, and other features of valley morphology along with a local history of erosion and sediment deposition. On the long journey from watershed divides to the oceans, all rivers must transport the erosional products of their source basins, while maintaining their own competence for self-perpetuation. As the drainage or watershed areas enlarge, so do the requirements for streamflow and sediment transport.

FUNDAMENTAL PRINCIPLES OF RIVER SYSTEMS

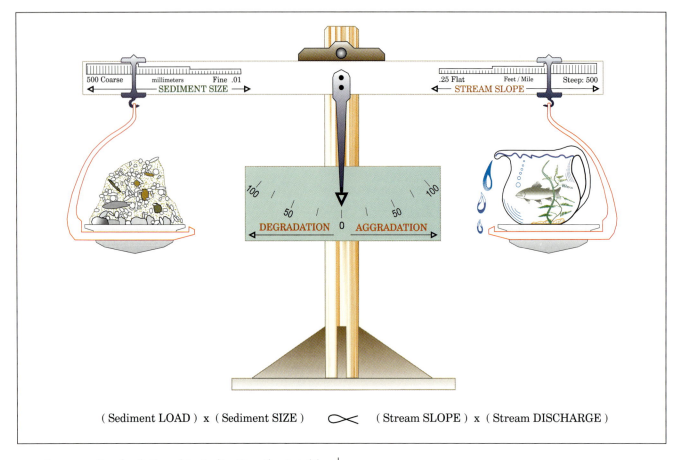

(Sediment LOAD) x (Sediment SIZE) ∝ (Stream SLOPE) x (Stream DISCHARGE)

FIGURE 2-1. Schematic of the Lane relationship for qualitative analysis. *(After Lane, 1955)*

A generalized relationship indicating the "stable channel balance" was proposed by Lane (1955) and is shown diagrammatically in ***Figure 2-1***. The relationship $Q_s\ D_{50} \sim Q\ S$ shows a proportionality between sediment discharge (Q_s), stream discharge (Q), particle size (D_{50}), and slope (S). A change in any one of these variables sets up a series of mutual adjustments in the companion variables with a resulting direct change in the characteristics of the river.

THE BANKFULL DISCHARGE

The term "bankfull" was originally used to describe the incipient elevation on the bank where flooding begins. In many stream systems, the bankfull stage is associated with the flow that just fills the channel to the top of its banks and at a point where the water begins to overflow onto a floodplain. The bankfull stage and its attendant discharge serve as consistent morphological indices which can be related to the formation, maintenance, and dimensions of the channel as it exists under the modern climatic regime.

Stream dimensions, patterns, and bed features associated with the longitudinal river profile are generally described as a function of channel width measured at the bankfull stage. Since streams are self-formed and self-maintained, it is important to relate measurable features one can identify in the field to a relatively frequent, corresponding bankfull discharge. This definition of bankfull, however, applies primarily to stream types that have an observable floodplain feature, no matter how wide. Floodplains can be quite small and inconspicuous in certain stream types, where they may be naturally indistinct or presently are being developed. Streams that are deeply entrenched in the landform do not exhibit significant changes in channel width as flood flows increase. With increasing stage, stream depth generally increases at a more rapid rate than the corresponding channel width. Bankfull stage can be observed and determined within the entrenched stream types by using a series of common stage indicators that may be situated along the boundary of the bankfull channel.

A commonly accepted and universally applicable definition of bankfull was provided by Dunne and Leopold (1978):

"The bankfull stage corresponds to the discharge at which channel maintenance is the most effective, that is, the discharge at which moving sediment, forming or removing bars, forming or changing bends and meanders, and generally doing work that results in the average morphologic characteristics of channels."

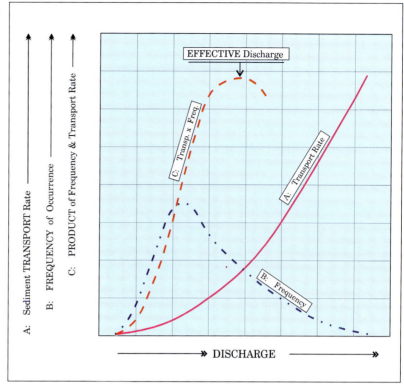

FIGURE 2-2. Relations between DISCHARGE, Sediment TRANSPORT Rate, FREQUENCY of Occurrence, and the PRODUCT of Frequency and Transport Rate. *(After Wolman and Miller, 1960)*

It is this discharge in concert with the range of flows that make up an annual hydrograph which govern the shape and size of the channel. Bankfull discharge is associated with a momentary maximum flow which, on the average, has a recurrence interval of 1.5 years as determined using a flood frequency analysis (Dunne and Leopold, 1978). Although great erosion and enlargement of steep, incised channels may occur during extreme flood events, it is the modest flow regimes which often transports the greatest quantity of sediment material over time, due to the higher frequency of occurrence for such events (Wolman and Miller, 1960). An example of the relationship between flow magnitude and frequency of flow occurrence is shown in *Figure 2-2*. The dominant, effective or bankfull discharge is associated with the peak of cumulative sediment transport for a given streamflow magnitude and frequency of occurrence. The majority of work over time is accomplished at moderate flow rates as shown in *Figure 2.2*. The effectiveness of bankfull discharge and a discussion of dominant or effective discharge theory was presented by Andrews (1980). Field methods to determine bankfull stage for a wide variety of stream types are presented in Chapter 5.

Recent analyses of peak flow data for gage stations on 47 rivers located in Ontario, Canada, indicated that their bankfull discharges have an average return interval of 1.6 years, with a range from 1.5 to 1.7 years (Annable, 1994). There exists in the literature, however, a range of return periods for "bankfull discharge" from 1 to 25 years (Williams, 1978), and (Nash, 1994). In the Williams study, however, there was not a clear distinction made between the elevations of the low terrace and the active floodplain, which serves as the indicator of bankfull stage. The low terrace, by definition, is an abandoned floodplain. The flows necessary to over-top the low terrace bank must be associated with a flood of large magnitude, much larger than the actual bankfull discharge. A low terrace feature is often mistaken for a floodplain by a field observer. In the summary provided by Nash, (1994), the "effective discharge" using a narrower definition, as the flow which attains a maximum sediment transport value over time, indicated a recurrence interval from one week to two decades. The computation used by Nash would indicate a poor correlation between flood frequency and the effective sediment transporting flows over time. This discrepancy can partly be explained by integrating different types of streams into a similar calculation. For example, stream types whose beds are composed of a lag deposit of boulders and large cobble, in low relief valleys, do not transport significant quantities of sediment associated with the 1.5 year return period discharge. Thus, larger, more infrequent flows may be more competent to move the bed material load than the more frequent "bankfull" flows. The definition presented by Dunne and Leopold (1978) also refers to the bankfull discharge that is "generally doing work that results in the average morphologic characteristics of channels." Streambank erosion that relates a given dimension of stream width to a corresponding return period discharge can often be verified in the field. For example, the observed width of a stream at a streamgage, whose bed is composed of boulders and cobble, does not correlate with the discharge that may mobilize this bed material. The observed widths are more closely related to the more frequent, normal high flows, rather than the widths corresponding to the rare, large floods. It is very difficult, indeed, to locate morphological features in the field that are related to the infrequently occurring flood flows. Morphological features can be located and described from field observations which are directly related to a more frequent recurrence interval discharge. The dimensions, pattern, and profile relations of rivers have been demonstrated to be proportionally related to the frequent rather than the infrequent discharges (Leopold et al. 1964 and Dunne and Leopold, 1978). Flood plains are often the "channels" associated with the infrequent, high magnitude, flood discharge.

Another problem using sediment computations to determine the recurrence interval of the bankfull or "effective" discharge is in the use of sediment relations vs discharge. Bedload is often computed rather than measured, and as a result, is often underestimated. Bedload and suspended sediment relations by stream types are described in Chapter 7. Bedload transport often becomes very significant at flows approaching the bankfull stage, where mobilization of the bed material occurs. In many computations bedload is shown to be less than 5 per cent of total load, although, for certain stream types, we have measured bedload that comprises over 75 percent of total load at the bankfull stage. The sediment computations used by Nash (1994) would possibly have been those more directly associated with suspended rather than measured bedload.

An analysis of return periods related to field determined bankfull discharge conducted by the author over the past 10 years and using data for gage stations located on rivers throughout North America indicates a range in return interval from 1.4 to 1.6 years. "Field determined bankfull" is the discharge and corresponding return period associated with the width, and cross-sectional area of the observed bankfull channel at a gaging station. Exceptions to the 1.4-1.6 year return period of the bankfull discharge have been observed in the highly developed, urban watersheds, where the bankfull discharge is closer to the 1.2 year return period, using the log-Pearson flood frequency analysis procedure.

Often, the U.S. Army Corps of Engineers field interpretation of "ordinary high water" and the bankfull stage are synonymous. Thus, one may conclude that the flow regime associated with bankfull discharge is a relatively frequent event. The USDA Forest Service Stream Systems Technology Center, has recently completed a video to assist in field identification of bankfull stage (USDA Forest Service 1995). Field procedures related to bankfull stage and discharge will be covered in more detail in the following chapters.

STREAM CHANNEL DIMENSIONS

Stream width is a function of streamflow occurrence and magnitude, size and type of transported sediment, and the bed and bank materials of the channel. Channel widths generally increase downstream as the square root of discharge (Leopold et al. 1964). Channel width can be modified by the following influences: direct channel disturbance such as channelization; changes in riparian vegetation that may alter the boundary resistance and susceptibility to streambank erosion; changes in streamflow regime due to watershed changes; and changes in sediment regime. The water surface width measurement at the bankfull stage often corresponds with the normal high water discharge as determined using a log-Pearson flood frequency analysis, and as presented earlier, is typically the discharge associated with the 1.5 year return period flow. Field methods to verify bankfull stage and the associated channel width are discussed in Chapter 5.

A channel can have a stable width even though the stream is migrating laterally at a constant annual rate. Stream width can remain relatively constant where the role of erosion on one bank is compensated with corresponding sediment deposition along the opposite bank. The adjustment processes of lateral accretion and associated point bar deposition associated with alluvial channels result in the building of floodplains as point bar areas are abandoned. Nonetheless, the bankfull width of alluvial channels remains relatively constant and thus becomes one of the most directly observable features used in correlations with selected streamflow magnitudes.

The mean depth of rivers varies greatly by individual reaches experiencing similar discharges due to the sequence of riffle and pool bed features. River morphology is influenced not only by the present streamflow and sediment regime, but also by the valley morphology, basin relief, and the nature of bed and bank materials.

Stream channel morphology is often described in terms of a width/depth ratio related to the bankfull stage cross-section. The width/depth ratio varies primarily with: the dimension of the channel cross-section for a given slope; the boundary roughness as a function of the streamflow and sediment regime; bank erodibility factors including the nature of streambank materials; degree of entrenchment (vertical containment); and the distribution of energy (boundary stress) in the stream channel.

The bankfull cross-sectional area of rivers (i.e., surface width times mean depth) is often correlated with streamflow and drainage area as an expression of channel size. Watershed research studies have confirmed that changes in bankfull channel dimensions correspond to changes in the magnitude and frequency of bankfull discharge. Such changes in the bankfull discharge can and do result from water diversions, reservoir regulation, forest clear cutting, vegetation conversion, roads, urban development, over-grazing, and other watershed changes. Streamflow and sediment regime changes directly alter the dimensions of stream channels.

STREAM CHANNEL PATTERNS

Streams are rarely straight for any appreciable distance, rather they tend to follow a sinuous course (Leopold et al. 1964). The planimetric view of various stream patterns, which may be qualitatively described as straight, meandering, or braided forms, will also exhibit specific geometric relationships that may be quantitatively defined through measurements of meander wavelength, radius of curvature, amplitude, and belt width. These relationship patterns are schematically illustrated in *Figure 2-3*.

Meander geometry is most often expressed as a function of bankfull width. An example of the relationships between bankfull width, meander wavelength and radius of curvature are shown in *Figure 2-4* (Leopold et al. 1964). These empirical relationships of channel dimensions with stream patterns are typically derived from analyses of meandering alluvial streams. For instance, the linear distance equal to one-half of a meander wavelength is related to the riffle/pool spacing or natural sequence of the steeps (riffle) and flatter slopes (pools), which are also then related to the bankfull width. Distance values ranging from 10 to 14 bankfull widths are common for an individual meander wavelength, while a linear distance of 5 to 7 bankfull widths are commonly noted for the spacing of riffle/pool features (Leopold et al. 1964).

Streamflow regimes not only influence bankfull channel widths but can also change stream patterns, depending on the magnitude and duration of flows.

FUNDAMENTAL PRINCIPLES OF RIVER SYSTEMS

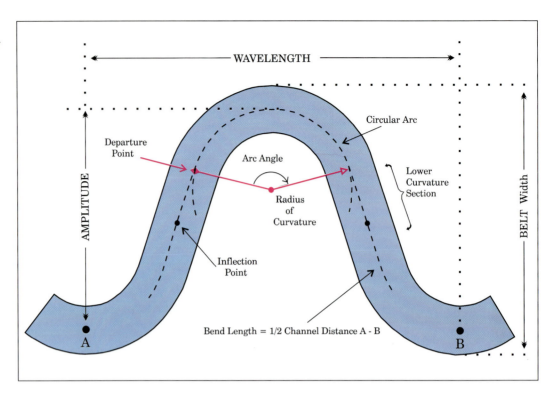

FIGURE 2-3. Schematic Meander Geometry Descriptions. *(After Williams, 1986)*

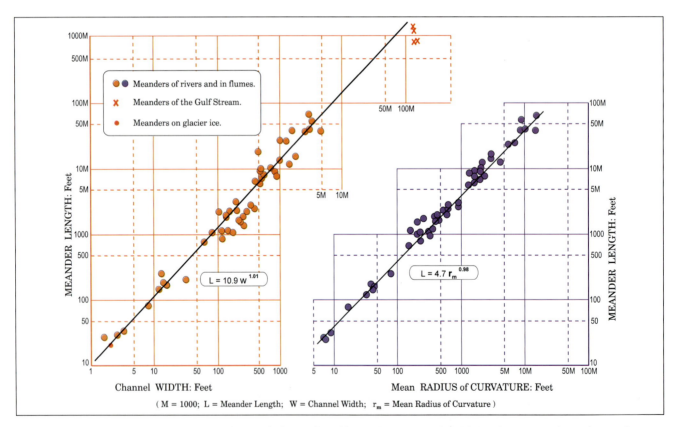

FIGURE 2-4. Relations between meander length and channel width, and meander length and mean radius of curvature. *(After Leopold, et al. 1964)*

FUNDAMENTAL PRINCIPLES OF RIVER SYSTEMS

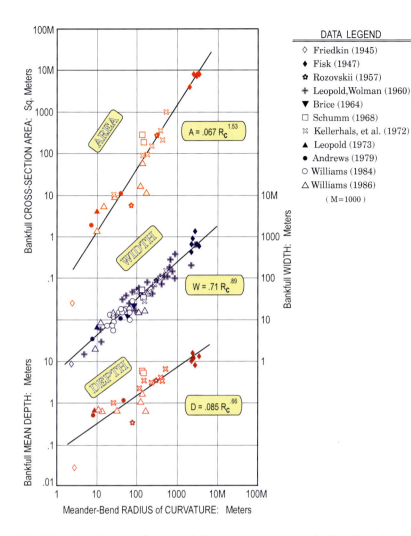

FIGURE 2-5. Relation of channel dimensions to meander-bend radius of curvature. (Data in meters.) *(After Williams, 1986)*

ture relationship is often used to evaluate channel resistance to erosion and bend or meander migration rates.

A study by Williams (1986) verified the relationships of Langbein and Leopold (1966) using a separate data base consisting of 79 sites. (*Figure 2-5*)

The patterns of rivers are naturally developed to provide for the dissipation of the kinetic energy of moving water, and the transportation of sediment. The meander geometry and associated riffles and pools within a river system adjust in such a way that the work expended on natural processes is minimized. Consequently, straightening stream channels ultimately leads to a state of disequilibrium or instability, often causing river entrenchment and corresponding changes in morphology and stability. One such study was conducted on the Mississippi River and some of its tributaries comparing the stability of reaches that were straightened versus those that were naturally meandering (Yang and Song, 1979). The more sinuous, meandering reaches were stable and maintenance free, while the reaches that had been straightened were highly unstable. Over a time period of many years a significant number of our rivers have been straightened, under the assumption that such treatments would enhance river functioning and stability. Even today, those kinds of river management philosophies and thought processes continue to result in a misdirection of much of the work done on rivers.

Extensive research has shown that natural rivers attempt to maintain a dynamic and continuing balance between sediment loads and the energy available from streamflow to perform work. The river pattern exhibits natural adjustments in sinuosity that result in maintaining a slope such that the stream system neither degrades nor aggrades. When the alignment of the river is changed, perhaps by

For example, as urban watersheds are developed, it should be of no surprise to observe widening of streams and changes in channel pattern. These system or channel adjustments brought on by an acceleration of streambank and bed erosion result in the noted increases in the width/depth ratio and corresponding bar development.

A sine-generated curve function was developed by Langbein and Leopold (1966) to describe the symmetrical meander paths of rivers. They derived the following relationship:

$$R_c = \frac{L_m K^{1.5}}{13(K-1)^{0.5}} \qquad (1)$$

in which R_c is radius of curvature, L_m is meander wavelength and K is sinuosity. The radius of curva-

FUNDAMENTAL PRINCIPLES OF RIVER SYSTEMS

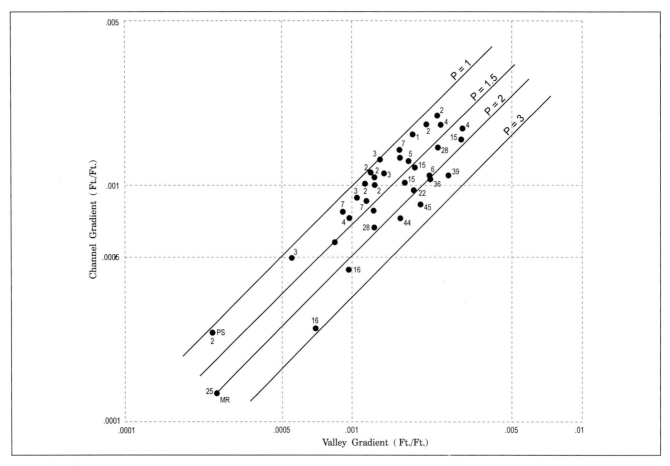

FIGURE 2-6. Relations between alluvial valley slope and channel gradient. Numbers beside points indicate percentage of silt-clay in channels (M). Four lines of equal sinuosity (P) are shown. *(After Shumm, 1968)*

reducing natural sinuosity, local reach slopes are changed and instability generally results.

River sinuosity (defined as the ratio of channel length/valley length, or approximated as valley slope/channel slope) indicates how a river has adjusted its slope to that of its valley. The degree of sinuosity is also related to channel dimensions, sediment load, streamflow, and the bed and bank materials. When the bankfull width/depth ratio increases, the stream length decreases, increasing local slope, setting in motion a series of channel adjustments. Streams that become entrenched (i.e., vertically contained) in their valley and abandon their floodplain are thus, also confined (i.e., laterally contained), normally resulting in a decrease in channel sinuosity.

A wide range of channel sinuosity values, (P) were plotted by Shumm (1968) and shown as a ratio of valley slope to channel slope. The data values were obtained from analyses of relatively flat valleys with a range in slopes of .0001 to .004, and attendant channel sinuosities ranging from 1 to over 2.7.

The ratio data were then stratified by the percentages of silt-clay materials found in the streambanks (*Figure 2-6*), with the highest percentage of silt-clay (45%) found in the banks of those channels in the 2.6 sinuosity range. This relation shows the inverse relation of sinuosity to slope.

STREAM CHANNEL PROFILE

Generally, channel gradient decreases in a downstream direction with commensurate increases in streamflow and a corresponding general decrease in sediment size. The relationship shown by Lane (1955), (*Figure 2-1*) shows that stream gradient is directly related to bed-material load and grain size and inversely related to streamflow. The shape of the longitudinal profile from the small order stream at the watershed divide to the lower valleys is generally concave. Along the extended profile, however, there exists significant differences in channel gradient, bed material size, and morphology which have been found to be independent of an associated drainage area, streamflow regime, stream order, and position in the watershed. A variety of different bed forms may exist, depending on the influence of local channel gradient and resident sediment materials.

Since steep gradient streams are relatively straight (low sinuosity), they dissipate energy along the longitudinal profile in relatively closed spaced features called steps and pools. Their spacing is inversely related to slope and proportional to the bankfull width. As gradient decreases, the nature of these alternating steep and flat components of a reach, change dramatically. The profile, corresponding to the gentle slopes, change into a series of riffles and pools, spaced farther apart as a function of the channel width. This change in profile is matched by a subsequent change in the pattern of the river as previously described.

APPLYING THE PRINCIPLES

River treatment applications that have not integrated fundamental principles of river behavior are shown in the following example where the natural dimension, pattern and profile of a river were changed in an attempt to "stabilize" the channel and control flooding. The river was over-widened, straightened, and adjacent levees were constructed to contain flood flows within one large channel. The objectives for channel alteration included the following:

1. Increase channel capacity by increasing the width of the channel.
2. Increase sediment transport capacity by straightening the river thus increasing channel slope and mean flow velocity.
3. Prevent flood-water encroachment of floodplain areas with the use of levees so that such areas may be commercially developed without risk of future flooding.
4. Increase the design channel shear stress (a depth-slope product) within the channel by increasing mean flow depths, and slope associated with flood flows, in order to assure sediment transport and maintain channel capacity.

In actuality the following process response results occurred:
1. The width/depth ratio of the *bankfull channel* was increased, thus, decreasing the bankfull stage shear stress. A change in velocity distribution occurred which then induced accelerated sediment deposition and channel aggradation.
2. The straightening of the river channel and the corresponding increase in energy changed the velocity distribution and increased the velocity gradient at the banks, as the stream was trying to regain its meander geometry and slope. Bank erosion occurred which required the installation of extensive rip-rap materials and subsequent maintenance.
3. The same flood stage was now associated with storms of less magnitude than the previous flood-magnitude storms. In effect, the flood stage was elevated due to bed aggradation, increasing the flood hazard for lower magnitude flows.
4. When a levee breach would occur, kinetic energy associated with flood waters was greater, and flood flows with higher than normal velocities would result in more extensive damage than would have occurred under the natural flood stages and velocities on the unaltered floodplain. Thus, the consequence of modification resulted in greater frequency and magnitude of flood impacts-just the opposite of the proposed objective.

Since the channel dimensions, river system pattern, and longitudinal profile are all interrelated and delicately adjusted to just carry the sediment load and flows of the watershed, purposeful and perhaps inappropriate changes in these variables result in predictable responses that are often ill-advised. Plan and section views of altered vs. natural channels are shown diagrammatically in *Figure 2-7* (Rosgen 1993b).

Streamflow and channel variables mutually interact to form the morphology of river systems. The unique dimensions, pattern, and profile of these river systems have evolved over time, such that they are able to effectively manage the disposition of

FUNDAMENTAL PRINCIPLES OF RIVER SYSTEMS

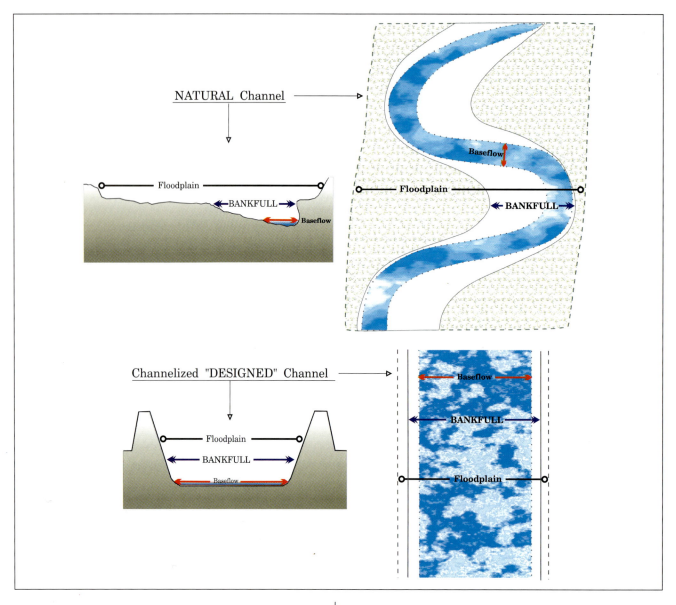

FIGURE 2-7. Comparison of "designed" channel dimensions and pattern with a natural channel. *(Rosgen 1993b)*

water and sediment originating within their watershed. Natural or imposed changes in any of the physical process variables will create a chain reaction of systematic adjustment that is often rapid and certainly significant. The stream classification system presented herein was primarily developed on the basis of measured morphological characteristics and combinations of "river formed" variables as described with field data collected from hundreds of actual river sites. A central concept serving the development of the stream classification system is to provide a mechanism to help others see and visualize the many unique river forms without having to intensively analyze all of the many contributing variables. Hopefully, use of the information-integration facilities available with a stream classification system will provide the basis for one to better understand the significance of the physical process interactions that occur within river systems. With this understanding, perhaps some of the mistakes that were made in the preceding examples may be avoided in the future.

CHAPTER 3
STREAM CLASSIFICATION

"Comparisons between streams of unknown characteristics must thus be of little help in understanding river process."
FAHNSTOCK AND BRADLEY (1973)

Classification can be defined as the ordering of objects into sets on the basis of their similarities or their relationships (Platts 1980). Classification enables us to infer attributes of individual objects on the basis of the characteristics used to set classification categories. The most effective classification systems are those that are based on objective, quantifiable criteria that permit consistent use of the classification system.

STREAM CLASSIFICATION REVIEW

The effort to classify streams is not new. Davis (1899) first divided streams into three classes based on relative stage of adjustment: youthful, mature, and old age. Additional river classification systems based on qualitative and descriptive delineations were subsequently developed by Melton (1936) and Matthes (1956).

Straight, meandering, and braided patterns were described by Leopold and Wolman (1957). Lane (1957) developed quantitative slope-discharge relationships for braided, intermediate, and meandering streams. A classification based on descriptive and interpretive characteristics was developed by Schumm (1963) where delineation was based, in part, on channel stability (stable, eroding, or depositing) and mode of sediment transport (mixed load, suspended load, and bedload). A descriptive classification was also developed by Culbertson et al. (1967) that utilized depositional features, vegetation, braiding patterns, sinuosity, meander scrolls, bank heights, levee formations, and floodplain

STREAM CLASSIFICATION

types. Thornbury (1969) developed a system based on valley types. Valley patterns were described as antecedent, superposed, consequent, and subsequent. Khan (1971) developed a quantitative classification for sand-bed streams based on sinuosity, slope, and channel pattern. The majority of these early classification systems relied heavily on qualitative interpretations of geomorphic features, resulting in inconsistent classification and limited predictive capabilities.

A descriptive classification scheme was developed for and applied on Canadian Rivers by Kellerhals et al. (1972, 1976) to cover a wider range of stream morphologies. The work of other Canadian researchers (Galay et al. 1973, and Mollard 1973) provides an excellent description and interpretation of fluvial features. The Canadian system has utility both for aerial photo delineation and for describing gradual transitions between classical river types and, to date, offers the most detailed and complete list of channel and valley features. The large number of possible interpretative delineations, however, makes the Galay-Mollard process quite complex for general planning objectives.

Schumm (1977) classified rivers in the Great Plains region using as delineative criteria, sediment transport, channel stability, and measured channel dimensions. Classifying stream systems solely on the basis of stability is often difficult because the qualitative criteria involved can be widely interpreted among observers, leading to inconsistencies in the classification. Similarly, data on the ratio of bedload to total sediment load, as needed in the Schumm classification, while useful, often is not readily available to those who need to classify streams. These variables are valuable for interpretation, but difficult for the initial stream type delineation.

Brice and Blodgett (1978) described four channel types: braided, braided point-bar, wide-bend point-bar, and equi-width point-bar. A descriptive inventory of alluvial river channels is well documented by Church and Rood (1983), and can be very useful for many purposes including the grouping of rivers based on similar morphological characteristics. Recent documentation by Selby (1985) showed a relationship between the form and gradient of alluvial channels and the type, supply, and dominant textures (particle sizes) of sediments. This relationship utilizes the Schumm (1977) classification in that an increase in the ratio of bedload to total sediment load with a corresponding increase in channel gradient leads to a decrease in stability causing channel patterns to shift from a meandering to braided channel form. Selby (1985) treats anastomosed and braided channel patterns similarly. However, the anastomosed rivers are not similar to braided rivers in slope, adjustment processes, stability, ratio of bed material to total sediment load, or width/depth ratios, as shown by (Smith and Smith 1980).

A recent classification, proposed by Montgomery and Buffington (1993) and developed in the North Cascades of Washington, delineates streams on the basis of source, transport and response reaches by slope position and stream bed features. Sediment supply and channel materials are indirectly inferred from bed features associated with the source, sediment transport, and response channel reaches.

Understanding the processes and characteristics of fluvial systems requires knowledge of their inherent hierarchical structure. The modern river reflects the effects of current climate, lithology, depositional and erosional history, and the mediating effects of broad vegetation zones. Because these factors change slowly, they establish the basic template for the characteristics of the fluvial system. For example, the interaction of climate, geology, and topographic relief drive the energy gradients within a fluvial system and determine subsequent erosional and depositional processes. Operating over extended time frames and great distances, these processes create valley morphologies and landforms.

The morphology of the modern river reflects not only the events of the past, but also the streamflow and sediment regime determined by climate and landform. The fundamental components of river morphology are its dimension, pattern, and profile. These components represent the integrated response of a river that enables it to be in balance with the prevailing energy gradients, sediment supply and sediment transport characteristics. Stream systems can be described with increasing detail at subsequent levels of organization by identifying the driving variables at finer scales of resolution.

OBJECTIVES OF STREAM CLASSIFICATION

Management strategies too often try to make all rivers meet all possible resource objectives and to force a river's pattern, dimension, and profile into a common homogeneous form. In so doing, we ignore the evolutionary sequence that created stream systems and their unique response potential. In contrast, the hierarchical assessment approach reflects the evolutionary history, and provides an organized procedure for determining the agents of formation and the processes utilized by different stream types for maintaining their form and function. Use of a hierarchical inventory method enhances the incorporation of companion analyses (e.g., studies of fisheries and riparian ecology) at levels compatible with the morphological information that determines the hydrologic and ecological response potentials of rivers.

The role of stream classification in bio-assessment as presented by Barbour, et al. (1991), stated

"One function of classification is to increase the resolving power or sensitivity to biological surveys to detect impairment by partitioning variation within selected environmental parameters or among sites. The importance of minimizing variation... Clearly it is easier to distinguish impairment if the parameters have low variability. Formal statistical tests (parametric and nonparametric) indicate greater resolution and power exist if there is low variance within elements being compared. Effective classification leads to improving resolving power by partitioning or accounting for variability. A coarse classification yields higher variance and therefore lower resolving power; vice versa for finer classifications."

The objective of classifying streams on the basis of channel morphology is to set categories of discrete stream types so that consistent, reproducible descriptions and assessments of condition and potential can be developed. Such assessments can then be extrapolated to similar stream reaches in other hydro-physiographic provinces. Field observations of hundreds of natural, stable channels provided the extensive data set that forms the basis and range of values for the delineative criteria selected to describe individual stream types. Analyses of the quantitative field measurements revealed that the values of the parameters used to describe stream morphology tend to cluster into definable groups and have predictable patterns of variation. The consistent patterns provide the breakpoints used for stratifying the continuum of channel morphology into discrete, recognizable units.

Specific objectives of the stream classification system, which are described in more detail in subsequent chapters of this book, include the following:

1. Predict a river's behavior from its appearance.
2. Develop specific hydraulic and sediment relationships for a given stream type and its state.
3. Provide a mechanism to extrapolate site-specific data to stream reaches having similar characteristics.
4. Provide a consistent frame of reference for communicating stream morphology and condition among a variety of disciplines and interested parties.

The stated objectives can be met through a hierarchical assessment of channel morphology. The advantage of hierarchical assessment is that it provides the physical, hydrologic, and geomorphic context for linking the driving forces and response variables at all scales of inquiry. The detail required at each level of assessment varies with the degree of resolution required.

HIERARCHY OF RIVER MORPHOLOGY

Combinations of morphological variables important for different scales of analysis from coarse to very fine resolutions were then used to create the

STREAM CLASSIFICATION

hierarchy of river morphology illustrated in *Figure 3-1*.

As shown in *Figure 3-1*, the hierarchy is comprised of four inventory or assessment levels that vary from a broad geomorphic characterization down to very detailed-specific description and assessment. The level of specificity required for the measurements taken when advancing from Level I to Level IV increases as do the associated interpretations of stream condition and response to disturbance. Since the assessment process is applied in a general manner, beginning at Level I and then advancing to more specific levels, it is not readily apparent that the procedure is process based. In fact, the classification system was developed by establishing morphological-process relations at reach specific levels and then developing methods to extrapolate these findings to other locations at broader levels of inquiry.

Level I describes the geomorphic characteristics that result from the integration of basin relief, landform, and valley morphology. The dimension, pattern, and profile of rivers are used to delineate geomorphic types at a coarse-scale. Many of the Level I criteria can be determined from topographic and landform maps, and from aerial photography. Even at this broad level, however, individual reaches are delineated and kept unique within the fluvial system.

Level II provides a more detailed morphological description of stream types extrapolated from field-determined reference reach information. The channel entrenchment, dimensions, patterns, profile, and boundary materials are quantified at this level and are described by discreet categories of stream types. This level provides a consistent, quantitative morphological assessment and provides a higher-resolution of information with utility for management applications.

Level III describes the existing condition or "state" of the stream as it relates to its stability, response potential, and function. At this level, additional field parameters are evaluated that influence the stream state (e.g., riparian vegetation, sediment supply, flow regime, debris occurrence, depositional features, channel stability, bank erodibility, and direct channel disturbances). Level III analyses are both reach-and feature-specific and are especially useful as a basis for integrating companion studies (e.g. fish habitat indices, and surveys of riparian communities).

Level IV is the level at which measurements are taken to verify process relationships inferred from preceding analyses. The objective is to establish empirical relationships for use in prediction (e.g., to develop Manning's "n" values from measured velocity; correlating bedload versus discharge by stream type to determine sediment transport relationships; or calculating hydraulic geometry from gaging station data). The developed empirical relationships are specific to individual stream types for a given state, and enable extrapolation to other similar reaches for which Level IV data is not available. Using relationships developed at Level IV, existing data from gage stations and research sites can be analyzed and extrapolated to similar stream types. Note, however, that without the geomorphic context provided by the preceding analyses, it is difficult to accurately extrapolate information obtained at Level IV. Thus, full use of the hierarchy from Level I to Level IV enables extrapolation and incorporation of existing data that could not otherwise be applied.

CONTINUUM OF STREAM MORPHOLOGY

The stream morphology classification uses discrete classes for a suite of morphologic parameters to set parameters or prescribe intervals for categorizing stream types. Discrete categories are necessary to create consistent, reproducible rules for inclusion of a stream within a stream type. The cutoff values for the category intervals were developed from frequency histograms for each of the morphologic parameters (these histograms are shown in Chapter 5).

In reality, however, stream morphology displays a continuum of form. The physical processes that create continuous adjustments in river morphology do not permit rigid, arbitrary boundaries to exist in nature. Thus, the stream morphology hierarchy also

STREAM CLASSIFICATION

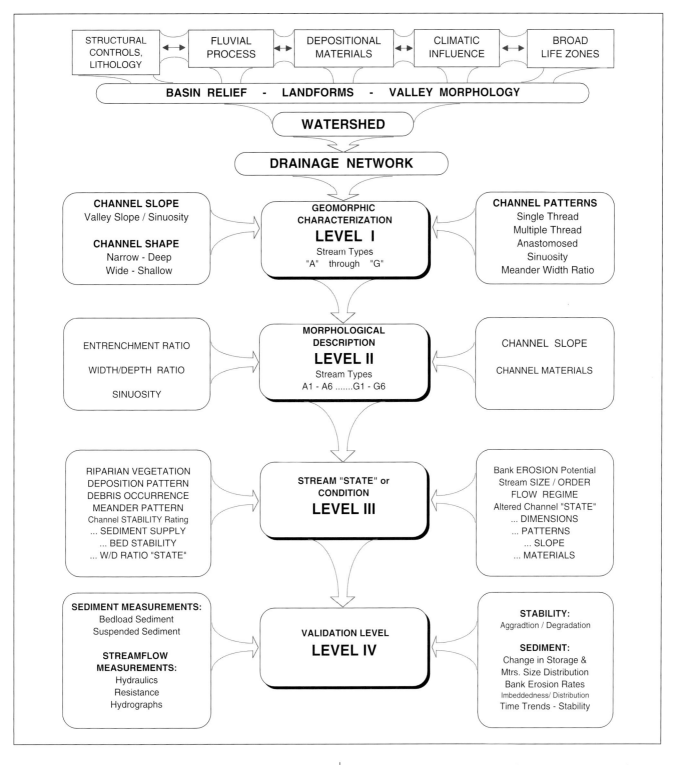

FIGURE 3-1. The hierarchy of river inventory and assessment.

uses a companion set of subclasses to accommodate a continuous range of parameter values. For example, entrenchment ratios of 1.4 to 2.2 are used to delineate B stream types. Slight deviations in entrenchment ratios (e.g., deviations of ± 0.2) do not signal a shift to a new morphology, but rather indicate that the observed value of the parameter is at the tail of its distribution for that morphological type.

Stream types are based on a minimum of six morphologic parameters. As adjustment processes occur, individual parameter values can shift from their central values (i.e., the average, "stable" value) to their extreme values. When several individual parameters shift sufficiently, the stream reach will express a new morphologic type. Guidelines describing limits in the range of adjustments to individual parameter values allowable within a given stream type category are presented in the discussion of individual parameters within subsequent chapters.

GENERAL APPLICATIONS OF THE STREAM CLASSIFICATION AND HIERARCHICAL INVENTORY

Geomorphologists have for many years investigated relationships between driving and response variables. A number of highly significant correlations have been documented and, while there can be limited exceptions, their strength is such that they become general guidelines for practicing geomorphologists. For example, discharge generally increases proportionally with increases in drainage area. The established quantitative relationships are essential for interpreting large quantities of data and for drawing conclusions.

In some cases, relationships between driving and response variables have yielded inconclusive results. For example, a general axiom is that streams get wider as drainage area increases. This is generally true; however, stream width can change in response to changes in width/depth ratio associated with changes in stream channel morphology independent of drainage area. Similarly, it is commonly believed that steep, low order, entrenched streams are sediment supply limited. This is generally true for streams having stable bed and bank materials (i.e., A1, A2, G1, and G2 stream types-see Chapter 5). However, sediment supply is not necessarily limited for streams of the same order and having similar gradients and entrenchment ratios, but which are incised in unstable, unconsolidated, heterogeneous material (i.e., A3, A4, A5, G3, G4 and G5 stream types, as described in Chapter 5). Often, it is the differences in stream channel morphology that explain the exceptions to these general axioms. Stream types which are stratified on the basis of discrete morphological characteristics can be used to refine the interpretation of previously determined geomorphological relationships and to define new relationships that previously could not be consistently observed. Examples of these relationships are presented in Chapters 4 and 5.

Communication

There are a diverse range of disciplines working with rivers. The complexities inherent in the study and understanding of river processes do not lead to obvious conclusions among such a diverse group. A presentation of a flow chart depicting the variables of the "alluvial channel system" is presented in ***Figure 3-2*** (Richards, 1982). All of these variables contribute to the alluvial channel system, but at a glance, one does not readily "see" or visualize an alluvial river. It is a challenge, then, to arrange many of the variables in a manner that describes the morphology of rivers. These variables interact with one another to provide an integrative, common form among certain rivers, but whose morphological forms vary among a large array of stream types.

In contrast to the very detailed, complex, and multiple variable descriptions, generalized terms are often used to simplify communication. Over-simplification, however, can lead to misunderstandings. An example of such a term in common usage among geologists, hydrologists and engineers is the "gravel bed river." In current usage, a "gravel bed river" encompasses a range of streams in gravel or in cobble that are wide and shallow, or narrow and deep, gentle to moderately steep, and include streams that have broad well developed floodplains to stream channels that are confined (laterally contained) and entrenched (vertically contained). Unfortunately, this generalization often conveys inconsistent interpretations and confusing communication, especially, since this one term covers such a wide range of stream types.

The physical and biological characteristics, interpretations, response, recovery potential, and other attributes of rivers are of common concern to many of the diverse groups working with rivers. A straight forward, simple, yet consistent communication among these workers is needed in order to share data, observations, and conclusions. Without a common, understandable description of a wide

STREAM CLASSIFICATION

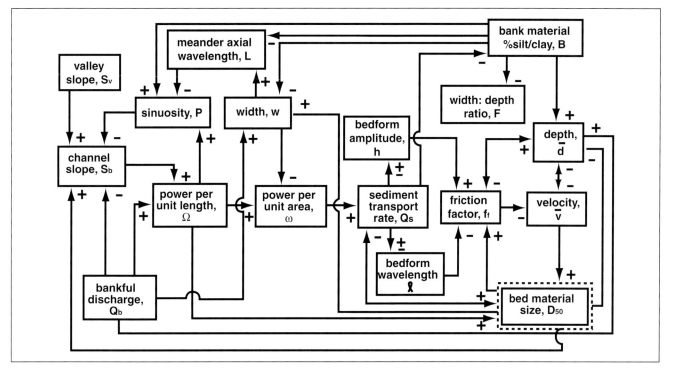

FIGURE 3-2. A flow chart of the "alluvial channel system" as presented by Richards, (1982)

range of river types, the complexities some of which are shown in Figure 3-2, do not lend to a common understanding. The attempt for this classification scheme was to integrate and arrange many of these same variables into an order that can be *visually* associated with the modern river.

General Applications

The stream classification system is based on extensive field observations and quantitative investigations of hundreds of stream systems over a period of twenty seven years. It has formed the basis for restoration designs for portions of more than 50 rivers, including the East Fork of the San Juan and Blanco Rivers of Colorado. The system presented here has been consistently revised to enhance its ability to discriminate between stream types and to improve its predictive capabilities. The procedure is fully operational and is designed for use in assessing channel condition, impact-use effects, and alternative management scenarios for either local or regional scale watersheds.

This book is designed to permit interested users to understand the basis of the morphological hierarchy and to use it to assess stream condition. Nonetheless, this system should not be considered static. Just as the USDA Soil Classification system has been refined and revised 12 times over 40 years, so should the database of morphologic parameters used in this hierarchy be expanded and revised as necessary. Readers are encouraged to conduct their own stream analyses to develop a local data base. Data collected by users at permanent stream gaging stations (such as those of the U.S. Geological Survey or state water resource agencies) would be especially helpful in refining estimates of the bankfull stage and corresponding stream type parameters and the development of local and regional relations.

Every day, however, decisions are made by resource managers and private landowners that affect the condition of watersheds. Some of these decisions are localized and incremental (e.g., should a road be widened or rerouted?). Others are more regional and comprehensive (e.g., how can ecosystem management principles be implemented to protect fisheries while allowing timber harvest or mining development?). Researchers say more research is needed, while managers say that we need more information on large areas and need to know how to apply the research findings. Both are correct.

STREAM CLASSIFICATION

The use of the stream hierarchical inventory is designed to:
- Stratify stream system inventories at appropriate levels.
- Provide an organization for integrating and analyzing information at selected levels.
- Assist in the assessment of cumulative watershed impacts.
- Provide a method to utilize sediment data, bank erosion and stability predictions.
- Provide a mechanism to integrate companion inventories such as fish habitat potential and riparian vegetation associations.

The following chapters will describe each of the various stream hierarchical inventory levels using the stream classification system and present examples for management applications.

CHAPTER 4
LEVEL I: GEOMORPHIC CHARACTERIZATION

"Eventually, all things merge into one, and a river runs through it.
The river was cut by the world's great flood
and runs over rocks from the basement of time.
It sings a song of wisdom and life far greater than man can hear."
NORMAN MACLEAN (1976)

The landscapes we observe in a watershed are formed by erosional and depositional processes resulting from a complex integration of climate, lithology, and vegetation patterns operating over extended time periods and great distances. The observed complexities of landform systems hardly inspire confidence that we can use these patterns in solving practical problems. A closer look at the pieces of the geomorphic maze, however, shows that landscape forms can be related to the processes responsible for their creation. Valley morphologies that have been shaped by erosional and depositional processes operating in various structural systems express certain features that can be extrapolated to other valleys having similar controls and exposed to similar processes, even though these valleys may be far distant.

Interpretations of drainage basin evolution and their resultant landscapes provide an understanding of the basin or watershed response to imposed

LEVEL I: GEOMORPHIC CHARACTERIZATION

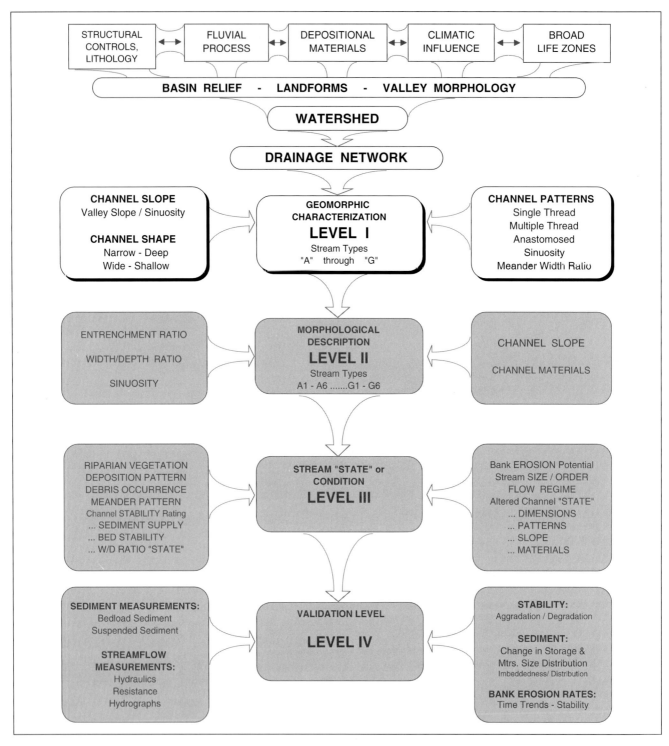

FIGURE 4-1a. Geomorphic characterization (Level I) for stream classification in relation to the hierarchical river inventory levels.

change. The influence of vegetative cover patterns, streamflow changes, erosional and depositional process are key variables that help shape the fluvial landscape. To help assemble the assortment of complex interrelated physical process variables into some meaningful form, it is essential to stratify the observed fluvial system, using a hierarchical inventory structure.

The concept of a hierarchical inventory system was presented in Chapter 3, *Figure 3-1* and will be utilized here to illustrate the Level I morphological description process for river systems (*Figure 4-1a*). A flowchart is presented in *Figure 4-1b* to depict

LEVEL I: GEOMORPHIC CHARACTERIZATION

the stream system variables utilized in the Level I hierarchical inventory. The integration of structural controls, fluvial process, depositional history, climate, and broad life zones are reflected as valley types, that when combined, result in the various landforms observable within a drainage basin. The valley types and their corresponding stream types can be delineated using the Level I inventory process.

Level I stream classifications serve three primary inventory functions:

(1) provide for the initial integration of basin characteristics, valley types, and landforms with stream system morphology.

(2) provide a consistent initial framework for organizing river information and communicating the aspects of river morphology. Mapping of physiographic attributes at Level I can quickly determine location and approximate percentage of river types within a watershed sub-basin, and/or valley type.

(3) assist in the setting of priorities for conducting more detailed assessments and/or companion inventories.

(4) correlate similar general level inventories such as fisheries habitat, river boating categories, and riparian habitat with companion river inventories.

However useful, the general information obtained about river systems using the Level I inventory process is, of course, the least specific. Chapters 5 through 7 discuss hierarchical levels II through IV, that deal with the development of more detailed, reach-specific information and interpretations.

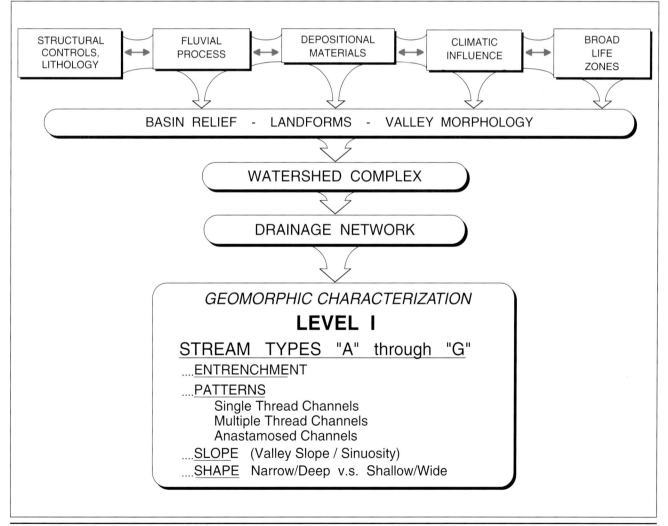

FIGURE 4-1b. Flow chart for delineation of the geomorphic characterization level stream classification (A through G).

LEVEL I: GEOMORPHIC CHARACTERIZATION

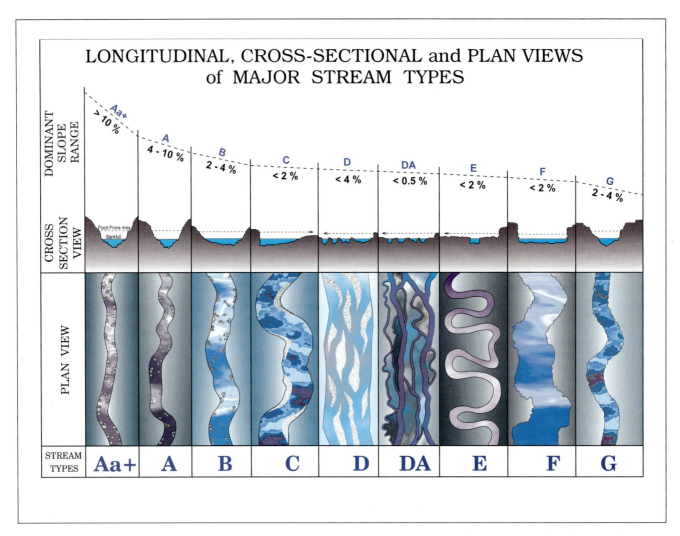

FIGURE 4-2. Broad level stream classification delineation showing longitudinal, cross-sectional, and plan-views of major stream types *(from Rosgen, 1994)*

Chapter 4 is presented in three parts: (1) a narrative description of stream types A through G, (2) an overview of the major associations between landforms and stream types that can be identified at a coarse-scale, and (3) delineation procedures to classify streams at Level I.

BROAD-LEVEL DESCRIPTIONS OF THE MAJOR STREAM TYPES

The advantage of a broad-level delineation is that it allows for a rapid initial morphological delineation of stream types and generally illustrates the distribution of these types that would be encountered within watersheds. The Level I classification and delineation process provides a general characterization of valley types and landforms, and identifies the corresponding major stream types, A through G. Illustrations depicting the broad-level morphological stream type groupings are shown in *Figure 4-2* and summarized in *Table 4-1*.

Stream Descriptions, "A" Through "G"

The following descriptions summarize each of the major morphological stream types:

The "Aa+" Stream Type

Stream type "Aa+" is very steep (>10%), well entrenched, has a low width/depth ratio, and is totally confined (laterally contained). The bedforms are typically a step/pool morphology with chutes, debris flows, and waterfalls. The "Aa+" stream types often occur in debris avalanche terrain, zones of deep deposition such as glacial tills and outwash

LEVEL I: GEOMORPHIC CHARACTERIZATION

Stream Type	General Description	Entrenchment Ratio	W/D Ratio	Sinuosity	Slope	Landform/Soils/Features
Aa+	Very steep, deeply entrenched, debris transport, torrent streams.	<1.4	<12	1.0 to 1.1	>.10	Very high relief. Erosional, bedrock or depositional features; debris flow potential. Deeply entrenched streams. Vertical steps with deep scour pools; waterfalls.
A	Steep, entrenched, cascading, step/pool streams. High energy/debris transport associated with depositional soils. Very stable if bedrock or boulder dominated channel.	<1.4	<12	1.0 to 1.2	.04 to .10	High relief. Erosional or depositional and bedrock forms. Entrenched and confined streams with cascading reaches. Frequently spaced, deep pools in associated step/pool bed morphology.
B	Moderately entrenched, moderate gradient, riffle dominated channel, with infrequently spaced pools. Very stable plan and profile. Stable banks.	1.4 to 2.2	>12	>1.2	.02 to .039	Moderate relief, colluvial deposition, and/or structural. Moderate entrenchment and W/D ratio. Narrow, gently sloping valleys. Rapids predominate w/scour pools.
C	Low gradient, meandering, point-bar, riffle/pool, alluvial channels with broad, well defined floodplains.	>2.2	>12	>1.2	<.02	Broad valleys w/terraces, in association with floodplains, alluvial soils. Slightly entrenched with well-defined meandering channels. Riffle/pool bed morphology.
D	Braided channel with longitudinal and transverse bars. Very wide channel with eroding banks.	n/a	>40	n/a	<.04	Broad valleys with alluvium, steeper fans. Glacial debris and depositional features. Active lateral adjustment, w/abundance of sediment supply. Convergence/divergence bed features, aggradational processes, high bedload and bank erosion.
DA	Anastomosing (multiple channels) narrow and deep with extensive, well vegetated floodplains and associated wetlands. Very gentle relief with highly variable sinuosities and width/depth ratios. Very stable streambanks.	>2.2	Highly variable	Highly variable	<.005	Broad, low-gradient valleys with fine alluvium and/or lacustrine soils. Anastomosed (multiple channel) geologic control creating fine deposition w/well-vegetated bars that are laterally stable with broad wetland floodplains. Very low bedload, high wash load sediment.
E	Low gradient, meandering riffle/pool stream with low width/depth ratio and little deposition. Very efficient and stable. High meander width ratio.	>2.2	<12	>1.5	<.02	Broad valley/meadows. Alluvial materials with floodplains. Highly sinuous with stable, well-vegetated banks. Riffle/pool morphology with very low width/depth ratios.
F	Entrenched meandering riffle/pool channel on low gradients with high width/depth ratio.	<1.4	>12	>1.2	<.02	Entrenched in highly weathered material. Gentle gradients, with a high width/depth ratio. Meandering, laterally unstable with high bank erosion rates. Riffle/pool morphology.
G	Entrenched "gully" step/pool and low width/depth ratio on moderate gradients.	<1.4	<12	>1.2	.02 to .039	Gullies, step/pool morphology w/moderate slopes and low width/depth ratio. Narrow valleys, or deeply incised in alluvial or colluvial materials, i.e., fans or deltas. Unstable, with grade control problems and high bank erosion rates.

TABLE 4-1 General stream type descriptions and delineative criteria for broad-level classification (Level I).

LEVEL I: GEOMORPHIC CHARACTERIZATION

terraces, or landforms that are structurally controlled or influenced by faults, joints, or other structural contact zones. Streamflow at the bankfull stage in the "Aa+" stream type is generally observed as a torrent or waterfall. The "Aa+" stream types can be associated with bedrock, and zones of deep deposition and/or be deeply incised in residual soils. The "Aa+" can often be described as high energy/high sediment supply systems due to their inherently steep channel slopes and narrow/deep channel cross-sections. "Aa+" stream types may also be found in alluvial landforms, where a change in the base level of the mainstem channel initiates a headward expansion of the tributary network through a channel rejuvenation process. Examples of rejuvenation may be observed where lower-slope position streams are deeply incised in over-steepened adjacent side-wall slopes, or older holocene terrace features that have cut their way through to the elevation of the existing mainstem river (***Figure 4-3a*** and ***Figure 4-3b***). The "Aa+" stream types are often found in valley types I, III, and VII (see Valley Morphology Related to Stream Types - page 4-11).

The "A" Stream Type

Stream type "A" is similar to the described "Aa+", in terms of associated landforms and channel characteristics. The exception being that channel slopes range from 4 to 10 percent, and streamflows at the bankfull stage are typically described as step/pools, with attendant plunge or scour pools. Normally, "A" stream types are found within valley types that due to their inherent channel steepness, exhibit a high sediment transport potential and a relatively low in-channel sediment storage capacity. Although a large number of "A" stream types occur as low-order streams, located at upper-slope positions, stream order for these stream types can range from 1st order up to 5th order or larger. Stream order referred to is that of Strahler (1952), where the incipient crenulation of a drainage way on the landscape is order 1 and the confluence of the first two drainage ways become order 2 and so on. The influx of large organic debris can play a major role in determining the bedform and overall channel stability of "A" stream types. Landforms associated with deeply incised fanhead troughs are associated with both "Aa+" and "A" stream types (***Figure 4-3c***). Valley types associated with the "A" stream types are I, III, and VII.

The "B" Stream Type

The "B" stream types exist primarily on moderately steep to gently sloped terrain, with the predominant landform seen as a narrow and moderately sloping basin. Many of the "B" stream types are the result of the integrated influence of structural contact zones, faults, joints, colluvial-alluvial deposits, and structurally controlled valley sideslopes which tend to result in narrow valleys that limit the development of a wide floodplain. "B" stream types are moderately entrenched, have a cross-section width/depth ratio (greater than 12), display a low channel sinuosity, and exhibit a "rapids" dominated bed morphology. Bedform morphology, which may be influenced by debris constrictions and local confinement, typically produces scour pools (pocket water) and characteristic "rapids." Streambank erosion rates are normally low as are the channel aggradation/degradation process rates. Pool-to-pool spacing is generally 4-5 bankfull widths, decreasing with an increase in slope gradient. Meander width ratios (belt width/bankfull width) are generally low which reflect the low rates of lateral extension. "B" stream types are usually found within valley types II, III, and VI.

The "C" Stream Type

The "C" stream types are located in narrow to wide valleys, constructed from alluvial deposition. The "C" type channels have a well developed floodplain (slightly entrenched), are relatively sinuous with a channel slope of 2% or less and a bedform morphology indicative of a riffle/pool configuration. The shape and form of the "C" stream types are indicated by cross-sectional width/depth ratios generally greater than 12, and sinuosities exceeding 1.2. The "C" stream type exhibits a sequencing of steeps (riffles) and flats (pools), that are linked to

LEVEL I: GEOMORPHIC CHARACTERIZATION

FIGURE 4-3a. Illustration of a main trunk river laterally eroding into an alluvial fan, causing tributary rejuvenation and adjustment of local base level.

FIGURE 4-3b. Photograph of the Green River (Utah) showing the location of oversteepened tributaries (Aa+ stream types) on lower slope positions due to tributary rejuvenation from main stem degradation.

FIGURE 4-3c. Illustration of a fan head trough (Aa+ stream type) creating high sediment supply.

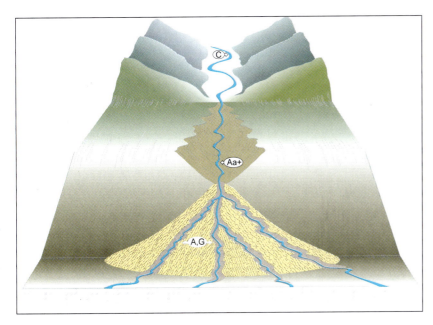

LEVEL I: GEOMORPHIC CHARACTERIZATION

the meander geometry of the river where the riffle/pool sequence or spacing is on the average one-half a meander wavelength or approximately 5-7 bankfull channel widths. The primary morphological features of the "C" stream type are the sinuous, low relief channel, the well developed floodplains built by the river, and characteristic "point bars" within the active channel. The channel aggradation/degradation and lateral extension processes, notably active in "C" stream types, are inherently dependent on the natural stability of streambanks, the existing upstream watershed conditions and flow and sediment regime. Channels of the "C" stream type can be significantly altered and rapidly de-stabilized when the effects of imposed changes in bank stability, watershed condition, or flow regime are combined to cause an exceedance of a channel stability threshold. "C" stream types may be observed in valley types IV, V, VI, VIII, IX and X. They can also be found on the lower slope positions of the very low gradient valley type III.

The "D" Stream Type

The "D" stream type is uniquely configured as a multiple channel system exhibiting a braided, or bar-braided pattern with a very high channel width/depth ratio, and a channel slope generally the same as the attendant valley slope. "D" type stream channels are found in landforms and related valley types consisting of steep depositional fans, steep glacial trough valleys, glacial outwash valleys, broad alluvial mountain valleys, and deltas. While the very wide and shallow "D" stream types are not deeply incised, they can be laterally contained in narrower or confined valleys. Bank erosion rates are characteristically high and meander width ratios are very low (**Figure 4-4**). Sediment supply is generally unlimited and bed features are the result of a convergence/divergence process of local bed scour and sediment deposition. The multiple channel features are displayed as a series of various bar types and unvegetated islands that shift position frequently during runoff events. Adjustments in channel patterns can be initiated with either natural or imposed changes in the conditions of the encompassing landform, contributing watershed area, or the existing channel system. Aggradation and lateral extension are dominant channel adjustment processes occurring within a range of landscapes from desert to glacial outwash plains. Typically, the runoff regime is "flashy," especially in arid landscapes with highly variable extremes of stage occurring on an annual basis which generates a very high sediment supply. Braided channel patterns can be found developing in very coarse materials located in valleys with moderately steep slopes, to very wide, flat, low gradient valleys containing finer materials. The "D" stream type may develop within valley types III, V, VIII, IX, X, and XI.

The "DA" (Anastomosed) Stream Type

The "DA" or anastomosed stream type is a multiple-thread channel system with a very low stream gradient and the bankfull width of each individual channel noted as highly variable. Stream banks are often constructed with fine grained cohesive bank materials, supporting dense-rooted vegetation species, and are extremely stable. Channel slopes are very gentle, commonly found to be at or less than .0001 (Smith, 1986). Lateral migration rates of the individual channels are very low except for infrequent avulsion. Relative to the "D" stream type, the "DA" stream type is considered as a stable system composed of multiple channels. Channel width/depth ratios and sinuosities may vary from very low to very high. The related valley morphology is seen as a series of broad, gently sloping wetland features developed on or within lacustrine deposits, river deltas or splays, and fine-grained alluvial deposits. The "DA" stream types make up a very small number of observed stream types, but are unique both in the process of their creation and maintenance. In certain locations operating at a "control" point within a valley, maintains the valley base level where a vertical balance exists between the rate of deposition and the rate of uplift (Smith and Putnam, 1980). The geologic processes responsible for development of the anastomosed river include subsidence of sedimentary basins in tectonically active forelands, valley base level rise at the basin outlet, regional basin tilting derived from glacial-induced differential isostatic rebound, and the uplifting of sea or lake bed levels (Smith 1986). The bedform features of the "DA" stream types are riffle/pool, similar to stream types "C" and "E." The streambanks and island surfaces between channels are well vegetated and constructed with either fine grained alluvium, or fine, cohesive depositional materials. The ratio of bedload to total sediment load is very low for these very stable stream types. The "DA" stream type normally occurs in valley types X and XI.

The "E" Stream Type

The "E" type stream channels are conceptually designated as *evolutionary* in terms of fluvial process and morphology. The "E" stream type represents the developmental "end-point" of channel stability and fluvial process efficiency for certain alluvial streams undergoing a natural dynamic sequence of system evolution. The "E" type system often develops inside of the wide, entrenched and meandering channels of the "F" stream types, following floodplain development on and vegetation recovery of the former "F" channel beds. The "E" stream types are slightly entrenched, exhibit very low channel width/depth ratios, and display very high channel sinuosities which result in the highest meander width ratio values of all the other stream types (*Figure 4-4*). The bedform features of the "E" stream type are predominantly a consistent series of riffle/pool reaches, generating the highest number of pools per unit distance of channel, when compared to other riffle/pool stream types (C, DA, and F). "E" type stream systems generally occur in alluvial valleys that exhibit low elevational relief characteristics and physiographically range from the high elevations of alpine meadows to the low elevations of coastal plains. While the "E" stream types are considered as highly stable systems, provided the floodplain and the low channel width/depth characteristics are maintained, they are very sensitive to disturbance and can be rapidly adjusted and converted to other stream types in relatively short time periods. The "E" stream type typically develops within valley types VIII, X, and XI.

The "F" Stream Type

The "F" stream types are the classic "entrenched, meandering" channels described by early day geomorphologists, and are often observed to be

FIGURE 4-4. Meander width ratio (belt width/bankfull width) by stream type categories.

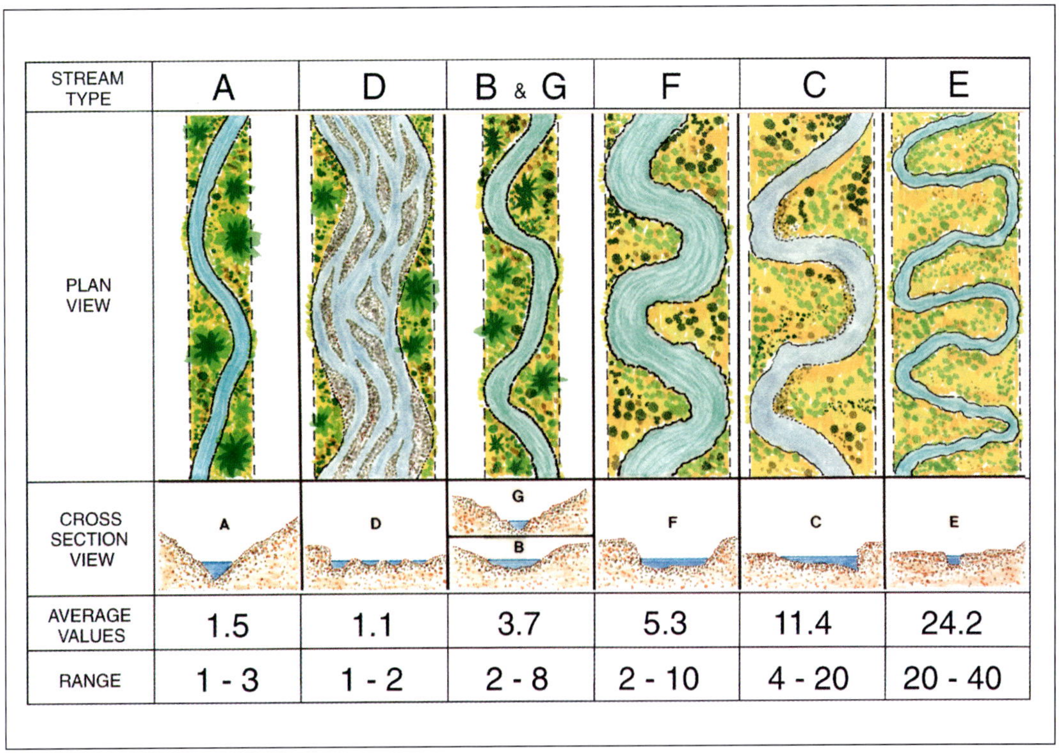

STREAM TYPE	A	D	B & G	F	C	E
PLAN VIEW						
CROSS SECTION VIEW						
AVERAGE VALUES	1.5	1.1	3.7	5.3	11.4	24.2
RANGE	1 - 3	1 - 2	2 - 8	2 - 10	4 - 20	20 - 40

LEVEL I: GEOMORPHIC CHARACTERIZATION

working towards re-establishment of a functional floodplain inside the confines of a channel that is consistently increasing its width within the valley. "F" stream types are deeply incised in valleys of relatively low elevational relief, containing highly weathered rock and/or erodible materials. The "F" stream systems are characterized by very high channel width/depth ratios at the bankfull stage, and bedform features occurring as a moderated riffle/pool sequence. "F" stream channels can develop very high bank erosion rates, lateral extension rates, significant bar deposition and accelerated channel aggradation and/or degradation while providing for very high sediment supply and storage capacities. The "F" stream types occur in low relief valley type III, and in valley types IV, V, VI, VIII, IX, and X

The "G" Stream Type

The "G" or "gully" stream type is an entrenched, narrow, and deep, step/pool channel with a low to moderate sinuosity. Channel slopes are generally steeper than .02, although "G" channels may be associated with gentler slopes where they occur as "down-cut" gullies in meadows. The "G" stream type channels are found in a variety of landtypes to include alluvial fans, debris cones, meadows, or channels within older relic channels. The "fanhead trench" which is a channel feature deeply incised in alluvial fans is typical of "G" type stream channels (*Figure 4-3c*). With the exception of those channels containing bedrock and boulder materials, the "G" stream types have very high bank erosion rates and a high sediment supply. Exhibiting moderate to steep channel slopes, low channel width/depth ratios and high sediment supply, the "G" stream type generates high bedload and suspended sediment transport rates. Channel degradation and sideslope rejuvenation processes are typical. The valley types supporting the "G" stream types are I, III, V, VI, VII, VIII, and X. The "G" stream type can also be observed in valley types II, VI, VIII and X, under conditions of instability or disequilibrium that are often imposed by watershed changes and/or direct channel impacts.

Morphological Descriptions

Considering the above general descriptions of the major categories of stream types it becomes apparent that development of a higher resolution information base is needed, that would allow for more detailed interpretations, predictions and applications. The morphological description along a river continuum was developed, wherein the natural range of individual parameter values for a given stream is determined from additional data collection and process-specific measurements taken at a higher resolution of inventory. The more detailed measurements primarily involve longitudinal profiles or channel slope, particle size distribution of channel bed and bank materials, and channel cross-sectional dimensions and form patterns which quantitatively describe and further define a particular stream type. These descriptions are presented in Chapter 5.

ASSOCIATIONS BETWEEN LAND-FORMS AND STREAM TYPES USED IN LEVEL I CLASSIFICATION

The nature and extent of erosional and depositional processes occurring in the various categories of landforms and valley types produce an array of fluvial and morphological features that can be correlated with stream channel types through the establishment of relationships involving the general features of dimension, pattern, longitudinal profile, and materials of the modern river.

Identification of valley types and related landforms can provide the basis for an initial indication of river morphology. For example, river breaklands and highly dissected fluvial slopes are indicative of steep, narrow, deeply incised, erosional A and G stream types. Narrow, confined canyons and deep gorges within landforms exhibiting low elevational relief characteristically contain "F" stream types. Broad alluvial valleys with well developed floodplains generally indicate the presence of "C" and "E" stream types. Cryoplanated uplands consisting of gentle terrain features, narrow valleys and colluvial slopes normally display relatively stable "B" stream types of both moderate width/depth ratio and

LEVEL I: GEOMORPHIC CHARACTERIZATION

entrenchment. Well vegetated lacustrine meadows with a gentle gradient typically contain the sinuous "E" and "C" stream types, and glacial outwash valleys or plains often exhibit well developed braided or "D" stream types.

The interpretation of stream type as derived from an analysis of landforms and related valley types are reliable. As with all coarse-scale inferences, however, such interpretative analyses should be verified by a "ground-truthing" process that utilizes additional information obtained with a sub-sample of representative landforms. An important supplemental benefit of a Level I classification is that the analysis process requires investigators to become familiar with the landforms and resident stream types occurring within the specific watersheds or river systems under study.

Several methodologies for describing land types and related fluvial features have been published. Wertz and Arnold (1972) developed a method for integrating information and data related to landforms, landtypes, and landtype associations. National Forest Service staff scientists on the Clearwater National Forest, Idaho are refining the Wertz-Arnold method by integrating landform and landtype associations with stream types (Wilson and Jones, personal communication 1993). A similar approach on the Helena National Forest, Montana combining landtype and landform information with stream types has provided an excellent water resource and riparian area inventory, (Stewart and Maynard, personal communication, 1994). Such efforts to combine soil, geomorphic, riparian, and stream type inventory data into an integrated data source have resulted in the development of a significant management and planning tool.

An important distinction of the stream classification hierarchy shown in *Figure 3-1* and *Figure 4-1* is that the determination of stream types are not solely based on the parameters of stream order or physiographic position within the drainage network. During the initial development of the stream type classification process, stream order (Strahler 1952), was considered as a delineative criteria, but was soon abandoned at the initial mapping level. Stream order was not consistent as a morphological indicator, given that a singular reach of a particular stream type may occur within the entire range of stream order values. Stream order, however is used in the Level III analyses, Chapter 6. The schematic landscape illustration, previously depicted in ***Figure 4-3a*** and ***Figure 4-3b***, illustrates how an "A" stream type can occur as a first-order stream at the headwaters and again as a larger order stream, deeply incised in a holocene terrace adjacent to the valley train.

The stream classification system and hierarchical inventory methodology is a physical process-based technique which can be applied to morphologically describe stream systems operating within states of dynamic equilibrium or quasi-equilibrium. The stream classification system contains the basic assumption that stream morphology is invariably fixed to landscape position.

The application of landform, landtype or valley type information to infer fluvial character, process and stream type with a Level I inventory is demonstrated with the following example. A stream type often associated with alluvial fans is the braided or multiple channel "D" stream type. The alluvial fan landform characteristically exhibit a high sediment supply originating from steep, entrenched, erosional drainages immediately upstream of the fan. The slopes of alluvial fans are generally much steeper than the slopes of the valley train they overrun, contain a heterogeneous mixture of unconsolidated sediments, are generally droughty, and thus are "severe" site conditions for riparian vegetation establishment. The "D" stream types located on alluvial fans experience high bank erosion, resulting from poorly established to non-existent streambank vegetation. During large runoff events, coarse sediment is often transported from the steeper, upslope (often an "A" stream type) channel system and dispersed over the fan surface. An accelerated sediment depositional or aggradation process serves to maintain the pattern of the braided channel, which frequently shifts laterally across the fan, in the presence of large quantities of coarse bedload sediment.

LEVEL I: GEOMORPHIC CHARACTERIZATION

When a stream channel becomes entrenched in an alluvial fan, the result is a moderately steep, single-thread, slightly meandering channel with actively eroding, rejuvenating stream banks, known as a "G" stream type. If the channel slope is steeper than .04 (4%), the stream type on the fan would be an "A." The morphology of "A" or "G" stream types, located on an alluvial fan, (*Figure 4-3c*) can be easily distinguished and directly in contrast to the "D" stream type, using appropriate aerial photos and selected topographic maps.

Valley Morphology Related to Stream Types

Valley Type I

A Type I valley is V-shaped, confined, and is often structurally controlled and/or associated with faults. Elevational relief is high, valley floor slopes are greater than 2%, and landforms may be steep, glacial scoured lands, and/or highly dissected fluvial slopes. Valley materials vary from bedrock to residual soils occurring as colluvium, landslide debris, glacial tills, and other similar depositional materials. Stream types commonly observed in valley Type I include types "A" and "G" which are typically step/pool channels with steeper channel slopes exhibiting cascade bed features. Stream channel erosional processes vary from very low and stable to highly erodible, producing debris torrents or avalanches. Often the "A" stream types in certain hydro-physiographic provinces are the starting or conveyance zones for snow avalanches. Examples of the Type I valley are shown in *Figures 4-5a* and illustrated in *Figure 4-5b*.

Valley Type II

Valley Type II exhibits moderate relief, relatively stable, moderate side slope gradients, and valley floor slopes that are often less than 4% with soils developed from parent material (residual soils), alluvium, and colluvium. Cryoplanated uplands dominated by colluvial slopes are typical of the landtypes that generally comprise Valley Type II in the northern Rocky Mountains. The stream type most commonly found in Valley Type II are the "B" types which are generally stable stream types, with a low sediment supply and bed features normally described as "rapids." Less common are "G" stream types that are observed generally under disequilibrium conditions. Examples of Valley Type II are shown with the photograph in *Figure 4-6a* and the illustration in *Figure 4-6b*.

Valley Type III

Valley Type III is primarily depositional in nature with characteristic debris-colluvial or alluvial fan landforms, and valley-floor slopes that are moderately steep or greater than 2%. Stream types normally occurring in Valley Type III are the "A," "B," "G," and "D" types. The "B" stream type which is less common on alluvial or colluvial fans occurs primarily on "non-building" stable fans and where riparian vegetation is well established along the drainage-way. The "G" stream type prevails where there is little established riparian vegetation in the presence of high bedload transport on actively "building" fans, similar to the multiple distributary channels of the "D" stream type. Examples of Valley Type III are shown with the photograph in *Figure 4-7a* and the illustration in *Figure 4-7b*.

Valley Type IV

Valley Type IV consists of the classic meandering, entrenched or deeply incised, and confined landforms directly observed as canyons and gorges with gentle elevation relief and valley-floor gradients often less than 2%. Valley Type IV is generally structurally controlled and incised in highly weathered materials. These stream types are also often associated with tectonically "uplifted" valleys. The "F" stream type is most often found in Valley Type IV, however, where the width of the valley floor accommodates both the channel and a floodplain, C channels are often observed. Depending on streamside materials, the sediment supply is generally moderate to high. Examples of Valley Type IV are shown with the photograph in *Figure 4-8a* and the illustration in *Figure 4-8b*.

Valley Type V

Valley Type V is the product of a glacial scouring process where the resultant trough is now a wide,

LEVEL I: GEOMORPHIC CHARACTERIZATION

FIGURE 4-5a. *Valley Type I,* "V" notched canyons, rejuvenated sideslopes (A and G stream types).

FIGURE 4-6a. *Valley Type II,* moderately steep, gentle sloping side slopes often in colluvial valleys (B stream types).

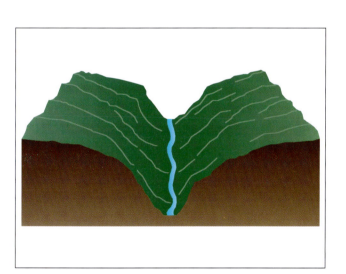

FIGURE 4-5b. *Valley Type I,* "V" notched canyons, rejuvenated sideslopes.

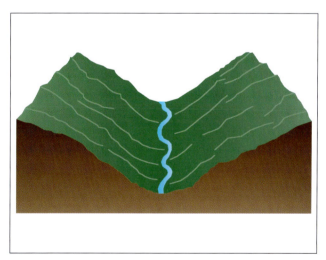

FIGURE 4-6b. *Valley Type II,* moderately steep, gentle sloping side slopes often in colluvial valleys.

LEVEL I: GEOMORPHIC CHARACTERIZATION

FIGURE 4-7a. *Valley Type III,* alluvial fans and debris cones (A, G, D and B stream types).

FIGURE 4-8a. *Valley Type IV,* gentle gradient canyons, gorges and confined alluvial valleys (F or C stream types).

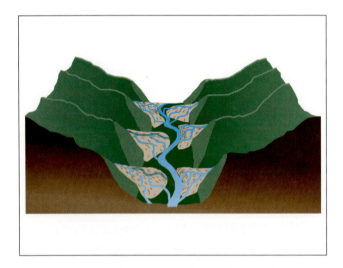

FIGURE 4-7b. *Valley Type III,* alluvial fans and debris cones.

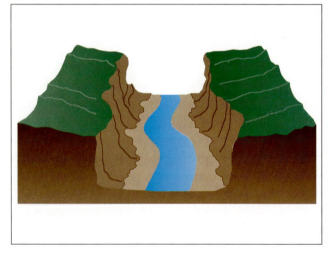

FIGURE 4-8b. *Valley Type IV,* gentle gradient canyons, gorges and confined alluvial valleys.

LEVEL I: GEOMORPHIC CHARACTERIZATION

FIGURE 4-9a. *Valley Type V,* moderately steep valley slopes, "U" shaped glacial trough valleys (D and C stream types).

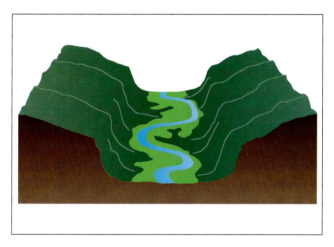

FIGURE 4-9b. *Valley Type V,* moderately steep valley slopes, "U" shaped glacial trough valleys.

"u"-shaped valley, with valley-floor slopes generally less than 4 percent. Soils are derived from materials deposited as moraines or more recent alluvium from the holocene period to the present. Landforms locally include lateral and terminal moraines, alluvial terraces, and floodplains. Deep, coarse deposition of glacial till is common, as are glacio-fluvial deposits, with the finer size mixture of glacio-lacustrine deposition above structurally controlled reaches. The stream types most often seen in Valley Type V are "C," "D," and "G." Examples of Valley Type V are shown with the photograph in *Figure 4-9a* and the illustration in *Figure 4-9b*.

Valley Type VI

Valley Type VI, termed a fault-line valley, is structurally controlled and dominated by colluvial slope building processes. The valley-floor gradients are moderate, often less than 4 percent. Some alluvium occurs amidst the extensive colluvial deposits and stream patterns are controlled by the confined, laterally controlled valley. Sediment supply is low. Stream types are predominantly "B" types with fewer occasions of "C" and "F" types in the wider and flatter valley reaches. Under disequilibrium conditions, "G" stream types are observed. Examples of Valley Type VI are shown in the photograph in *Figure 4-10a* and the illustration in *Figure 4-10b*.

Valley Type VII

Valley Type VII consists of a steep to moderately steep landform, with highly dissected fluvial slopes, high drainage density, and a very high sediment supply. Streams are characteristically deeply incised in either colluvium and alluvium or in residual soils. The residual soils are often derived from sedimentary rocks such as marine shales. Depositional soils associated with these highly dissected slopes can often be eolian deposits of sand and/or marine sediments. This valley type can be observed over a variety of locations, from the provinces of the Palouse Prairie of Idaho, the Great Basin or high deserts of Nevada and Wyoming, the Sand Hills of Nebraska, to the Badlands of the Dakotas. The majority of stream types found in Valley Type VII are the "A" and "G" types which are channels that have moderate to steep gradients, are entrenched (deeply incised), confined, and unstable due to the active lateral and vertical accretion processes. Examples of Valley Type VII are shown in the photograph, *Figure 4-11a* and the illustration in *Figure 4-11b*.

Valley Type VIII

Valley Type VIII is most readily identified by the presence of multiple river terraces positioned laterally along broad valleys with gentle, down-valley elevation relief. Alluvial terraces and floodplains are the predominant depositional landforms which produce a high sediment supply. Glacial terraces can

LEVEL I: GEOMORPHIC CHARACTERIZATION

FIGURE 4-10a. *Valley Type VI*, moderately steep, fault controlled valleys (B, G and C stream types).

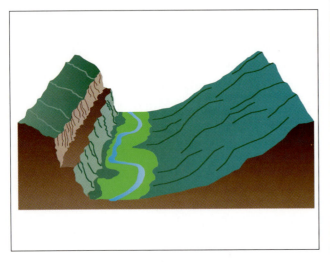

FIGURE 4-10b. *Valley Type VI*, moderately steep, fault controlled valleys.

also occur in these valleys, but stand much higher above the present river than the alluvial (Holocene) terraces. Soils are developed predominantly over alluvium originating from combined riverine and lacustrine depositional processes. Stream types "C" or "E," which have slightly entrenched, meandering channels that develop a riffle/pool bed-form, are normally seen in the Type VIII valley. However, "D," "F," and "G" stream types can also be found, depending on local stream and riparian conditions. Examples of Valley Type VIII are shown in the photograph, *Figure 4-12a* and the illustration in *Figure 4-12b*.

Valley Type IX

Valley Type IX is observed as glacial outwash plains and/or dunes, where soils are derived from glacial, alluvial, and/or eolian deposits. Due to the depositional nature of the developed landforms, sediment supply is high, and the commonly occurring "C" and "D" stream types are associated with high rates of lateral migration. Examples of Valley Type IX are shown in the photograph, *Figure 4-13a* and the illustration in *Figure 4-13b*.

Valley Type X

Valley Type X is very wide, with very gentle elevation relief and is mostly constructed with alluvial materials originating from both riverine and lacustrine deposition processes. Soils are primarily alluvium, and while less common, may also be derived from eolian deposition. Landforms commonly observed as Valley Type X are coastal plains, broad lacustrine and/or alluvial flats, which may exhibit peat bogs and expansive wetlands. Stream types "C," "E," and "DA" are the most commonly observed, although in many instances, where streams have been "channelized," or the local base level has been changed, "G" and "F" stream types are found. Examples of Valley Type X are shown in the photograph, *Figure 4-14a* and the illustration in *Figure 4-14b*.

Valley Type XI

Valley type XI is a unique series of landforms consisting of large river deltas and tidal flats constructed of fine alluvial materials originating from riverine and estuarine depositional processes. The Type XI valleys or delta areas are often seen as freshwater and saltwater marshes, natural levees, and crevasse splays. There are four morphologically distinct delta areas, initially described by Fischer, et al. (1969), which produce different stream types or patterns and include: the elongated, high-constructive delta (*Figure 4-15*); the lobate, high constructive delta (*Figure 4-16*); the wave-dominated, high-destructive delta (*Figure 4-17)*; and the tide-dominated, high-constructive delta (*Figure 4-18*). An additional delta landform is presented

LEVEL I: GEOMORPHIC CHARACTERIZATION

FIGURE 4-11a. *Valley Type VII,* steep, highly dissected fluvial slopes (A and G stream types).

FIGURE 4-12a. *Valley Type VIII,* wide, gentle valley slope with a well developed floodplain adjacent to river terraces.

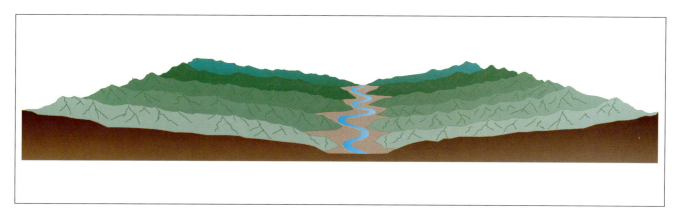

FIGURE 4-11b. *Valley Type VII,* steep, highly dissected fluvial slopes.

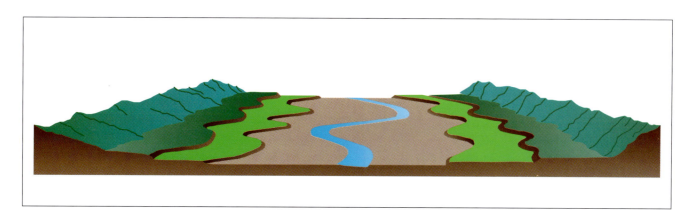

FIGURE 4-12b. *Valley Type VIII,* wide, gentle valley slope with well developed floodplain adjacent to river terraces.

LEVEL I: GEOMORPHIC CHARACTERIZATION

FIGURE 4-13a. *Valley Type IX,* broad, moderate to gentle slopes, associated with glacial outwash and/or eolian sand dunes (Predominately D and some C stream types).

FIGURE 4-14a. *Valley Type X,* very broad and gentle slopes, associated with extensive floodplains - Great Plains, semi-desert and desert provinces; coastal plains and tundra.

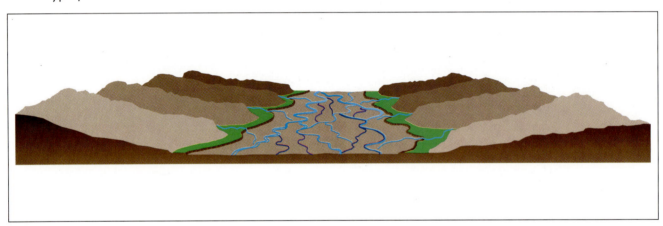

FIGURE 4-13b. *Valley Type IX,* broad, moderate to gentle slopes, associated with glacial outwash and/or eolian sand dunes.

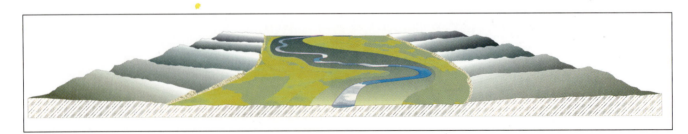

FIGURE 4-14b. *Valley Type X,* very broad and gentle slopes, associated with extensive floodplains - Great Plains, semi-desert and desert provinces; coastal plains and tundra.

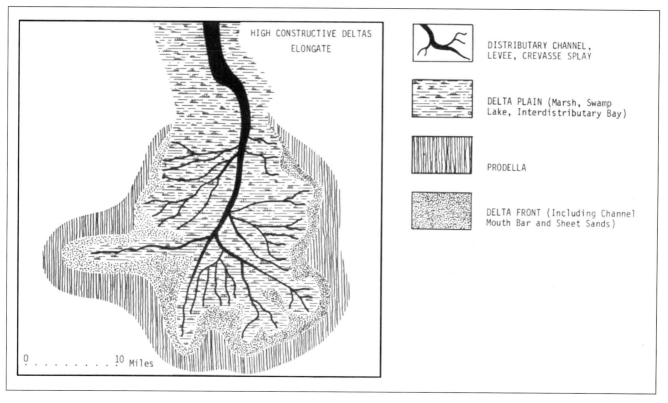

FIGURE 4-15. *Valley Type XI,* Deltas - elongated, highly constructive delta with a distributary channel system. (Fisher et al., 1969)

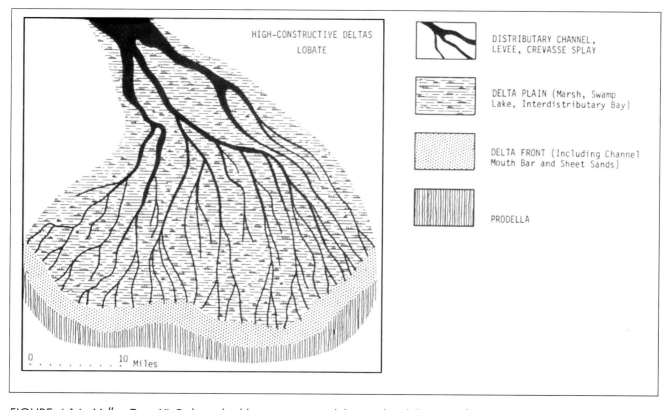

FIGURE 4-16. *Valley Type XI,* Deltas - highly constructive deltas with a lobate configuration and distributary channel system (Fisher et al., 1969)

LEVEL I: GEOMORPHIC CHARACTERIZATION

here, representative of extensive wetlands, peat, and cohesive sediments with multiple, stable channels typical of the "DA" (anastomosed) stream type (*Figure 4-19*).

The corresponding stream types found in delta areas are primarily the distributary channels of stream type "DA," or the multiple channel systems of the "D" stream type, along with occasional "C" and "E" stream types. The "DA" stream type is more common to the delta landforms shown in *Figure 4-19*, which are the tide-dominated, stable deltas with numerous wetland islands, and the base level of the channel system controlled by either lake or sea levels.

DELINEATION METHODS

Classification and delineation of stream types is most effective if one organizes the overall analysis process to first map and identify the origin and character of landforms, and then to overlay the drainage systems of interest. It is also important to recognize that for information about depositional features to be useful, the elevation of existing terraces must be located with respect to elevations of the valley floor. Differences in terrace elevations are used to separate geomorphic features associated with Pleistocene, Holocene and modern depositional features, the products of different climatic regimes. Stream channels incised within landforms of different origins can be observed to have different materials. The following examples of landform/soil relations help in this initial delineation: glacial terraces are typically constructed of large cobble and gravel; glacial outwash terraces normally contain sand and small gravel; Holocene terraces are composed of a stratification of a variety of alluvial materials sizes that range from cobble and gravel/sand to gravel and sand-with the strata material composition varying by the elevation of the particular terrace above the valley floor. Lacustrine valleys characteristically contain finer soils stratified into alternating layers that may range from gravel/sand to silt/clay, with the layers occasionally containing "varves" or lenses of fine silt/clay and organic materials.

The morphological context of the planform, profile, and shape of the river is observed at a broad scale by overlaying a delineation of the system on the fluvial landscape which initially provides: (a) an indication of the general channel slope; (b) an interpretative base for describing channel bed features in terms of a step/pool versus riffle/pool configuration; (c) an estimate of channel shape (i.e., general width/depth ratio categories); (d) an assessment of the channel patterns and profiles that denote the presence of a flood plain, or express apparent entrenchment of the stream within a surrounding landform; (e) a determination of plan-view patterns related to a single versus multiple channel system; and (f) a determination of the general degree of confinement or lateral containment of the river.

Level I stream type classifications are based on geomorphic features that can be interpreted from aerial photography, topographic maps, geologic maps, soil maps, and a strong individual familiarity with the stream systems and landforms within the watershed of interest.

Delineation of Valley Types and Landforms

Delineation of the pertinent Valley Types provides a logical stratification with which to identify the potential number of stream types within a broad area. Identification of the various landforms such as alluvial fans, glacial and/or fluvial terraces, floodplains, hanging valleys, or other erosional/depositional features within the valley type further reduces the possible stream types or stream type combinations. Based on landform and valley type analyses, one can then proceed to develop interpretations as to the stream types present, and their morphological character.

Plan-view Morphology

Plan-views of river patterns are grouped as: ***relatively straight*** ("A" stream types); ***low sinuosity*** ("B" stream types); ***meandering*** ("C" and "F" stream types); ***torturously meandering*** ("E" stream types); and the ***complex stream patterns*** which are the multiple channel, braided ("D" stream types), and anastomosed ("DA") stream types.

LEVEL I: GEOMORPHIC CHARACTERIZATION

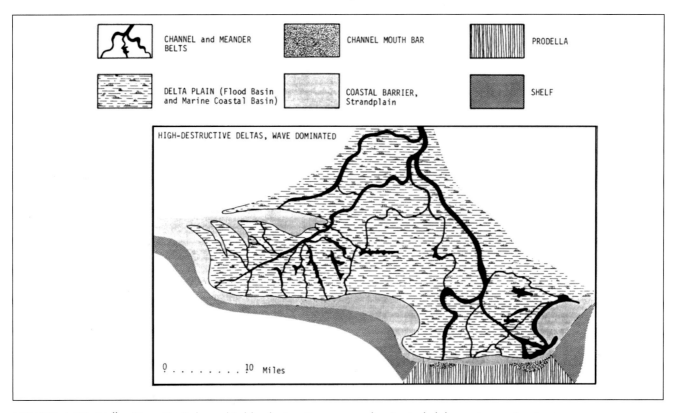

FIGURE 4-17. *Valley Type XI,* Deltas - highly destructive, wave dominated delta (Fisher, et al., 1969)

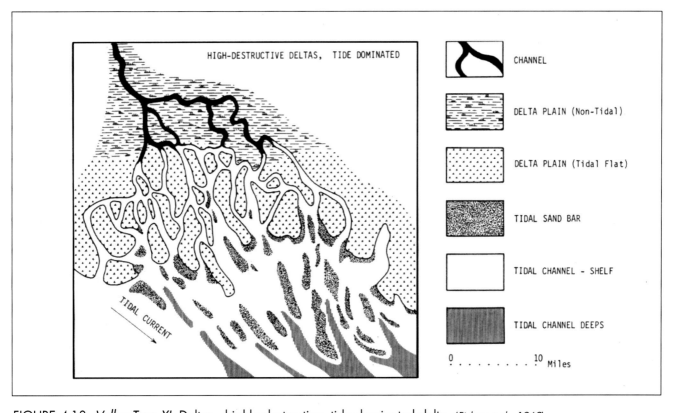

FIGURE 4-18. *Valley Type XI,* Deltas - highly destructive, tide dominated delta. (Fisher et al., 1969)

LEVEL I: GEOMORPHIC CHARACTERIZATION

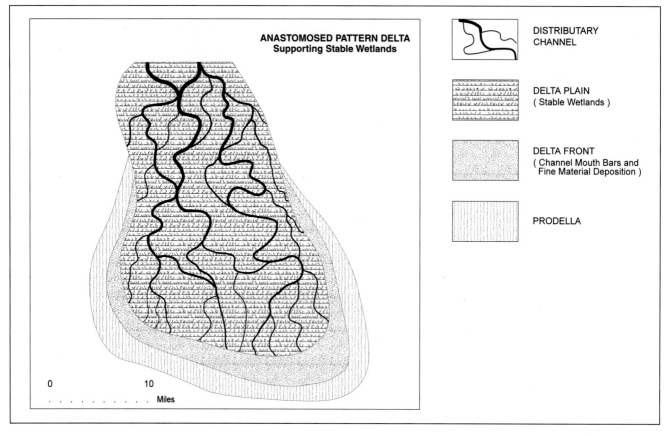

FIGURE 4-19. *Valley Type XI*, Deltas - anastomosed river delta pattern with supporting stable wetlands and channels.

A planform geometry index that was initially developed by Inglis (1942) and later discussed by Lane (1957) was modified and applied to assist in delineating the various stream types at a broad level of inventory. Known as the "meander width ratio," the index is defined as meander belt width divided by bankfull channel width, as defined previously in *Figure 2-3* and *Figure 4-4*. Originally calculated without regard to stream classification, the meander width ratio provided an objective criteria, but had little interpretive value due to the great range of values observed among rivers. Later, the meander width ratio values were stratified by stream type (Rosgen, 1994). The stream type stratifications then allowed the ratio values to serve as an effective index to the degree of channel confinement (lateral containment) and thus improve the process of stream type delineation and interpretation.

Cross-section Morphology

The Level I stream type classification is also based on interpretations as to channel cross-sectional shapes that would distinguish narrow, deep streams from wide, shallow stream types. Channel system characteristics important at Level I are: the degree of channel incision within the parent valley; the location and extent of floodplains; the occurrence and position of terraces; the prevalence of colluvial slopes; the presence of structural control features; the degree of channel confinement and entrenchment; and the overall valley versus channel macro-dimensions.

For example, as shown in *Figure 4-2* and *Table 4-1*, Type "A" streams are narrow, deep, steep, confined, and entrenched. The width of Type "A" channels and their valleys are similar. In contrast, Type "C" streams have a wide, shallow channel with well-developed floodplains established in broad, gently sloped valleys. Type "D" streams are associated with the highest width/depth ratio of all stream types for similar discharges. Often the channel width may occupy the full width of a valley.

The "DA" stream type, however, is associated with a very wide range of cross-section morphology ranging from very low width/depth ratios to very high. The multiple channels which are very stable are often delineated due to this wide range of stable cross-sectional morphology. Type "E" streams have narrow, deep channels and generally occur in wider valleys with well-developed floodplains. Type "F" streams have entrenched wide, shallow, and meandering channels, with little or no adjacent floodplain. Type "G" channels have low width/depth ratios similar to Type "E" streams, except that Type "G" streams are well entrenched, steeper, and less sinuous. The interpretation of patterns and shapes of rivers is of prime assistance to the initial Level I classification of a stream reach, into a particular morphological type.

Channel Sinuosity

Channel sinuosity is the ratio of stream channel length to down-valley distance, or also is the ratio of valley slope to channel slope, and best measured from aerial photographs. With the exception of large rivers, if sinuosity is computed for smaller stream systems using topographic maps, significant underestimations will generally be encountered since the actual channel sinuosity is not accurately transposed in the process of map construction. Topographic maps, however, can be appropriately used to obtain valley slope.

Channel sinuosity which is a primary indicator of stream type, is also an indication of how the stream channel slope is adjusted to that of the valley slope. When measured accurately from aerial photographs, channel sinuosity may also be used to estimate channel slope by dividing valley slope by sinuosity.

Channel Slope

Since the Level I stream channel classification process often does not provide for extensive field measurements (other than for "ground-truthing"), certain morphologic features such as channel slope are estimated on the basis of aerial photograph and topographic map interpretations. For instance, determining valley slope from topographic maps, measuring sinuosity from aerial photographs, and developing the ratio of valley slope/sinuosity provides a generalized estimate of channel slope.

Bed Features

The longitudinal profile, which can be generally inferred from topographic maps, changes with channel sinuosity in the down-valley direction, and serves as the basis for delineating stream reaches into general slope categories that reflect channel profile morphology. Bed features are related to channel slope, and both parameters are integral to delineating stream types. Examples of channel slope categories stratified by stream type are shown in *Figure 4-2* and *Table 4-1*. For example, "Aa+" stream types are very steep (greater than 10% slope), have frequently spaced, vertical-drop, scour-pool bed features, that tend to transport high debris loads and have waterfalls. Type "A" streams are less steep (i.e., 4-10% slope) and have cascading, step/pool bed features. Type "B" streams, with slopes less than 4%, are dominated by riffles, often have a "rapids" configuration with infrequently spaced scour-pools at bends or constricted areas. The "C," "E," and "F" stream types have gentle channel gradients (less than 2%) and have riffle/pool bed features with a spacing sequence proportional to the channel slope and bankfull width. The "DA" stream types also exhibit riffle/pool features, with channel slopes less than 0.5%. Type "G" streams or gullies generally have channel slopes greater than 2% and are typically step/pool channels. Finally, the "D" stream types are the braided or multi-channel systems, with channel slopes of generally less than 4%. Some "D" stream types on alluvial fans can often be steeper than 4%. The braided channel stream type develops convergence and divergence bed features, which tend to have localized, unstable, frequently spaced local scour, and depositional bed forms. Examples of some major stream types within a watershed and

LEVEL I: GEOMORPHIC CHARACTERIZATION

their associated bed features are shown in *Figure 4-20.*

Grant et al. (1990) described bed features such as pools, riffles, rapids, cascades, and steps as a function of bed-slope gradient, illustrated in *Figure 4-21*. The Grant relationship was modified by plotting the same slope categories by stream types "A" through "G." (Rosgen, 1994) As shown in *Figure 4-21*, stream types grouped by slope: riffle/pool stream types "C," "E," and "F" at less than 2% slope; "rapids" dominated stream types "B" and "G" at 2-4% slope; cascading stream type A at 4-10% slopes; and step/pool stream types "A" and "Aa+" at slopes of 10-40%. In common usage, "step/pool" channels cover a wide range of stream types including Aa+, A, B and G. Using the relations shown in Figure 4-21, a more specific bedform description such as "rapids dominated" for B stream types may be more consistent.

Examples of Broad Level Delineation

A broad level characterization of stream types by watershed is shown in the illustration, *Figure 4-22.* This delineation depicts the integration of stream types within valley types, landtypes and landtype associations. To demonstrate procedures and examples of a broad Level I stream channel classification and delineation of the "A" through "G" stream types, a U.S. Geological Survey (7.5 min 1:24000 scale) quadrangle map was obtained for an area along the Colorado and Fraser Rivers in Colorado (*Figure 4-23*). Evident on the maps are landforms found within a wide alluvial valley which include the features of Holocene terraces, floodplains, alluvial fans, and colluvial slopes. Valley morphology varies from the narrow, steep, "v-notch" Valley Type I to the wide, gentle relief of a Valley Type VIII. The Level I mapping can be done readily, with interpretative verification using aerial photography and "ground-truth" (field site validation). The "D," "C," and "E" stream types, as delineated in *Figure 4-24*, are separated primarily on the basis of channel pattern (meandering or braided), sinuosity, and meander width ratio. The observed "G" stream type in a landtype often termed a "fanhead trench," is deeply incised within an alluvial fan. The more stable "B" stream types are seen with colluvial side slopes and moderate channel slopes in narrower valleys (Valley Type II). The "A" stream types are noted as the steep, entrenched channels in the "v-notched," narrow and steep valleys (Valley Type I).

An aerial photograph can significantly provide for additional interpretations through the illustration of river patterns more accurately than topographic mapping particularly for the smaller stream reaches. An example of delineation of stream types is shown with the aerial photo in *Figure 4-24*. A geologic "nick point" in the valley created fine material deposition headward, a process associated with glacio-lacustrine or lacustrine deposits.

These wide, alluvial valleys on gentle gradients are indicative of "E" stream types where the channel plan-view pattern shows a well developed floodplain, very high meander width ratios, and high sinuosity. Downstream of this stream type, the valley confinement changes abruptly as indicated by the change in channel pattern and valley width. The materials in this location would still be depositional, but the gradient is steeper as evidenced by the very low sinuosity. This low sinuosity also suggests an entrenched channel. These variables would lead to the delineation of a "G" stream type.

As the valley narrows, the channel is influenced by colluvial soils as contributed from both valley sides. The gradient, river pattern, and valley morphology indicates a "B" stream type. The "B" stream type continues until the valley morphology, channel pattern (sinuosity), gradient, and nature of deposition change. The new stream morphology associated with these changes is an alluvial, riffle/pool, "C" stream type.

The reach of the "B" stream type is a comparatively stable channel and is an efficient sediment transport reach, where in-channel sediment storage is limited and channel stability is mostly determined by the "structural" control features. The "G" type streams located in the center of the photo in *Figure 4-24* are deeply incised channels in alluvial fans associated with "fanhead troughs." The "G" and a

LEVEL I: GEOMORPHIC CHARACTERIZATION

Waterfalls, chutes and steep vertical steps. (*Stream Type Aa+, Valley type I*)

Cascading, step/pools (*Stream Type A, Valley Type I*)

Riffle Pool (*Stream Type C, Valley Type X*)

Rapids (*Stream Type B, Valley type II*)

Convergence/divergence (*Stream Type D, Valley Type IX*)

FIGURE 4-20. Examples of broad level valley type and stream type delineations and associated bed features of their channels.

LEVEL I: GEOMORPHIC CHARACTERIZATION

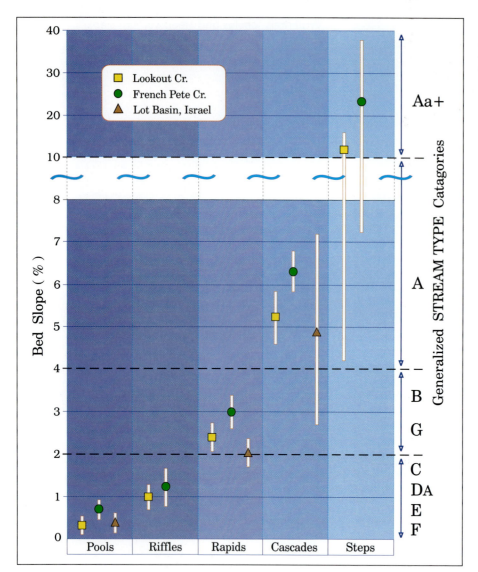

FIGURE 4-21 Relationship of bed slope to bed forms (*from Grant et al., 1990*) for various stream types. (*Rosgen, 1994*).

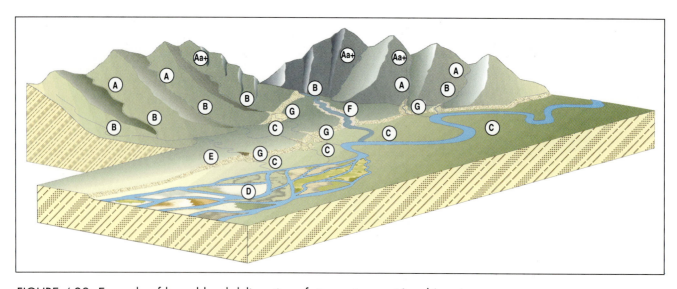

FIGURE 4-22. Example of broad level delineation of stream types at Level I.

LEVEL I: GEOMORPHIC CHARACTERIZATION

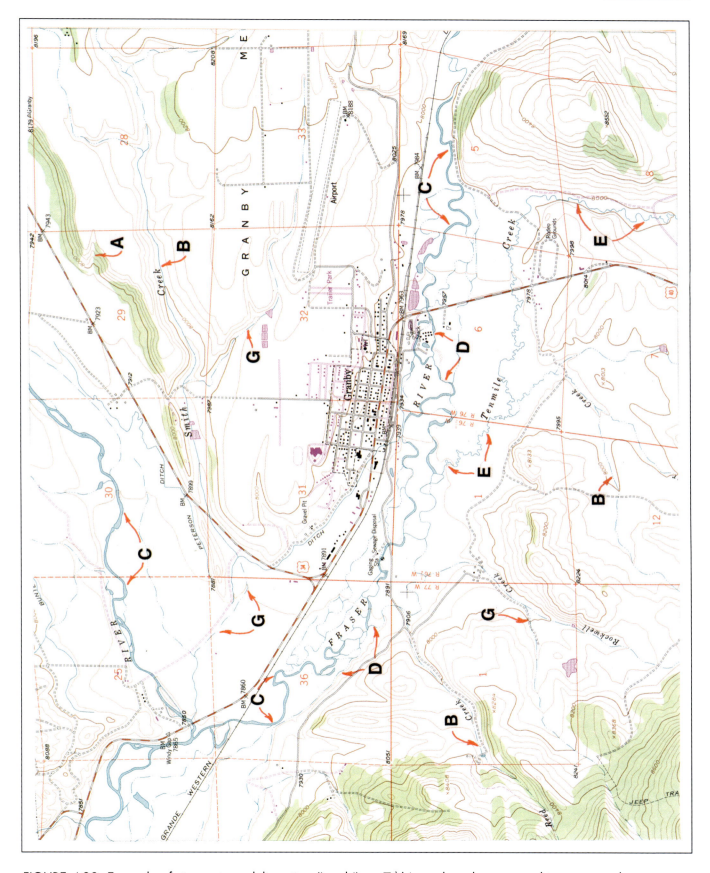

FIGURE 4-23. Example of stream type delineation (Level I) on 7-1/2' quadrangle topographic maps on the upper reaches of the Colorado and Fraser Rivers - Colorado.

LEVEL I: GEOMORPHIC CHARACTERIZATION

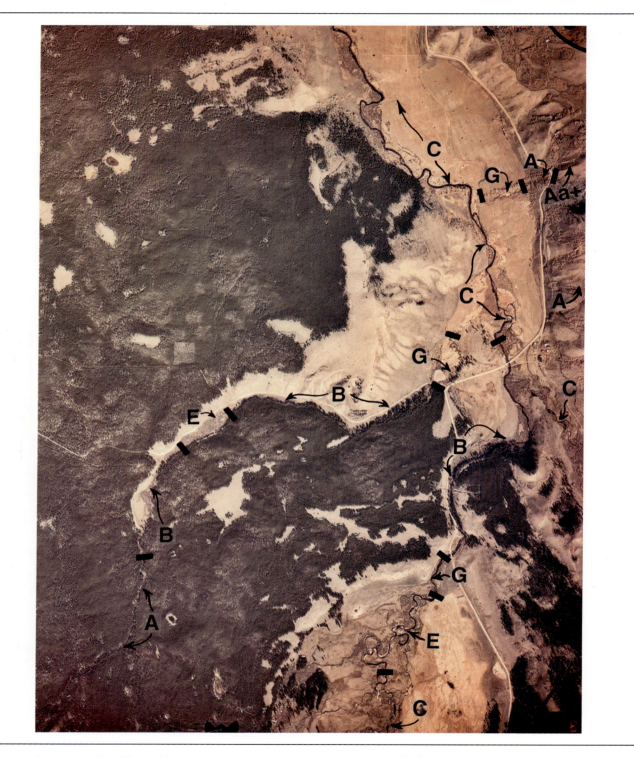

FIGURE 4-24. Example of broad level stream type delineation using aerial photography.

few of the "A" stream types are channels incised in the high, middle, or low Holocene terraces along the mainstem river. The "C" stream types in the alluvial valley are channels with moderate to high sinuosities, well developed floodplains, and gentle channel slopes. The "A" and "Aa+" stream types are observed as tributary rejuvenation channels, partly originating from a change in base-level or lowering of the valley floor over time.

An additional example of a Level I inventory and stream type delineation is presented for an area near Wolf Creek pass in the West Fork San Juan River Basin, Colorado (*Figure 4-25*, USGS Topographic. Map). The area selected exhibits a range of landform slopes and different geology representing volcanism, alpine glaciation, landslides and fluvial erosion and deposition, resulting in the building of a series of depositional terraces and more recent alluvial features related to the modern climate.

The geomorphic characterization delineation from this topographic map produced the following summary: Valley type II, or narrow colluvial valley on slopes less than 4% associated with "B" stream types (aerial photo analysis associated with field verification assists in the accuracy and efficiency of integrating valley types and landforms with the broad level delineation of stream types at the geomorphic characterization level); Valley type III, or alluvial fans and landslide debris associated with "A" and "G" stream types; Valley type V, "U" shaped glacial valleys associated with "C" and "D" stream types; and Valley type VIII, or wide alluvial valleys associated with holocene terraces and a well developed floodplain associated with "C" stream types.

Discussion

Interpretations as to the probable mode of channel adjustment, either in a vertical or lateral direction, or both, and a process oriented perspective of available energy and its distribution can be inferred for the major categories of stream types. Many of the individual classification parameters that are not discrete delineative variables integrate with a Level I inventory to produce an observable morphology. An example is the influence of a deep sod-root mass that normally comprises the floodplain surface of type "E" streams, producing a low channel width/depth ratio, gentle gradient, and a high meander width ratio. Riparian vegetation is not singled out for mapping in the Level I inventory, but its presence and effect is implicit in the resulting interpreted morphology. If the existing, densely rooted mat vegetation is changed, the width/depth ratio and other features may result in adjustments to the higher width/depth ratio of the type "C" stream morphology. Detailed riparian vegetative information is obtained at the "channel state" level (Chapter 6), to provide more detailed information related to the existing channel condition.

Delineating stream types provides an initial sorting of types within large basins and allows a general level of interpretation. Field checking the "remote sensing" mapping effort that utilizes aerial photographs and topographic maps, can lead to proper interpretations. Delineation of stream types at this broad level leads to data organization and the ability to develop a set of analysis priorities for the next, more detailed level of stream classification inventory.

LEVEL I: GEOMORPHIC CHARACTERIZATION

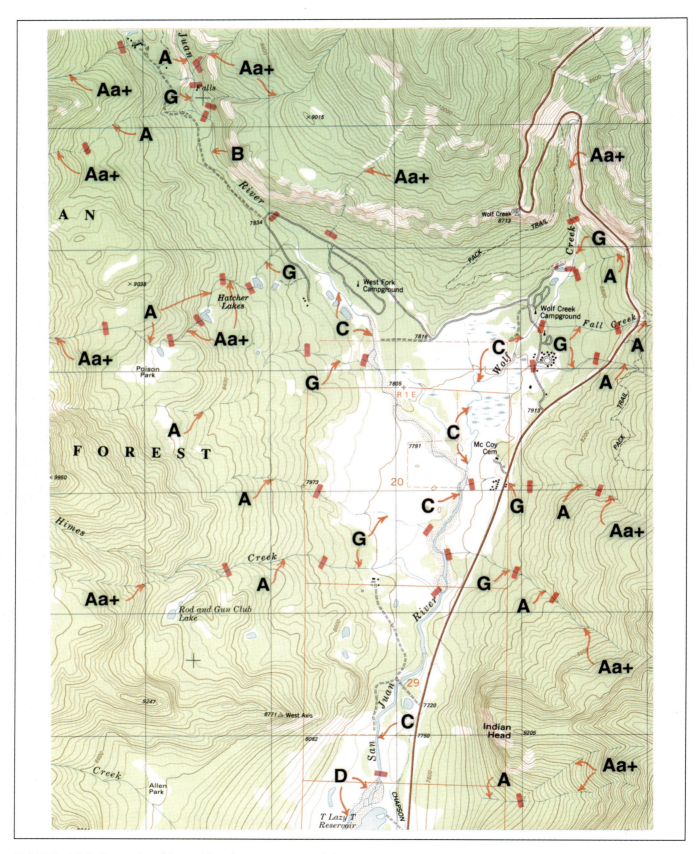

FIGURE 4-25. Example of broad level I stream type delineation using USGS topographic maps in mountainous terrain, Wolf Creek and West Fork San Juan River - Colorado.

4-30

CHAPTER 5
LEVEL II: THE MORPHOLOGICAL DESCRIPTION

"The river is the carpenter of its own edifice."
LUNA B. LEOPOLD (1994)

River and stream morphology is determined by the interplay of the forces acting to create channel dimensions versus the forces resisting that action. River bed features, dimensions, and patterns for rivers influenced by structural controls are naturally different from those systems influenced by alluvial patterns of deposition. While Level I stream types are distinguished primarily on the basis of the valley landforms and channel dimensions observable on aerial photos and maps, Level II stream types are determined with field measurements from specific channel reaches and fluvial features within the river's valley.

The Level II classification processes employ more finely resolved criteria in order to address questions of sediment supply, stream sensitivity to disturbance, potential for natural recovery, channel response to changes in flow regime, and fish habitat potential. These questions require, at a minimum, interpretations based on data and information developed at least to the resolution of a Level II classification.

This chapter is presented in 6 parts: (1) Level II delineative criteria; (2) the channel reference reach; 3) the role of bankfull discharge; 4) the principle of continuity in channel form; (5) field methods for

LEVEL II: THE MORPHOLOGICAL DESCRIPTION

stream type delineation; and (6) morphological descriptions and examples of specific stream types.

LEVEL II DELINEATIVE CRITERIA

Level II stream type delineation criteria are based on the following characteristics of channel cross-section, longitudinal profile, and planform features as measured and computed from collected field data.

Cross-section

Entrenchment Ratio: A computed index value which is used to describe the degree of vertical containment of a river channel (width of the flood prone area at an elevation twice the maximum bankfull depth/bankfull width).

Width/depth Ratio: An index value which indicates the shape of the channel cross-section (ratio of bankfull width/mean bankfull depth).

Dominant Channel Materials: A selected particle size index value, the D50, representing the most prevalent of one of six channel material types or size categories, as determined from a channel material size distribution analysis.

Longitudinal Profile

Slope: Slope of the water surface averaged for 20-30 channel widths.

Bed Features: Secondary delineative criteria describing channel configuration in terms of riffle/pools, rapids, step/pools, cascades and convergence/divergence features which are inferred from channel plan form and gradient.

Plan-form (pattern)

Sinuosity: Defined as stream length/valley length or valley slope/channel slope).

Meander Width Ratio: A secondary delineative criteria defined as meander belt width/bankfull width that describes the degree of lateral channel containment, and is primarily used in assisting aerial photo delineation of stream types.

The hierarchy of river inventory shows the morphological characterization of Level II, covered in this chapter in relation to the other levels (*Figure 5-1a*). The flowchart in *Figure 5-1b* shows how these selected criteria are used to conduct a Level II stream type classification. *Figure 5-2* illustrates the representative channel cross-section configurations, channel materials, and primary morphologic criteria for the full set of 41 stream types. As shown in *Figure 5-2*, the nine Level I, or major stream types are refined by the addition of six categories of channel materials (bedrock through silt and clay); and by quantitative criteria for entrenchment, sinuosity, width/depth ratio, and water surface slope.

It may at first seem unnecessary and difficult to distinguish an individual sub-category stream type from among the total of forty-one stream types. To assist in the delineation, *Figure 5-3* provides a dichotomous key for efficient "typing" of stream reaches illustrating how the array of morphological criteria can be used to quickly identify the appropriate stream types for further consideration. For example, as shown in *Figure 5-3,* a single thread channel that is entrenched can only be a sub-category of an "A," "G," or "F" stream type; and the other 23 potential Level II types need not be considered. Moreover, Level II stream type distinctions are essential for interpreting those significant differences in stream condition and stream response to disturbance, that are partially masked at the Level I classification.

THE REFERENCE REACH

The morphological variables used to define stream types at Level II can and do change within short distances along a river channel, due to changes in geology and tributary influence. A Level II classification may apply to a reach that is only a few tens of meters in length, or to a reach distance of several kilometers. It is important to note that data from individual channel reaches are not averaged over entire basins. Extrapolations necessary to describe the variety of stream types that may exist within a broad area are instead based upon Level II field measurements taken from selected "reference" reaches. Interpretations developed on the basis of data and analysis related to the reference reach can then be extrapolated to other similar reaches, where such detailed data is not readily available. Level II classification data can be easily extrapolated from reference reaches to other areas identified as having

LEVEL II: THE MORPHOLOGICAL DESCRIPTION

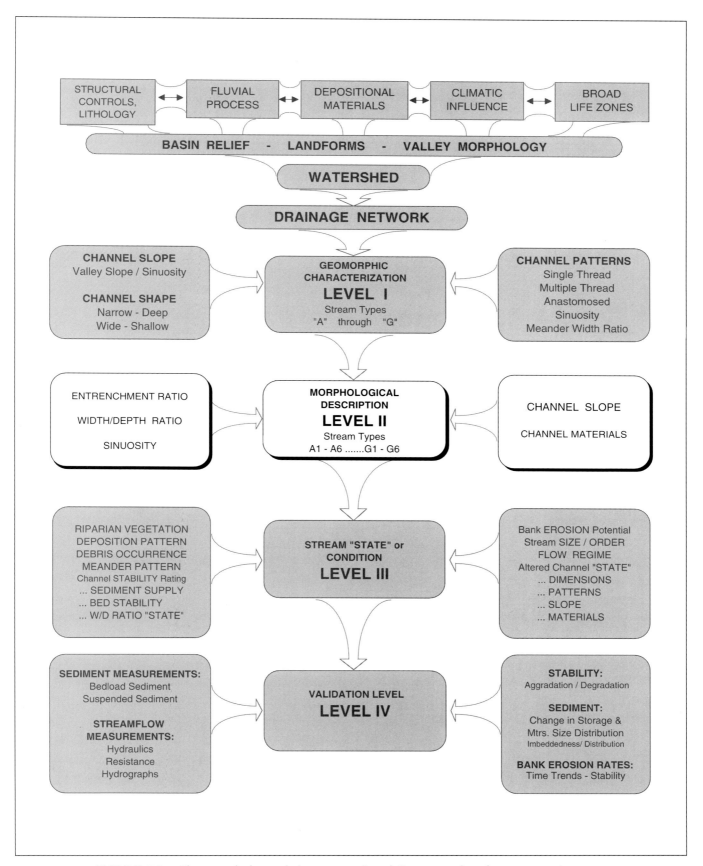

FIGURE 5-1a. The morphological description (Level II) stream classification in relation to the hierarchical river inventory levels

LEVEL II: THE MORPHOLOGICAL DESCRIPTION

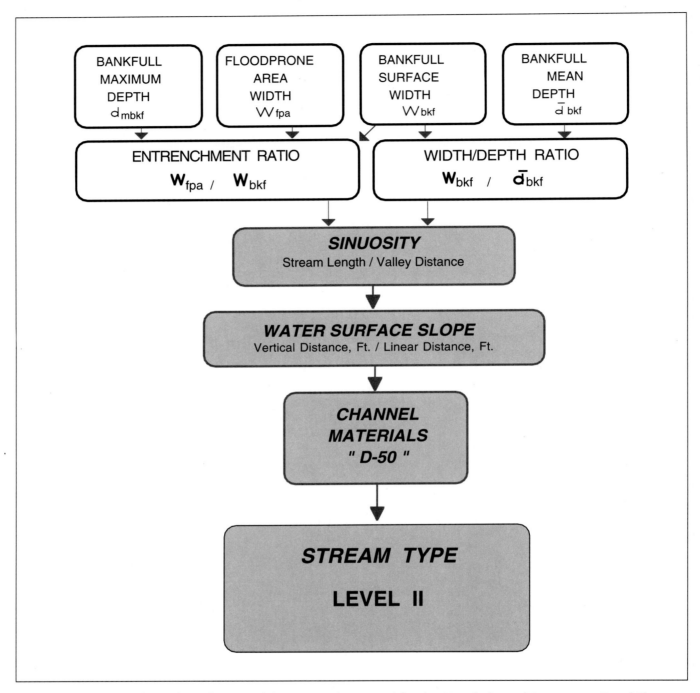

FIGURE 5-1b. Flow Chart showing delineative criteria used for the Morphological Description (Level II).

LEVEL II: THE MORPHOLOGICAL DESCRIPTION

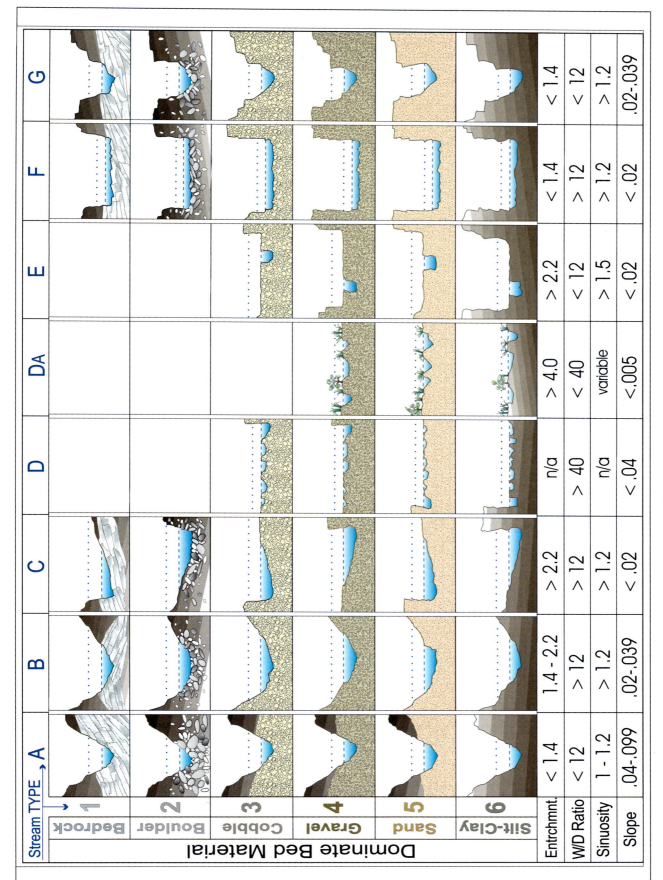

FIGURE 5-2. Primary delineative criteria for the major stream types.

LEVEL II: THE MORPHOLOGICAL DESCRIPTION

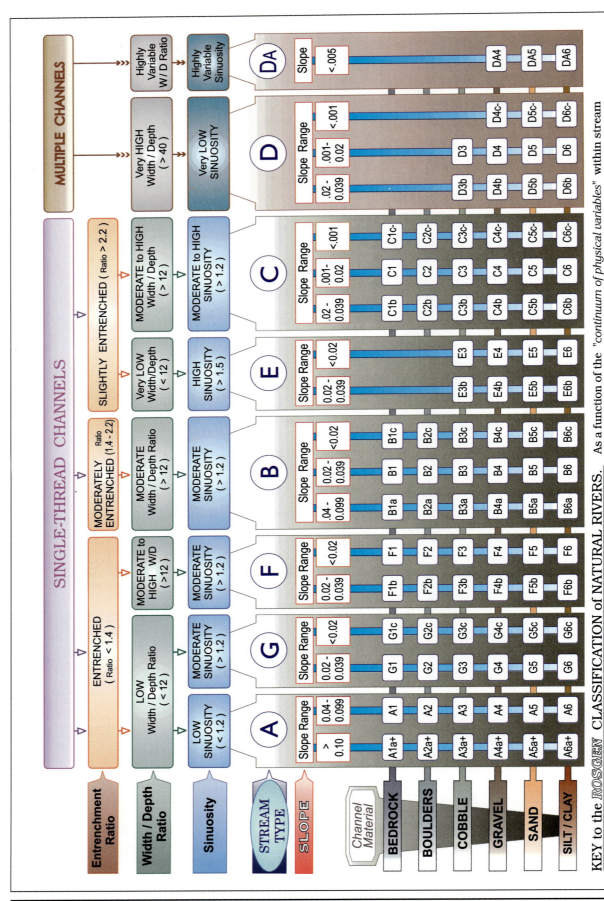

FIGURE 5-3. Classification key for natural rivers.

LEVEL II: THE MORPHOLOGICAL DESCRIPTION

similar valley and lithological types through the use of aerial photography or topographic, and geologic maps.

The number and location of reference reaches are determined initially by information obtained through a Level I survey. There should be at least one Level II reference reach established for each Level I stream type identified. In practice, multiple reference reaches are usually required, one for each change in channel material encountered for a given Level I stream type e.g., establish a reference reach for a "C3" type (D50 particle size is cobble) and for a "C4" type (D50 particle size is gravel).

The use of a reference reach concept enables Level II stream type classifications to be completed for other similar areas without requiring extensive on-site data collection. It is essential, however, to: (1) calibrate the bankfull channel dimensions at gaging stations for representative stream types within the particular hydro-physiographic regime; (2) use aerial photography and topographic maps for preliminary identification of stream types; and (3) measure and analyze representative or reference reaches to verify the initial interpretive classifications. It is especially critical to verify and calibrate the Level I stream type classifications with additional data from the selected reference reaches, and to subsequently re-calibrate stream type interpretations with field data as variations in landform and associated new stream types are observed. Not only will the accuracy and speed of delineation skills improve, but the reference reach database for future classifications and validation monitoring will be greatly enhanced.

THE CONTINUUM OF CHANNEL FORM

All Level II stream types can be referenced to a continuum of channel form. *Figure 5-3* shows the range of typical values for each delineative criteria. A change in stream type is generally signaled when the measured values for the delineative criteria fall outside the range of typical values. Exceptions occur infrequently, where the value for one criterion falls outside the range characteristic of a particular stream type.

Occasional deviations of measured values from the average values are described by simple subclasses. For example, *Figure 5-3* shows that slope values for "A" stream types are not lumped together in the range of 4 to >10% as with the broader Level I classes, but are described as the following subclasses: a+, for slopes steeper than 0.10; a, for slopes of 0.04-0.10; b, for slopes of 0.02 to 0.039; c, for slopes less than 0.02; and c-, for slopes less than 0.001. Adjustment values to accommodate the continuum of possible values for channel entrenchment, sinuosity, and width/depth ratio are also shown in *Figure 5-3*. Measured values for entrenchment and sinuosity can vary by ± 0.2 units; while values for width/depth ratios can vary by ± 2.0 units.

The continuum adjustment allows the initial Level I classes to propagate systematically throughout the inventory and classification hierarchy without a break in stream type (i.e., from Level I to Level IV). The characteristic of process consistency is helpful for subsequent interpretations of stream response, since date obtained at finer resolutions can then be extrapolated to coarse-level delineations, e.g., Level II. The continuum adjustment concept accommodates all of the stream types that have been observed and measured by independent users of this stream classification procedure.

THE ROLE OF BANKFULL DISCHARGE

The stage or elevation of bankfull discharge is the single most important parameter used in Level II classifications. As discussed in Chapter 2, the stage of bankfull discharge is related to channel dimensions such as width, and channel patterns such as meander length, radius of curvature, belt width, meander width ratio and amplitude. Moreover, the bankfull channel width is required to estimate two of the five primary Level II criteria (i.e., entrenchment ratio and width/depth ratio). Correct and reliable interpretations of the interrelationships between dimension, pattern, profile, and streamflow depends upon the correct field identification of bankfull stage and the related discharge. Bankfull

LEVEL II: THE MORPHOLOGICAL DESCRIPTION

discharge as described in Chapter 2 represents the upper level of the range of channel-forming flows which transport the bulk of the available sediment over time (Wolman and Miller 1960). It is also the flows that take care of the annual maintenance of transporting sediment supplied from upstream sources, forming and removing bars, forming or changing bends, and generally doing work that results in the average morphologic characteristics of the channel (Dunne and Leopold, 1978). Similarly, the bankfull discharge has a frequent recurrence interval that can be consistently determined using geomorphic evidence and data obtained in the field. Often that morphologic feature described as "top-of-bank" is confused with the elevation of the bankfull stage, where low terraces (abandoned floodplains) are present and closely adjacent to the channel.

Field Determination of Bankfull Stage

The most consistent bankfull stage determination is obtained from identification of the top of the floodplain. This elevation is where incipient flooding begins for those flows that extend above the bankfull stage. Many floodplains are constructed as the river moves laterally, away from established point bars. The elevation of the top of point bars and the bankfull stage thus share a common elevation that is directly related to the development of floodplains within the valley, given the current climate regime. It is important for the field observer to recognize the physical and morphological difference between a low terrace and floodplain, since alluvial channels with well developed floodplains can often have low elevation river terraces (abandoned former floodplains) adjacent to the channel. For those stream types that exhibit a well-developed floodplain, - such as the "C," "D," "DA," and "E" types - the bankfull stage is easily and reliably identified as the elevation of the floodplain.

Where floodplains are not well developed, the identification of bankfull stage must be determined by field stage indicators that may be combined as corroborating evidence for an indication of a consistent and common elevation. The appropriate use of any or all of the bankfull stage indicators requires adherence to four basic principles:

(1) Seek indicators in the locations appropriate for specific stream types.
(2) Know the recent flood and/or drought history of the area to avoid being mislead by spurious indicators (e.g., colonization of riparian species within the bankfull channel during drought, or flood debris accumulations caught in willows that have rebounded after flood flows have receded).
(3) Use multiple-indicators wherever possible for reinforcement of a common stage or elevation.
(4) Where possible, calibrate field determined bankfull stage elevations and corresponding bankfull channel dimensions to known recurrence interval discharges at gaged stations. This procedure can verify the difference between the floodplain of the river and a low terrace.

There are several visual or physical indicators of the bankfull stage that enable field determination of this important parameter for areas where streamflow records are not available. The bankfull stage indicators vary in their importance and discriminating power for different stream types. A partial listing of these indicators follows:

(a) The presence of a floodplain at the elevation of incipient flooding.
(b) The elevation associated with the top of the *highest* depositional features (e.g., point bars, central bars within the active channel). These depositional features are especially good stage indicators for channels in the presence of terraces or adjacent colluvial slopes.
(c) A break in slope of the banks and/or a change in the particle size distribution, (since finer material is associated with deposition by overflow, rather than deposition of coarser material within the active channel).
(d) Evidence of an inundation feature such as small benches.
(e) Staining of rocks.
(f) Exposed root hairs below an intact soil layer indicating exposure to erosive flow.
(g) Lichens and - for some stream types and locales - certain riparian vegetation species.

LEVEL II: THE MORPHOLOGICAL DESCRIPTION

Before selecting a reference reach within a channel for the purpose of obtaining bankfull cross-section data, it is considered essential to survey a longitudinal profile for at least 20 bankfull channel widths, both upstream and downstream of the reach of interest, to determine the nature and presence of representative bankfull stage indicators for that reach. When corroborating, features and/or representative stage indicators have been identified, then consistent measured values for bankfull channel width can be determined. Identification of corroborating bankfull stage indicators on a reach basis is a mark of good field technique, and greatly improves the reliability of resulting field determined bankfull stage estimates. Again, it is important to first measure bankfull cross-sections and longitudinal profiles at gaged station reaches. This allows one to calibrate and compare the interpretations of geomorphic bankfull stage features to known streamflow and the corresponding return period.

Appropriate locations for the determination of bankfull channel widths are summarized for riffle/pool and step/pool channels, as shown in *Figure 5-4* and *Figure 5-5*, respectively. In general, for all stream types, the best location to measure bankfull channel width is within the narrowest segment of the selected reach, where the channel can freely adjust its lateral boundaries under existing streamflow conditions. The plan form position of the narrowest reach segment varies by stream type. For example, the best locations for determining bankfull channel dimensions are at the riffle or "cross-over" reach of "C," "E," and "F" stream types (Figure 5-4); within the middle of the "rapid" reach for "B" stream types; and the narrow width of the transition reach as it extends from the "step" into the head of the pool for "Aa+," "A," and "G" stream types (*Figure 5-5*). Deflectors such as rocks, logs, other debris, nickpoints or unusual constrictions that make the stream especially narrow, or that create exceptionally wide backwater conditions must be avoided. Additional examples of field methods for locating and measuring the bankfull stage of riffle/pool channels are summarized by Lowham (1976), and described for step/pool channels by Osterkamp (1994 pers. commun.).

Many species of riparian plants are widely distributed, occurring across a variety of hydro-physiographic provinces. Using vegetation to identify bankfull stage must be done cautiously, since some species and age classes can establish themselves on suitable substrate well within the boundaries of the bankfull channel. The elevation of bankfull stage is frequently underestimated when determined solely on the basis of vegetation, and such unilateral determinations should be avoided. Nonetheless, some common riparian species can be reliably used as indicators of bankfull stage, such as certain mature species of birch (Betula spp.), dogwood (Cornus spp.), and alder (Alnus spp.), which tend to consistently colonize and become established at levels very close to the bankfull stage. Similarly, mature alders generally are not found within the bankfull channel, unless the supporting bank materials have become undercut and "slumped" into the channel area. Conversely, smaller woody plants, grasses, and forbs can colonize on suitable substrate areas within the bankfull channel, especially during periods of drought or low flow as noted with certain species and age classes of willows (Salix spp.). Such species should not be used as reliable indicators of the bankfull stage. In cases where there is little choice but to use riparian vegetation as a stage indicator, it is best to seek the advice of a riparian ecologist familiar with the study area and to verify the relations of these species to stream stage at gaged sites within the same hydro-physiographic province.

Calibrating Bankfull Stage to Known Streamflows

A common error in the Level II classification process is the failure of field observers to calibrate the elevations of appropriate field indicators of bankfull stage to known streamflows. Such calibration is essential until one gains sufficient field experience in a given locale to be sure of the proper interpretation of those indicator features representing the stage or elevation of the bankfull discharge.

LEVEL II: THE MORPHOLOGICAL DESCRIPTION

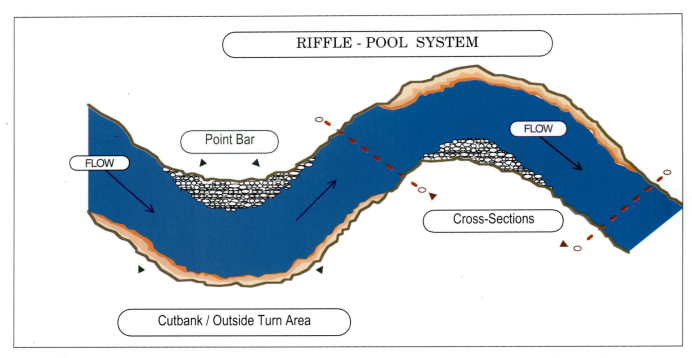

FIGURE 5-4. Recommended cross-section locations for bankfull stage measurements in "riffle/pool" systems.

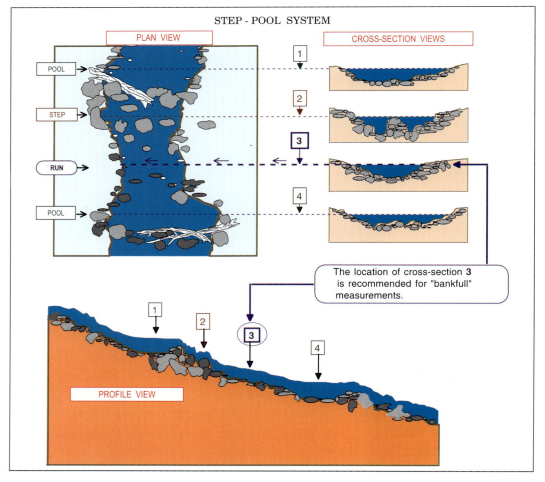

FIGURE 5-5. Recommended location for measurement of bankfull stage in "step/pool" systems.

LEVEL II: THE MORPHOLOGICAL DESCRIPTION

TABLE 5-1. Checklist of recommended Procedure @ USGS Gage or other streamflow measurement locations.

A. Describe Site
1. Geomorphic Setting - Valley Types I through XI. See chapter 4 for descriptions.
2. Channel Materials (Pebble Count) (D16, 35, 50, 84, 95)
   a. Bed Material - Pebble Count
   b. Bank Material - Pebble Count and core sample
   c. Bar Material - core sample
3. Locate on Topographic Map
4. Photo Document - Up/Downstream
5. Compute percentage of Watershed Hydraulically Impacted

B. Longitudinal Profile
1. Measure Average Water Surface Slope
   a. Riffle Slope
   b. Pool Slope
2. Measure Valley Slope
3. Sequence of Riffle/Pool or Step/Pool as a function of Bankfull Width
4. Locate Bankfull Stage along Longitudinal Profile

C. Plan View
1. Measure Sinuosity (SL/VL) (VS/CS), where: SL=stream length; VL=valley length; VS=valley slope; and CS=channel slope
2. Meander Geometry
   a. Meander Length (LM)
   b. Belt Width (BW)
   c. Radius of Curve (RC)
   d. Meander Arc Length (ML)
   e. Meander Width Ratio

D. Cross-Section (Dimension)
1. Cross-section of Channel + Valley Features - Terrace/Floodplain (to be identified on cross-section plot).
   a. Bankfull Width (W_{bkf})
   b. Bankfull Mean Depth (d_{bkf})
   c. Bankfull Maximum Depth (d_{mbkf})
   d. Flood Prone Area Width (W_{fpa})
   e. Entrenchment Ratio (W_{fpa}/W_{bkf})
   f. Bankfull Cross-sectional Area (A_{bkf})
   *g. Bankfull Velocity (U_{bkf})
   h. Estimated Bankfull Discharge (Q_{bkf})

Estimate from various sources.

2. Calibrate Bankfull Estimates
   a. Survey Estimated Bankfull Stage
   b. From gage plate, extrapolate stage reading associated with estimated "Bankfull"
   c. Read Discharge from Rating Curve @ Gage (Stage/Discharge Relation)
   d. Determine recurrence interval in years from flood frequency curves at station. (Should be 1-2 years or average of 1.5 year Q).
   e. Analyze hydraulic geometry data from 9-207 forms (discharge notes) for width, depth, velocity and cross-sectional area vs. stream discharge. Plot data on logalog paper and run a regression to obtain slope and intercept values for each variable.
   f. Develop dimensionless hydraulic geometry relations. This is to be applied for extrapolation purposes to rivers of the same stream type, but for various sizes. W/W_{bkf} vs Q/Q_{bkf} (Complete for depth, velocity and cross-sectional area).

LEVEL II: THE MORPHOLOGICAL DESCRIPTION

The recommended procedure for calibrating field identified bankfull stage with known streamflows and return period is as follows:

(1) Locate all current and discontinued stream gaging stations within the study basin and/or in nearby similar basins.

(2) Make a field visit to each station to collect supplemental data which will be needed to interpret existing hydrologic records at each station. Note that these field visits are not an unnecessary extravagance, nor are they likely to be a major time encumbrance.

FIGURE 5-6a. Jefferson Creek, Oregon - @ gaging station (27.8 sq. mi.) Entr. ratio: 2.2, W_{bkf}: 32.5', d_{bkf}: 2.2', W/D: 14.8, Sinuosity: 1.25, Channel materials: D-15,12$_{mm}$, D-34,28$_{mm}$, D-50,48$_{mm}$, D-84,94$_{mm}$, Stream Type C4.

Investigators will be fortunate if they can find a half-dozen gaging stations within a selected study area, and often it may be necessary to travel outside the area of interest to obtain representative data for extrapolation.

(3) The use of information in *Table 5-1* will serve as a check list for procedures to be performed at the gaged site. A portion of the data to be collected is not entirely necessary for stream classification, but is necessary to perform the sediment and hydraulic analyses described in Chapter 7.

(4) *Table 5-2* provides a form for recording gaged and field data. The primary uses of these data involve the calibration of field-estimated bankfull stage to a corresponding measured record). To this available published data, one must add their own collected field-data, describing the reference reach slope, particle size distribution, bankfull stage and channel characteristics, hydraulic geometry, and, of course, stream type.

All gaging stations normally will have a permanent benchmark installed for the purpose of maintaining an elevation control and referencing to a corresponding discharge measurement cross-section. The established gage station cross-section should be re-surveyed and expanded laterally to include the active floodplain, low terraces, and other valley features of interest.

Two examples of field identification of bankfull stage and subsequent verification using gaging station data are illustrated in *Figure 5-6a* and *Figure 5-6b*. The bankfull stage line in *Figure 5-6a*, for Jefferson Creek, Oregon, was determined by two coincident indicators: (1) the presence of a stand of mature alders, and (2) a major break in slope. The bankfull stage elevation was surveyed on the gage station staff plate to obtain the corresponding discharge as determined from the stage/discharge relationship developed for that gage station. The recurrence interval for the field-determined bankfull stage and attendant bankfull discharge was estimated using data developed with a Log-Pearson type flood frequency analysis. This comparative analysis procedure indicated that the selected bankfull stage produced a bankfull discharge that had a 1.3 year return period, which is a

LEVEL II: THE MORPHOLOGICAL DESCRIPTION

TABLE 5-2. Sample form for recording gage station and field data.

SUMMARY of USGS GAGE DATA/RECORDS for STREAM CHANNEL CLASSIFICATION

Station NAME: _____   Station NUMBER: _____
LOCATION; _____
Period of RECORD: _____ Yrs.   Percentage of Watershed Hydraulically Impacted _____ %
Drainage AREA: _____ Ac. _____ Sq.Mi.   Drainage MEAN ELEVATION: _____ Ft.
Reference REACH SLOPE: _____   STREAM TYPE: _____

BANKFULL CHARACTERISTICS

Determined by FIELD MEASUREMENT
Bankfull WIDTH _____ Ft. (W_{bkf})
Bankfull MEAN DEPTH _____ Ft. (d_{bkf})
Bankfull STAGE _____ Ft.
Bankfull Xsec. AREA _____ Sq. Ft. (A_{bkf})
Wetted PERIMETER _____ Ft. (WP)
Est. Mean VELOCITY _____ Ft./Sec. (V_{bkf})
Est. Bankfull DISCHARGE _____ Cfs (Q_{bkf})

Determined by GAGE DATA Analyses
Bankfull WIDTH _____ Ft. (W_{bkf})
Bankfull MEAN DEPTH _____ Sq. Ft. (d_{bkf})
Bankfull STAGE _____ Ft.
Bankfull Xsec. AREA _____ Sq.Ft. (A_{bkf})
Wetted PERIMETER _____ Ft. (WP)
Bankfull Mean VELOCITY _____ Ft./Sec. (V_{bkf})
Bankfull DISCHARGE _____ Cfs (Q_{bkf})

Bankfull DISCHARGE associated with *"field determined"* Bankfull STAGE: _____ Cfs. (Q_{bkf})
(From Gage Height reading at Staff Plate and tabular Stage-Discharge Curve Data)

Recurrence Interval (Log-Pearson) associated with *"field determined"* Bankfull Discharge
R.I.= _____ Years

From the <u>Annual Peak Flow Frequency Analysis</u> data for the <u>Gage Station,</u> determine:
    1.0 Year R.I. Discharge= _____ CFS
    1.5 Year R.I. Discharge= _____ CFS
    2.0 Year R.I. Discharge= _____ CFS

MEANDER GEOMETRY

Determined by FIELD MEASUREMENT

Meander Length (L_m) _____ Ft.   Radius of Curvature (R_c) _____ Ft.
Belt Width (W_{blt}) _____ Ft.   Meander Width Ratio (W_{blt}/W_{bkf}) _____

HYDRAULIC GEOMETRY

Based on <u>USGS Discharge Summary Notes</u> data (Form 9-207) and regression analyses of measured discharge (Q) with the hydraulic parameters of Width (W), Area (A), Mean Depth (d), Mean Velocity (V); determine the *intercept coefficient* (a) and the <u>slope exponent</u> (b) values for a power function of the form $Y = aX^b$; when Y is one of the selected hydraulic parameters, and X is a given discharge value (Q).

	Width (W)	Depth (d)	Area (A)	Velocity (V)
Coefficient (a)				
Slope Expn. (b)				

Hydraulic Radius (R=A/WP) _____ Ft.   Manning's "N" (Rough. Coeff.) at Bankfull Stage _____.
"N"=1.486 / Q_{bkf} [(Area) (Hydraulic Radius ^2/3) (Slope ^1/2)]

5-13

LEVEL II: THE MORPHOLOGICAL DESCRIPTION

FIGURE 5-6b. East Fork River, Colorado - @ gaging station
Entr. ratio: 6.0, W_{bkf}: 50', d_{bkf}: 2.5', W/D: 20. Channel materials: D-15,12$_{mm}$, D-34,28$_{mm}$; D-50,58$_{mm}$, D-84,120$_{mm}$. Stream Type C4.

A field guide for bankfull stage determination and conducting a stream channel survey was recently published by the USDA Forest Service, (Harrelson et al, 1994). This guide is very helpful in describing stream survey methods, bankfull stage surveys, pebble counts, and other channel inventory methods. A video produced by the USFS (1994) is also very helpful in selecting bankfull stage.

reasonable estimate of a frequently occurring flow, confirming the reliability of the bankfull indicators at this site.

The bankfull stage was determined for the East Fork of the San Juan River (***Figure 5-6b***) by the direct observation of the elevation of the active floodplain. The starting elevation of the floodplain was also linked to the same elevation as a resident stand of mature alders, a break in slope, and the top of the point bar. These "indicators" could be used for bankfull stage determination for a stream type where a floodplain may not be present. The recurrence interval associated with the bankfull discharge corresponding to the bankfull indicator and feature elevations of the East Fork San Juan River, as verified by gage data was 1.45 years. It is important to note the similarities in recurrence intervals for bankfull flow in these two very different hydro-physiographic provinces.

Examples of bankfull stage are shown in ***Figure 5-7*** and ***Figure 5-8*** for a variety of stream types in locations that range from Texas to Alaska. In these examples, field evidence that supports the location of bankfull stage is described for each photograph.

Drainage Area vs. Bankfull Channel Dimensions by Stream Type

Bankfull channel dimensions of cross-sectional area, width, mean depth, and the related streamflow velocities tend to increase linearly with increases in drainage area (Leopold et al, 1964). When stratified by stream type, plots of bankfull channel dimensions prove even more useful for estimating similar channel dimensions for ungaged areas. For example, selected average bankfull channel dimensions for four hydro-physiographic regions are shown in ***Figure 5-9*** (Dunne and Leopold, 1978). The majority of these streams are alluvial channels. The bankfull channel dimensions that have been collected and field-calibrated at stream gages should be plotted similarly in order to build a supporting database useful for refining estimates of bankfull channel dimensions for ungaged areas, (Rosgen, 1993a and 1994). As one constructs local relationships and curves from the gage data, the plot of bankfull channel dimensions by drainage area should not only be integrated or stratified for a hydro-physiographic province, but also by stream type.

LEVEL II: THE MORPHOLOGICAL DESCRIPTION

a) Note staining on rocks which correspond to brush/rock interface at the bankfull stage - B3 Stream Type, Lake Creek, Alaska.

b) Note top of point bar and rock "line" on opposite right bank delineating the bankfull stage - F4 Stream Type, Duchesne River, Utah. (Photo by J. Winston)

c) Note top of point bar indicating bankfull stage in entrenched G4 Stream type - Rito Blanco, Colorado.

d) Top of Point bar and willows indicating bankfull stage in C4 Stream Type - Upper Willow Creek, Colorado.

FIGURE 5-7. Photographs of various stream types depicting bankfull stage and corresponding indicators.

FIELD METHODS FOR STREAM TYPE DELINEATION

This section explains the significance of each of the Level II inventory criteria, and describes methods for their field determination. The criteria are presented in an order that corresponds to their use in the dichotomous key presented in *Figure 5-3*; and not necessarily in order of their hydrologic or geomorphic significance. Comments regarding the relative importance of the criteria are described in the text. A "reference reach" field data form designed for the purpose of documenting measured values for the delineative criteria is provided in *Table 5-3*.

Entrenchment

Significance

Entrenchment describes the relationship of the river to its valley and landform features. Entrenchment is qualitatively defined as the vertical containment of a river and the degree to which it is incised in the valley floor (Kellerhalls et al. 1972).

LEVEL II: THE MORPHOLOGICAL DESCRIPTION

TABLE 5-3. Reference reach field data form for stream classification.

REFERENCE REACH FIELD FORM
STREAM CHANNEL CLASSIFICATION LEVEL II

STREAM TYPE: _____

STREAM NAME: _____ DRAINAGE AREA: _____ BASIN NAME: _____
OBSERVERS: _____ DATE: _____
LOCATION: _____ Twp. _____ Rge. _____ Sec. _____ Qtr. _____

Bankfull WIDTH _____ Ft. (W_{bkf})   Bankfull MAX> DEPTH _____ Ft. (d_{max})   Channel SLOPE _____ Ft/Ft _____ %
Bankfull Mean DEPTH _____ Ft. (d_{bkf})   Flood Prone Area WIDTH _____ Ft. (W_{FP})   Valley SLOPE _____ Ft/Ft _____ %
WIDTH/DEPTH Ratio _____   ENTRENCHMENT Ratio _____   SINUOSITY (Stream Dist/Valley Dist.) _____

Channel MATERIALS: (Pebble Count)   D15 ____ mm   D34 ____ mm   D50 ____ mm   D84 ____ mm   D95 ____ mm

photo

photo

LEVEL II: THE MORPHOLOGICAL DESCRIPTION

a) Conifers as indicators of bankfull stage, B4 Stream type.

b) Top of point bar corresponding to woody species indicating the bankfull stage. Floodplain on right. C5 Stream type.

c) Highest central bars coincide with bank on right, D3 stream type - Blanco River.

d) Base of alders and change in bank slope in Stream type B3 - Leche Creek.

e) C5 Stream type showing top of point bars and riparian plant indicators of the bankfull stage.

f) Top of bank and "bankfull" at same level, E3 Stream type - O'Neill Creek.

FIGURE 5-8. Photographs of various stream types depicting bankfull stage and corresponding indicators.

LEVEL II: THE MORPHOLOGICAL DESCRIPTION

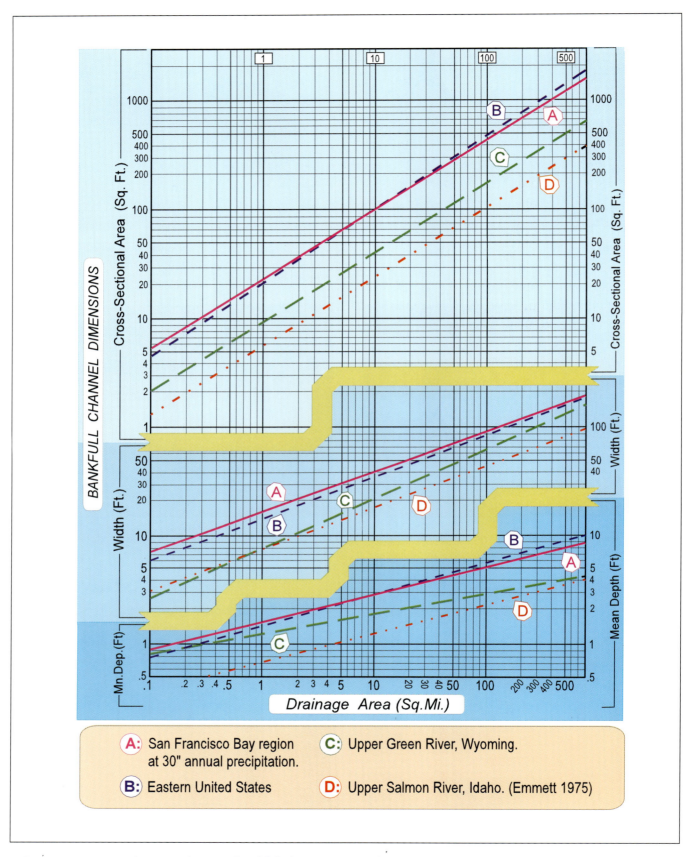

FIGURE 5-9. Regional curves showing bankfull dimensions vs. drainage areas for various hydro-physiographic provinces (Dunne and Leopold, 1978).

When determining entrenchment, it is important to distinguish whether the flat adjacent to the channel is a frequent floodplain, a terrace (abandoned floodplain), or is well outside of the flood-prone area. Recall that the term "flood-prone" is used to avoid having to determine or map an arbitrary or political flow level such as the "100 year floodplain." Such a designation is normally made for socio-economic descriptions and has no particular geomorphic significance. The flood-prone area generally includes the active floodplain and the low terrace.

Methods

The term "entrenchment ratio," which is the vertical containment of the river, has been quantitatively defined (Rosgen 1994) to provide a consistent method for field determination. The entrenchment ratio is the ratio of the width of the flood-prone area to the surface width of the bankfull channel. The flood-prone area width is measured at the elevation that corresponds to twice the maximum depth of the bankfull channel as taken from the established bankfull stage.

Figure 5-10 shows representative entrenchment ratios from cross-sections of various stream types. The degree of entrenchment described by these ratios is based on empirically derived relationships between flood-prone areas and bankfull channel cross-sections, developed from hundreds of stream channel measurements representing a majority of the described stream types. Ratios of 1-1.4 represent entrenched streams; 1.41-2.2 represent moderately entrenched streams; and ratios greater than 2.2 indicate rivers only slightly entrenched in a well-developed floodplain.

Figure 5-11 illustrates the steps needed to determine entrenchment. To measure the width of the flood prone area, select the elevation that corresponds to twice the maximum bankfull channel depth as determined by the vertical distance between bankfull stage and the thalweg of a riffle. Field observations show that for most stream types,

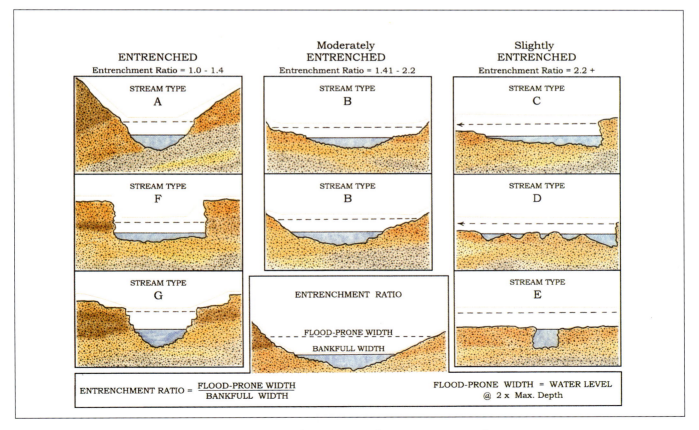

FIGURE 5-10. Representative entrenchment ratios for cross-sections of various stream types.

LEVEL II: THE MORPHOLOGICAL DESCRIPTION

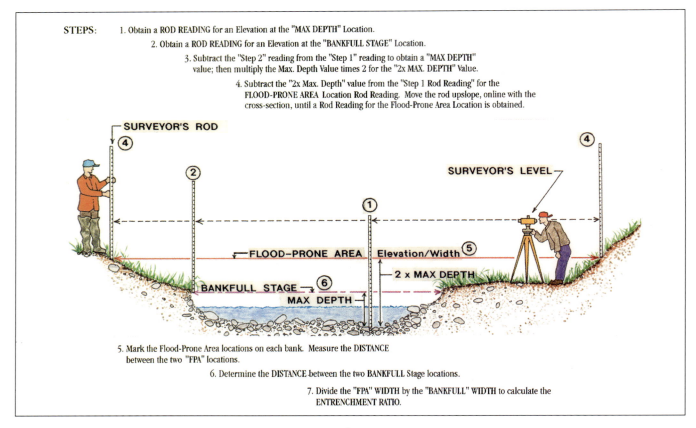

FIGURE 5-11. Determining a Flood-Prone Area elevation/width for calculation of Entrenchment Ratio.

this elevation is associated with a < 50 year return period flood, rather than with a very rare flood.

The entrenchment ratio is based in part on dimensionless rating curves that plot mean depth to bankfull depth (d/d_{bkf}) against the corresponding discharge ratios (Q/Q_{bkf}) (Dunne and Leopold 1978). *Figure 5-12* illustrates this relationship for streams in the eastern U.S. and for Idaho rivers (Emmett 1975). As shown, the 50 year flood has approximately a 1.7 ratio value, indicating that the 50 year flood would be about 1.7 times deeper than the mean bankfull channel depth as determined from gage station data.

Figure 5-13 extends the use of a dimensionless rating curve to illustrate the relationship between stream morphology and entrenchment for two streamflows. The figure was developed by simulating two discharges: bankfull flow (represented by 50 cfs) and a 50-year recurrence interval flow (represented by 200 cfs). These flow stages were then related to the characteristic stream morphologies for Level I stream types, A through G. The graphic relationship enabled a comparison by stream type of the ratio of flood depth to bankfull depth. As shown in Figure 5-13, the 50-year flood is distributed differently within the channel and on the adjacent flood prone area for different stream types. The 50-year recurrence interval flood discharge yields a dfpa/dbkf ratio of between 1.3 to 2.7 across all stream types, with an average ratio value of 2.0. Obviously, this value is also a function of the total width of the flood prone area available to the river. Field verification of hundreds of cross-sections, is the basis for using the value of 2 times the maximum depth at the bankfull stage to estimate the elevation and extent flood-prone area in the field. The average ratio value selected is 2.0 rather than the stratified Level II ratios of 1.3 to 2.7. When mean values were used rather than maximum values, the low terrace elevations in the C stream types would not have been included in the flood-prone area, contrary indicators from the gage data/flood stage data. Based on these observations, twice maximum bankfull depth was selected as an

LEVEL II: THE MORPHOLOGICAL DESCRIPTION

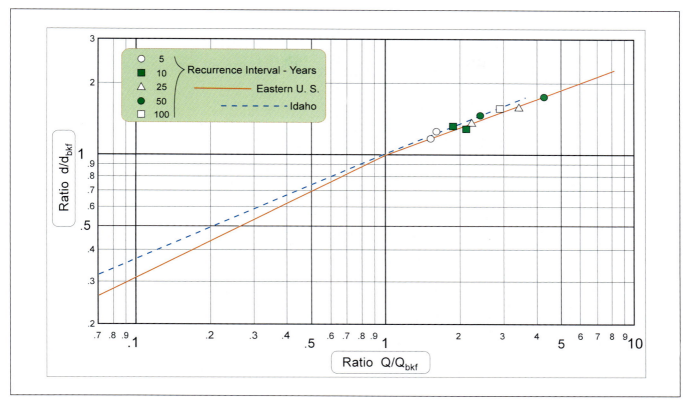

FIGURE 5-12. Dimensionless rating curve for two regions, eastern United States and Idaho.
(Idaho data from Emmett, 1975.) From Dunne and Leopold, 1978

index, to represent an estimate of the representative flood prone area elevations.

For stream types that are only slightly entrenched (e.g., stream types "C," "D," "DA," and "E"), flows greater than the bankfull stage overtop their streambanks and extend onto their floodplain. This natural phenomena does not hold true for deeply entrenched channels (e.g., stream types "A," "F," and "G") where the actual top-of-bank elevations are much higher than the bankfull stage. For entrenched channels, streamflows greater than bankfull increase in depth much faster than in width, as discharge increases. In entrenched channels, the "flood-prone" area increases only marginally in width with an increasing flow-stage above bankfull elevations. The distribution of shear stress (a product of the specific weight of water, hydraulic mean depth and channel slope) and of very high velocity gradients (high boundary stress) on both high banks of an entrenched channel during floods is partly responsible for the associated high bank erosion rates observed in the "A," "F," and "G" stream types. Graphic relationships between bank-

full stage and bank height is shown with the illustrations in *Figure 5-2* and *Figure 5-10*.

Dimensionless rating curves simulated for a range of stream types are compared on the curves developed by Dunne and Leopold (1978), for the eastern U.S (*Figure 5-14*). The illustration indicates the possible variation in dimensionless ratio values for depth, by stream type for bankfull and flood discharges. The importance of a simple method for consistent quantitative determination of entrenchment in the field is that it enables the identification of flood-prone areas that are associated with relatively frequent, rather than rare flood events.

Width/depth Ratio
Significance

The width/depth (W/D) ratio is defined as the ratio of the bankfull surface width to the mean depth of the bankfull channel. The width/depth ratio is key to understanding the distribution of available

LEVEL II: THE MORPHOLOGICAL DESCRIPTION

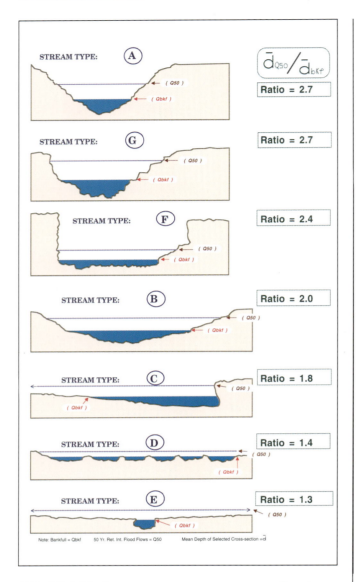

FIGURE 5-13. Representative cross sections of various stream types depicting two flows including bankfull (50 cfs) and a 50 year flood discharge (200 cfs).

energy within a channel, and the ability of various discharges occurring within the channel to move sediment. Of the Level II criteria, the W/D ratio is the most sensitive and positive indicator of trends in channel instability. Determination of the width/depth ratio provides a rapid, visual assessment of channel stability, which, of course, must be verified with field measurements as described in the Level III inventory process (Chapter 6).

Measurement of the width/depth ratio is also valuable for describing channel cross-section shape and comparison of ratio values can be used to interpret shifts in channel stability following disturbances to channels or watersheds. Osterkamp et al. (1983) derived power functions used to predict streamflow using W/D ratios as a surrogate for channel sediment characteristics and the distribution of shear stress. Osborn and Stypula (1987) used W/D ratios to characterize channel hydraulics by describing channel boundary shear as a function of channel shape.

The distribution of energy within channels having high W/D ratios (i.e., shallow and wide channels) is such that stress is placed within the near bank region. As the W/D ratio value increases (i.e, the channel grows wider and more shallow), the hydraulic stress against the banks also increases and bank erosion is accelerated. The accelerated erosion process is generally the result of high velocity gradients and high boundary stress, as mean velocity, stream power, and shear stress decrease in the presence of an increase in width/depth ratio values. Increases in the sediment supply to the channel develop from bank erosion, which - by virtue of becoming an over widened channel - gradually loses its capability to transport sediment. Deposition occurs, further accelerating bank erosion, and the cycle continues.

Figure 5-3 shows that for all stream types, except multiple thread channel, a W/D ratio of 12 is the high end value for "A," "G," and "E" stream types and the low end value for "B," "C," and "F" stream types. Analyses of field data showed a width/depth ratio continuum value of 10, but a ratio value of 12 was selected as the most frequently observed value (see frequency distribution data by stream type later in this chapter). The W/D ratio value of 12 is empirically derived, but is related to physical processes governing the distribution of energy and resultant sediment transport. Channel dimension, pattern, profile, and corresponding stream types change when the width/depth ratio is significantly altered. Evolutionary sequences of adjustments in channel morphology are discussed more fully in Chapter 6.

LEVEL II: THE MORPHOLOGICAL DESCRIPTION

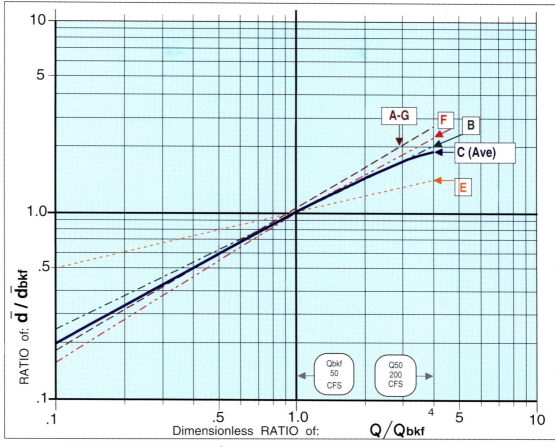

FIGURE 5-14. Dimensionless ratios for various stream types for a Bankfull Discharge of 50 CFS, and a 50 year flood of 200 CFS.

Methods

Width/depth ratios should be determined on selected reference reaches using the same criteria for determining bankfull stage indicators and bankfull channel measurements. That is, the selected reach should be free to adjust its boundaries in response to frequent flows and should not have excessive deposition, nor constrictions. Criteria for selecting representative channel reaches are shown in *Figure 5-4* and *Figure 5-5*. Wherever possible, at least one or more permanent cross-section should be established to permit resurveying so that changes in state or stability of a stream may be effectively determined. Permanent benchmarks for use as reference elevations can be installed using either cemented or driven metal rods. Width/depth ratios are determined from measured channel cross-sections as shown in *Figure 5-15*.

Examples of bankfull channel cross-sections and typical W/D ratios for various stream types are shown in *Figure 5-16*. Mean values and ranges in width/depth ratios for each major stream type are described in the data summaries presented at the end of this chapter. Width/depth ratios can occasionally vary by ± 2 units without necessarily indicating a change in stream morphology or type.

Sinuosity

Significance

Sinuosity is the ratio of stream length to valley length. It can also be described as the ratio of valley slope to channel slope. Meander geometry characteristics are directly related to sinuosity, consistent with the principle of minimum expenditure of energy. Langbein and Leopold (1966) suggested that a sine generated curve describes symmetrical meander paths, permitting the estimation of a radius of curvature for meander bends, as developed from data analyses related to meander wavelength and channel sinuosity. Applying the same analyses techniques to an expanded data set, Williams (1986) found highly significant relations between predicted vs. observed values for radius of curvature and channel sinuosity for 79 streams.

LEVEL II: THE MORPHOLOGICAL DESCRIPTION

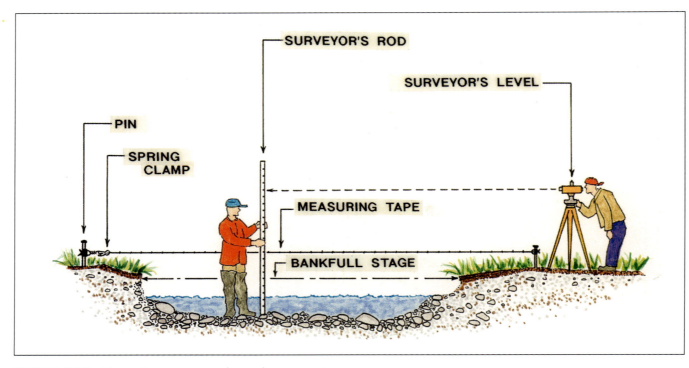

FIGURE 5-15. Measuring a stream channel cross-section.

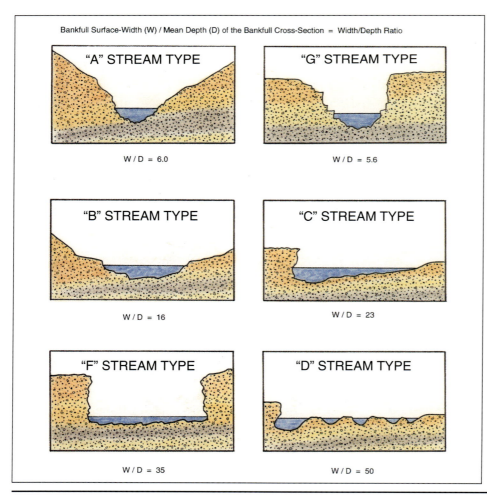

FIGURE 5-16. Examples of various Width/Depth Ratios for selected stream types.

Sinuosity, however, carries the least weight of all the criteria used to delineate Level II morphologies. It is used primarily as a channel plan-form and mapping indicator for Level I classification, since it is based on easily observable and interpreted patterns. Sinuosity can be modified by bedrock control, roads, channel confinement (lateral containment), and vegetation. As channel gradient and dominant particle size decreases, there is generally a corresponding increase in sinuosity.

Methods

Sinuosity can best be measured using aerial photography as previously described in Level I classification.

Channel Materials
Significance

While channel bed and bank materials influence the cross-sectional form, plan-view, and longitudinal profile of rivers, they also determine the extent sediment transport and provide the means of resistance to hydraulic stress. Additionally, an assessment of the nature and distribution of channel materials is critical for interpreting the biological function and stability of rivers. Knowledge of channel materials is essential for Level II classification. When inventories of channel materials are obtained, there is often some confusion as to what and how to sample. For stream channel classification purposes, channel material refers primarily to the surface particles that make up both the bed and banks within the bankfull channel. The surface particles are referred to as the "pavement" of the channel. A diagram is presented in *Figure 5-17* which depicts the coarser surface particles and finer subpavement particles. The subpavement is indicative of the range of sizes of sediment that are likely to be mobilized when streamflows are approaching or are at bankfull discharge levels. As the surface particles decrease in size, the ratio of pavement particle size to subpavement particle size decreases. Additional information on sediment data will be presented in Chapter 7.

Methods

Wolman (1954) first developed the "pebble-count" method for field determination of the particle size distribution of channel materials. Wolman's method, since modified to account for both bank and in-channel material, for sands and smaller particle sizes, and for bedrock is used for Level II stream classification. The modified Wolman method (Rosgen 1993a) uses a stratified, systematic sampling method based on the frequency of riffle/pools and step/pools occurring within a channel reach that is approximately 20-30 bankfull channel widths in length (or two meander wavelengths). The modified method adjusts the material sampling locations so that various bed features are sampled on a proportional basis along a given stream reach.

For example, assume that across two meander wavelengths, 70% of the channel reach length is composed of riffles and 30% is composed of pools. A minimum of 70 observations should be taken within riffle areas and 30 observations should then be taken within pool areas, such that the minimum sample size totals 100. In addition, the particles should be sampled at systematic intervals along a given transect or sample route. To avoid potential bias, the actual particle picked up for measurement must be selected on the "first blind touch," rather than seen and then selectively picked up.

This procedure, shown in *Figure 5-18*, ensures that a representative pebble count may be obtained through proportional sampling of riffles, steps, and pools. Representative sampling is essential, since the generally flatter, deeper pools contain finer sized materials relative to the coarser materials found in steeper, shallower riffles and steps. The designed proportional procedure is necessary to counter the observer's natural tendency to sample only the shallow areas that promotes easy wading and to selectively sample the more obvious, larger particles.

The intermediate axis of the particle is measured (as described by Wolman, 1954) such that the particle size selected would be retained or pass a standard materials sieve of fixed opening. Particles sampled along transects in the pools are tallied separately from those sampled in the riffles on the form

LEVEL II: THE MORPHOLOGICAL DESCRIPTION

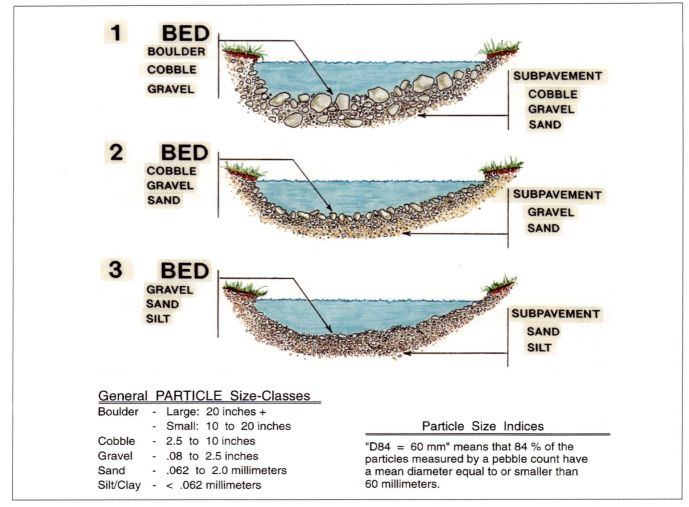

FIGURE 5-17. General categories of channel bed and sub-pavement materials.

provided in *Table 5-4*. The segmented particle size data is then added together for a composite total for stream classification purposes. However, the separate data sets are used to provide additional data for biological and sediment source/transport evaluations. The data for the riffles, pools, and composite totals can be effectively plotted for individual interpretations on log-normal graph paper, shown as *Figure 5-19*. In some bi-modal particle distributions the indexed D-50 (50 per cent of the sampled population is equal to or finer than the representative particle diameter) particle may not even be present. In this case, the *dominant* particle size should be determined as that particle size which has provided the largest number of observations. The Y axis to the right of the log-normal graph is used to plot the number of particles rather than percent. These data related to individual particles will also provide additional biological and sediment source/yield interpretations for individual bed features or reach sediment characteristics.

To obtain the representative median particle diameter or the D-50 index diameters, the pebble count data is then plotted on a log-normal graph as a cumulative percent. The plot shown in *Figure 5-19* procedures contains both the cumulative per cent plot as well as the numbers of individual particles by sizes.

Pebble count sampling accuracy can be further improved by using a tape line with equally spaced intervals to assist in determining an appropriate location for selecting in-channel particles for measurement. Pebble count data can be tallied separately for riffles, runs, glides, and pools, and then be summed to obtain a distribution of the total

LEVEL II: THE MORPHOLOGICAL DESCRIPTION

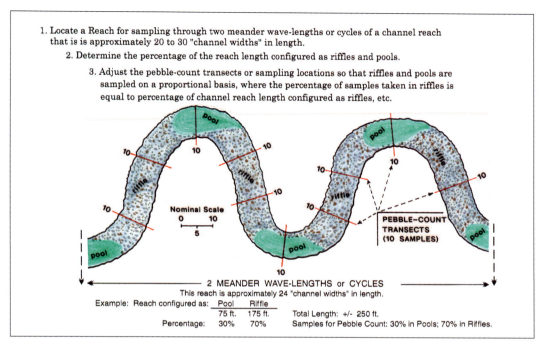

FIGURE 5-18. Representative Pebble-Count procedures.

number of particles sampled. Beyond the use prescribed for Level II stream classification, such data is also valuable for assessing fisheries habitat, sediment studies, and impacts to the watershed and/or trends in river stability. ***Table 5-4*** illustrates a field form for recording and analyzing pebble count data related to three separate reaches or separate bed features at the same sampling location. To estimate bank erodibility, similar pebble-count data for stream bank materials may be obtained separately and used in erodibility calculations as described in Chapter 6.

Slope
Significance

The slope of the water surface is a major determinant of river channel morphology, and of the related sediment, hydraulic, and biological functions. A longitudinal profile surveyed along a selected channel reach is recommended to be established for slope determinations. When integrated with previous stream classification data, the longitudinal profile survey can also provide slope facet information that can be related to depth and particle sizes for specific bed features. Sediment transport as well as designs for river restoration or fish habitat structures often require such individual slope facet data. With a sufficient array of longitudinal profile data, specific characteristics of riffles, runs, glides, and pools can be compared between each feature and between features of other stream types. It is desirable to describe riffles or pool dimensions in terms of ratios of average slope, depth, and particle size. The existing quality of the surveyed bed features can be related to their response potential by reference reach comparisons.

Methods

Water surface slope is determined along the longitudinal profile of the channel by measuring the difference in water surface elevation per unit stream length. Slope measurements should be taken through a channel reach that is a minimum of 20 channel widths in length or for a distance equal to two meander wavelengths. Field methods for proper measurement of water surface slope are shown in ***Figure 5-20***. As shown, the water surface slope should be measured by taking the difference in elevation from the one bed feature to the same bed feature either upstream or downstream. For riffle/pool channels, it is generally best to measure from the top of one riffle to the top of another. It is important to measure the actual stream length rather than estimating the length or running a tape that does not follow the actual channel alignment to avoid overestimating slope. Instruments should be

LEVEL II: THE MORPHOLOGICAL DESCRIPTION

INCHES	PARTICLE	MILLIMETER		Particle Count	TOT #	ITEM %	% CUM	TOT #	ITEM %	% CUM	TOT #	ITEM %	% CUM
	Silt/Clay	< .062	S/C										
	Very Fine	.062 - .125	S										
	Fine	.125 - .25	A										
	Medium	.25 - .50	N										
	Coarse	.50 - 1.0	D										
.04 - .08	Very Coarse	1.0 - 2	S										
.08 - .16	Very Fine	2 - 4											
.16 - .24	Fine	4 - 6	G										
.24 - .31	Fine	6 - 8	R										
.31 - .47	Medium	8 - 12	A										
.47 - .63	Medium	12 - 16	V										
.63 - .94	Coarse	16 - 24	E										
.94 - 1.26	Coarse	24 - 32	L										
1.26 - 1.9	Very Coarse	32 - 48	S										
1.9 - 2.5	Very Coarse	48 - 64											
2.5 - 3.8	Small	64 - 96	C										
3.8 - 5.0	Small	96 - 128	O										
5.0 - 7.6	Large	128 - 192	B										
7.6 - 10	Large	192 - 256	L										
10 - 15	Small	256 - 384	B										
15 - 20	Small	384 - 512	L										
20 - 40	Medium	512 - 1024	D										
40 - 160	Lrg-Very Lrg	1024 - 4096	R										
	BEDROCK		BDRK										
				TOTALS									

TABLE 5-4. Field form for documentation and analysis of Pebble Count Data

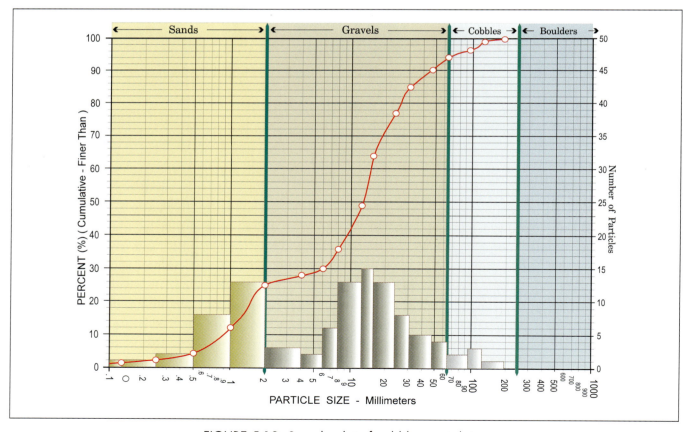

FIGURE 5-19. Sample plot of pebble-count data.

LEVEL II: THE MORPHOLOGICAL DESCRIPTION

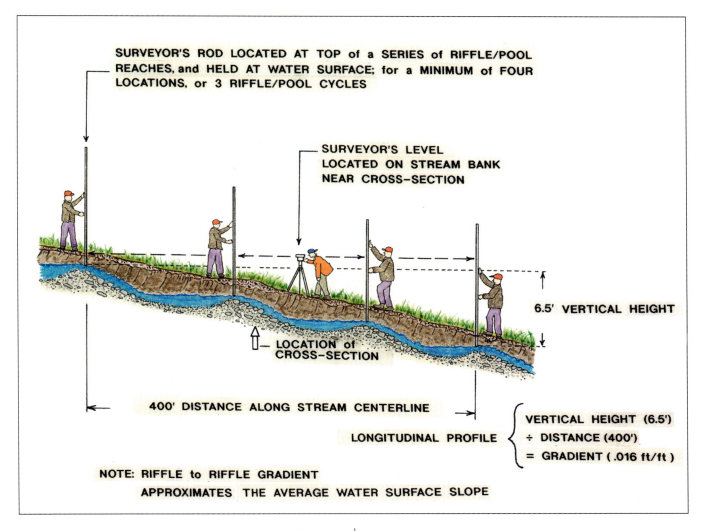

FIGURE 5-20. Measuring stream gradient through a typical riffle/pool sequence.

used for elevation change which are appropriate to accurately measure 0.1 feet change in 100 feet distance. The use of clinometers should be discouraged for measuring stream slope, except for very steep streams over 10 percent. On lower gradient reaches, clinometers tend to consistently over-estimate water surface slope. Laser levels are often preferred, as one person can accurately obtain all of the survey data needed for a stream reach inventory. Hand levels, with zoom scopes, properly used in conjunction with survey rods have given sufficiently accurate slope readings.

The Level II stream classification key shown in *Figure 5-3* lists the most frequently observed slope categories by stream type. Stream type categories having slope values greater or lesser than those shown in *Figure 5-3* for a given stream type are denoted with subscripts. The subscripts and their associated slope values are: a+ = >0.10; a = 0.04 to 0.1; b = 0.02 to 0.04; c = < 0.02; and c- = < 0.001. Where possible, valley slope should also be measured to establish relationships between slope for a given valley and channel materials. Valley slope measurements also provide a check on sinuosity measurements (sinuosity = valley slope/channel slope). In addition, longitudinal profiles are often obtained in conjunction with slope in order to describe the frequency of bed features as a function of bankfull width.

MORPHOLOGICAL DESCRIPTIONS AND EXAMPLES

Data developed from Level II classification of stream types that have been field calibrated and ver-

LEVEL II: THE MORPHOLOGICAL DESCRIPTION

ified at reference reaches permit the extrapolation of appropriate channel data to other similar locations. To extrapolate pertinent data to other areas, similarities between the calibrated reference reach and the desired extrapolation locations must be determined. Objective comparisons of landform, soils, and channel patterns, are obtained from aerial photographs, topographic, geologic, and soil area maps.

Remote Sensing Methods for Level II Delineation

An example of Level II classification inferred from aerial photography is shown in *Figure 5-21*, for a reach along the East Fork San Juan River in southwestern Colorado. As shown here, the reach was classified using observed valley types, landforms, and depositional features. The use of sinuosity as an indicator of channel pattern and associated stream type is also illustrated. The Level II classifications for various reaches are shown on a topographic map of the watershed in *Figure 5-22*.

Figure 5-23 illustrates a Level II classification inferred from oblique photographs of the North Fork of the South Platte River, Colorado. Close-ups of the different channel morphologies associated with the stream types in this watershed are shown in the photo-insets with an E4 stream type at the watershed divide, and continuing into a B4 and an A2 stream type. Field verification is needed where stream types change or where variations are observed. Oblique imagery was similarly used to classify a tributary to the Fall River, in the Rocky Mountain National Park, after the Lawn Lake flood, as shown in **Figure 5-24**. The flood created a deeply incised channel in glacial till and created a large alluvial (debris) fan at the mouth of the tributary. The stream types delineated here are "A3a+," " A3," "A2," " G3," and "D4."

Helicopter surveys can also be used to rapidly classify stream types at the Level II inventory once individual observer experience at field-calibrated sites in similar hydro-physiographic settings has been well established. Helicopter surveys permit low overflights for direct observation of stream reaches, as well as occasional on-site landings for collecting certain reference reach data. With the assistance of Lee Chavez, and Bo Stewart, USDA Forest Service Hydrologists, the author conducted a helicopter flight where approximately 3,500 miles of first to sixth order streams were classified and mapped in approximately 14 days (USDA Forest Service, 1989b). *Figure 5-25* displays the frequency distribution of stream types mapped at Level II within the surveyed watersheds.

To record and compute the field data for stream classification at Level II, a field form shown in ***Table 5-3*** is provided. This form should be completed for each representative stream type in the watershed being studied. Once the appropriate data is obtained, the field person then becomes an experienced observer with an ability to classify reaches in the same watershed without requiring a detailed reach inventory.

This chapter concludes with a summary for each level II stream type, characteristic morphologies, channel materials, and associated attributes of energy and sediment supply. Example photographs of similar stream types from different areas, schematic drawings, and summary statistics for each delineative criteria are also presented.

LEVEL II: THE MORPHOLOGICAL DESCRIPTION

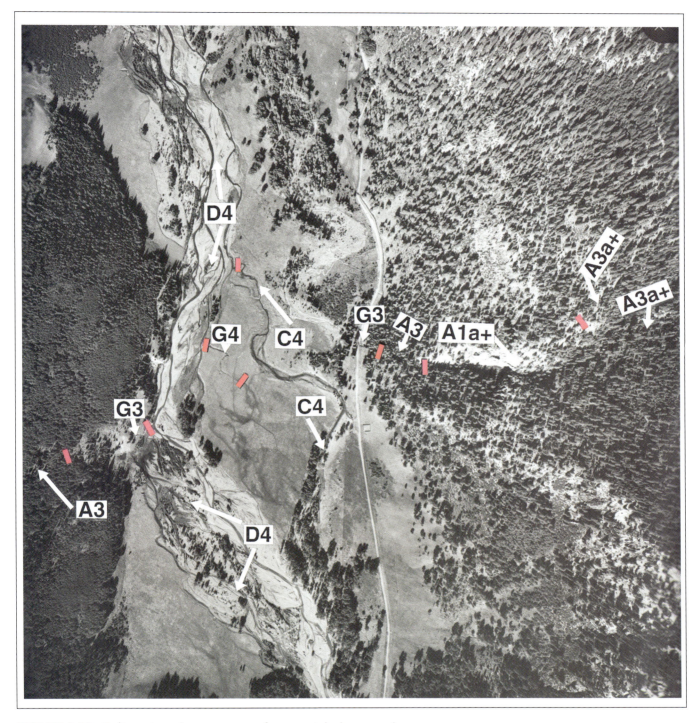

FIGURE 5-21. Delineation of stream types from aerial photography.

LEVEL II: THE MORPHOLOGICAL DESCRIPTION

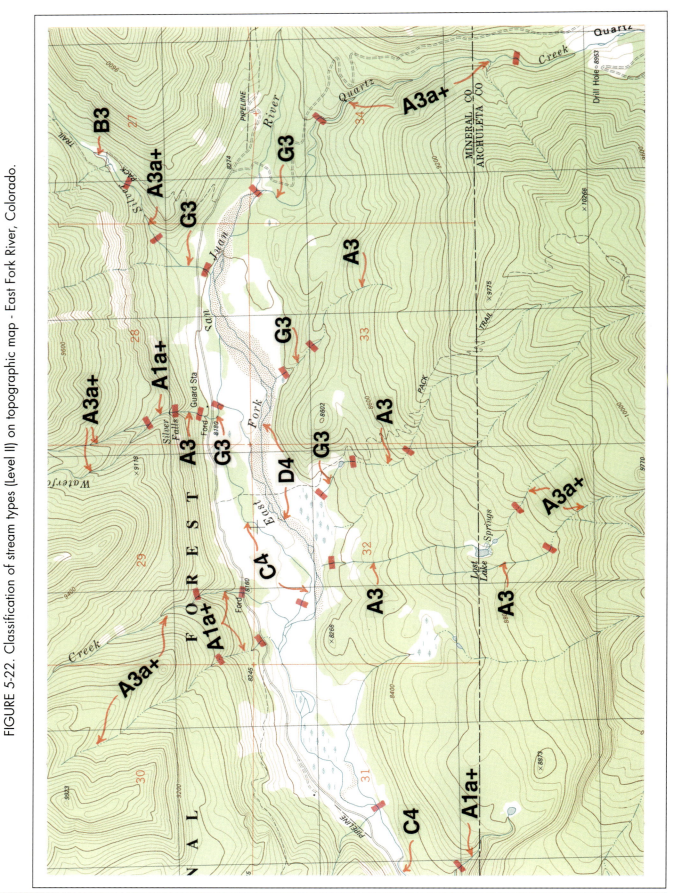

FIGURE 5-22. Classification of stream types (Level II) on topographic map - East Fork River, Colorado.

LEVEL II: THE MORPHOLOGICAL DESCRIPTION

FIGURE 5-23. Delineation of stream types from aerial oblique photo and associated cross-sections.

LEVEL II: THE MORPHOLOGICAL DESCRIPTION

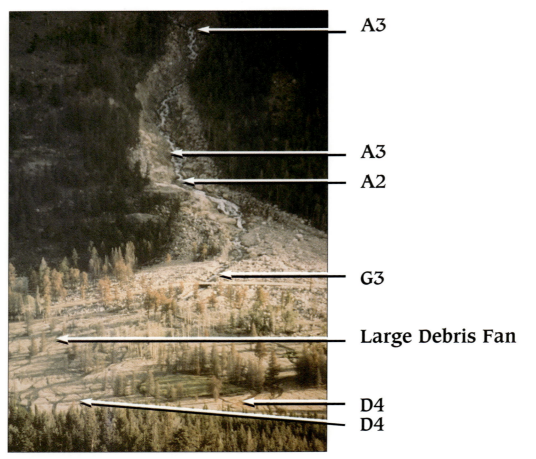

FIGURE 5-24. Aerial oblique - Lawn Lake flood depicting stream types onto alluvial fan.

FIGURE 5-25. Distribution of stream types as mapped by the author in several watersheds of the South Platte River Basin, Colorado. (*Helicopter Aerial Survey, 3200 miles, USDA Forest Service, 1989.*)

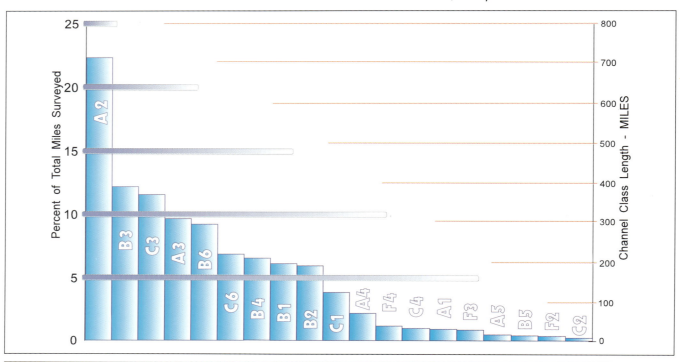

5-34

LEVEL II: THE MORPHOLOGICAL DESCRIPTION

MORPHOLOGICAL DESCRIPTION AND EXAMPLES OF STREAM TYPES

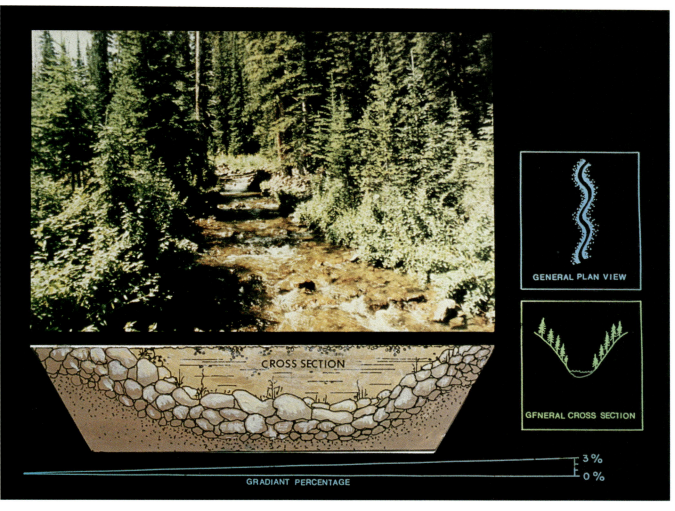

LEVEL II: THE MORPHOLOGICAL DESCRIPTION

MORPHOLOGICAL DESCRIPTION AND EXAMPLES OF STREAM TYPES

A1 Stream Type

Idaho

The A1 stream type is a steep, entrenched and confined channel in bedrock, that is associated with faults, scarps, folds, joints and other structurally controlled drainage ways. The A1 stream type is typically found in a valley where stream slopes can range from 4-10%. The stream bed is seen as a cascading, step/pool morphology with irregularly spaced drops and deep scour pools. Channel gradients greater than 10% are categorized as A1a+ and may produce stream- flow described as torrents or with torrential flows, waterfalls, and bed forms described as "chutes," or a series of vertical drops. Pool spacing is highly irregular and is controlled by bedrock and large, woody organic debris. As gradient increases, the steps become more closely spaced. The width/depth ratios are characteristically low; however, some observations indicate width/depth ratios greater than 12, due to bedrock sills. The A1 stream type is relatively straight with a sinuosity less than 1.2. Channel materials are principally bedrock although boulders and, in smaller amounts, cobbles and gravels may be included.. The dimension, pattern, and profile of this stream type is little influenced by the modern flows and sediment regime. The A1 stream type is a high energy-low sediment supply stream system.

LEVEL II: THE MORPHOLOGICAL DESCRIPTION

DELINEATIVE CRITERIA (A1)

Landform/soils: Bedrock controlled, steep slopes and channel. Glacial scoured slopes, faults, folds and joints. Soils predominantly bedrock and coarse colluvium.

Channel materials: Bedrock with lesser amounts of boulders, cobble and gravel.

Slope Range: .04 - .10 (A1a+ > .10)   **Entrenchment Ratio:** < 1.4

Width/depth Ratio: < 12   **Sinuosity:** < 1.2

LEVEL II: THE MORPHOLOGICAL DESCRIPTION

A1. Washington

A1. Colorado

A1. Colorado

A1. Wyoming

LEVEL II: THE MORPHOLOGICAL DESCRIPTION

A1. Idaho

A1. Colorado

A1. Wyoming

A1. Wyoming

5-39

LEVEL II: THE MORPHOLOGICAL DESCRIPTION

MORPHOLOGICAL DESCRIPTION AND EXAMPLES OF STREAM TYPES

A2 Stream Type

Colorado

The A2 stream type is a steep, deeply entrenched and confined stream channel associated with faults, scarps, folds, joints, and other structurally controlled drainageways. Land forms supporting A2 stream types include canyons, steep side slopes, talus fields, glacial moraines, lag deposits, and coarse colluvial deposition. The A2 stream type is normally situated in valley types I and II. Slope ranges are 4-10 %, producing channels that exhibit step/pool bed features. However, the A2 stream type also occurs on slopes greater than 10 %, (A2a+), which promote cascades and "chutes". The sinuosity is low (<1.2) as is the width/depth ratio (<12). Width/depth ratios greater than 12 can occur where larger boulders contribute to channel bed and bank stability. Stream types are incised in predominantly boulder-sized channel material with lesser amounts of cobble and gravel materials present. The A2 stream type is a high energy and low sediment supply stream type, with corresponding low bedload transport rates. The channel bed and stream banks are normally stable and contribute little to sediment supply.

LEVEL II: THE MORPHOLOGICAL DESCRIPTION

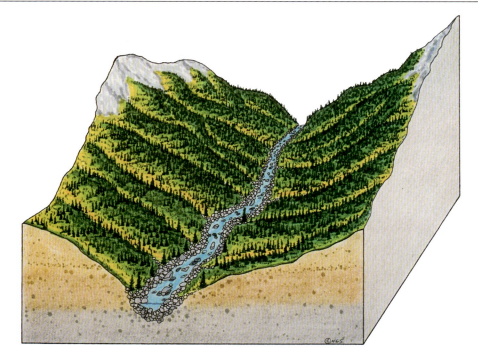

DELINEATIVE CRITERIA (A2)

Landform/soils: Steep, structural controlled slopes with colluvial deposition in narrow and confined valleys.

Channel materials: Predominantly boulders, with lesser amounts of cobble, gravel and sand. Some bedrock sporadically spaced.

Slope Range: .04 - .10 (A2a+ > .10) **Entrenchment Ratio:** < 1.4

Width/depth Ratio: < 12 **Sinuosity:** < 1.2

LEVEL II: THE MORPHOLOGICAL DESCRIPTION

STREAM TYPE A2

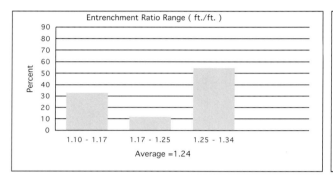
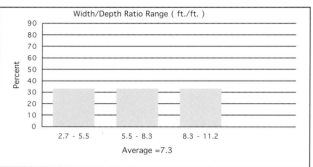
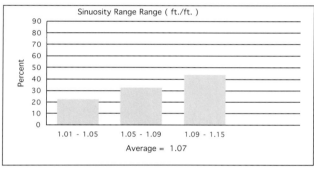
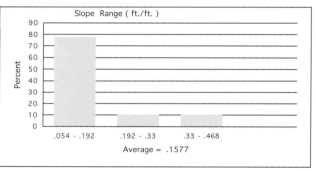
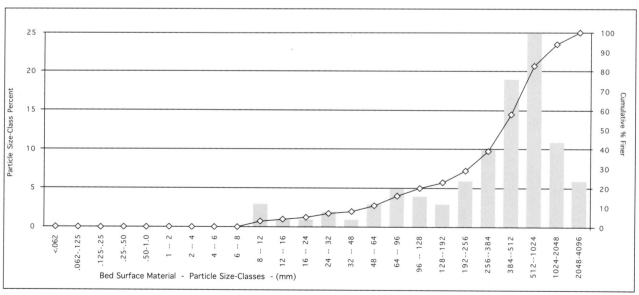

LEVEL II: THE MORPHOLOGICAL DESCRIPTION

A2. Colorado

A2. Colorado

A2. Colorado

A2. Colorado

LEVEL II: THE MORPHOLOGICAL DESCRIPTION

MORPHOLOGICAL DESCRIPTION AND EXAMPLES OF STREAM TYPES

A3 Stream Type

Colorado

The A3 stream type is a steep, deeply entrenched, and confined channel that is incised in coarse depositional soils. Landforms in which A3 types may be found are noted as glacial moraines or tills, alluvial fans, lag deposits, slide debris, and other coarse textured colluvial deposition. The A3 stream type can also be situated as lateral tributary streams intersecting the valley train, and deeply incised in glacial and Holocene terraces adjacent to the river. The A3 stream type is observed in valley types I, III, and VII. The width depth ratios are low (< 12), with low sinuosity (< 1.2). The channel materials are typically unconsolidated, heterogenous, noncohesive materials, dominated by cobbles but also containing some small boulders, gravel, and sand. The A3 stream types develop both high energy (high stream power and shear stress values) and a high sediment supply, with corresponding very high bedload sediment transport rates. The A3 streams are generally unstable, with very steep, rejuvenated channel banks that contribute large quantities of sediment. The A3 bedform occurs as a step/pool, cascading channel which often stores large amounts of sediment in the pools associated with debris dams. The A3a+ stream types (slopes > 10%) are generally found in landforms associated with slump/earthflow and debris torrent/debris avalanche erosional processes. Characteristic stream bank erosional processes for the A3a+ stream type are fluvial entrainment, collapse (mass wasting), dry ravel, freeze/thaw, and debris flow scour.

LEVEL II: THE MORPHOLOGICAL DESCRIPTION

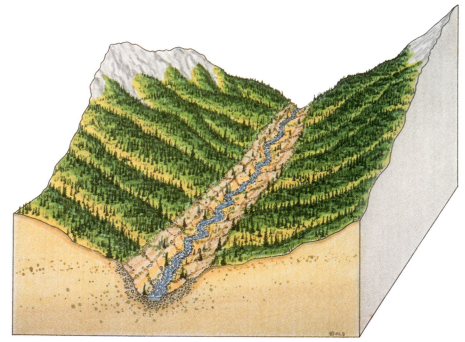

DELINEATIVE CRITERIA (A3)

Landform/soils: Steep, narrow depositional slopes typical of glacial moraines and debris slides associated with unconsolidated, heterogeneous and non-cohesive materials.

Channel materials: Predominantly cobble with a mixture of boulders, gravel and sand.

Slope Range: .04 - .10 (A3a+ > .10) **Entrenchment Ratio:** < 1.4

Width/depth Ratio: < 12 **Sinuosity:** < 1.2

LEVEL II: THE MORPHOLOGICAL DESCRIPTION

STREAM TYPE A3

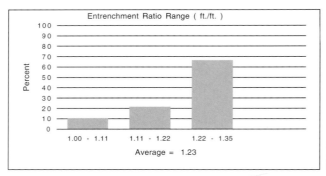
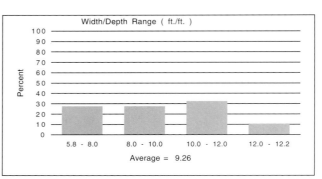
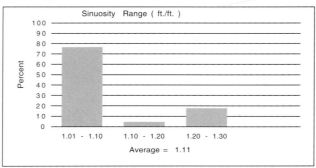
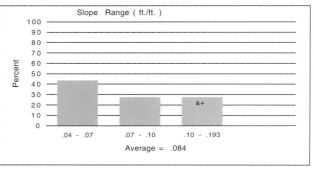
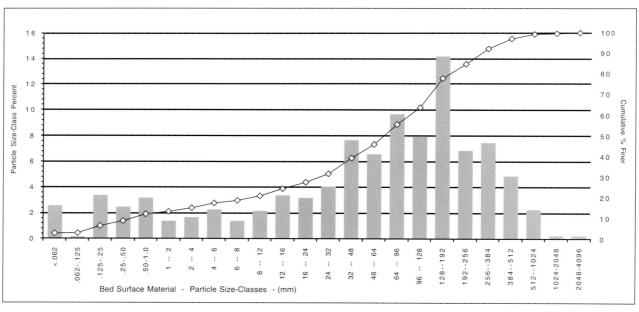

5-46

LEVEL II: THE MORPHOLOGICAL DESCRIPTION

A3a+ - Colorado

A3 - Colorado

A3 - Colorado

5-47

LEVEL II: THE MORPHOLOGICAL DESCRIPTION

MORPHOLOGICAL DESCRIPTION AND EXAMPLES OF STREAM TYPES

A4 Stream Type

The A4 stream type is a steep, deeply entrenched and confined, channel that is incised in coarse depositional materials. Landforms in which the A-4 types develop are primarily glacial moraines or glacial tills, alluvial fans, lag deposits, landslide debris, and other coarse textured colluvial deposition. The A4 types can also be situated as lateral tributary streams intersecting the valley train and deeply incised in glacial and Holocene river terraces. The A4 stream type appears in valley types I, III, and VII, in which some of the soils are residual and associated with highly weathered rock, such as grussic granite. The width depth ratios are low (< 12), with low sinuosity, (< 1.2). The A4 channel materials are typically unconsolidated, heterogenous, noncohesive materials, dominated by gravel, but also containing small amounts of boulders, cobble, and sand. The A4 channel bed features may be described as a step/pool or cascading channel, that is often influenced by the occurrence of organic woody debris that form debris dams, behind which are stored significant amounts of sediment in the pools. The A4 stream types typically have a high sediment supply which is combined with high energy streamflow to produce very high bedload sediment transport rates. The A4 stream types are generally unstable, with very steep, rejuvenated banks that contribute large quantities of sediment. Characteristic stream bank erosional processes for the A4 type are fluvial entrainment, bank collapse, dry ravel, freeze/thaw and lateral scour from debris flows. The A4a+ stream types (slopes > 10%) are usually located in slump/earthflow land forms and are often associated with debris avalanches and debris torrent erosional processes.

LEVEL II: THE MORPHOLOGICAL DESCRIPTION

DELINEATIVE CRITERIA (A4)

Landform/soils: Steep, confined, depositional slopes in glacial and landslide debris. Soils heterogeneous mixture of unconsolidated colluvial and alluvial deposition.

Channel materials: Predominantly gravel with lesser amounts of boulders, cobble, and sand.

Slope Range: .04 - .10 (A4a+ > .10) **Entrenchment Ratio:** < 1.4

Width/depth Ratio: < 12 **Sinuosity:** < 1.2

LEVEL II: THE MORPHOLOGICAL DESCRIPTION

STREAM TYPE A4

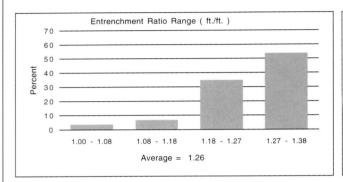
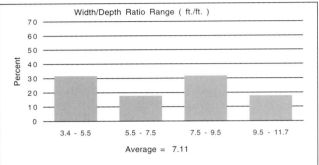
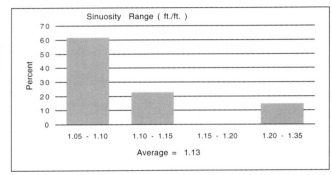
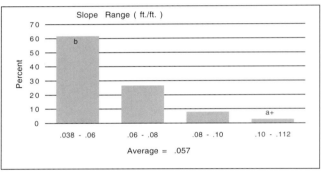
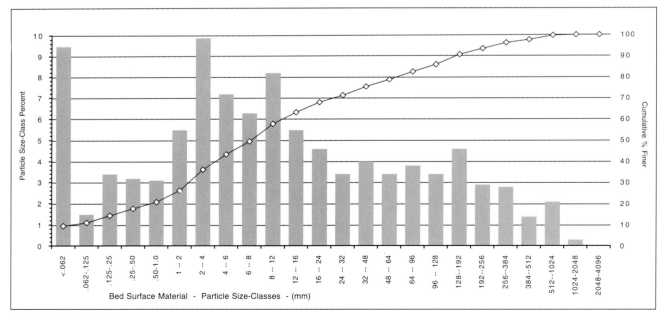

LEVEL II: THE MORPHOLOGICAL DESCRIPTION

A4 - Colorado

A4 – Colorado

A4a+ - Colorado

A4a+ - California

LEVEL II: THE MORPHOLOGICAL DESCRIPTION

MORPHOLOGICAL DESCRIPTION AND EXAMPLES OF STREAM TYPES

A5 Stream Type

The A5 stream types are steep, entrenched and confined channels, incised in predominantly sandy materials that are frequently intermixed with gravels. The A5 types are situated in both depositional landforms and residual soil areas that have developed from grussic granite and highly weathered sedimentary rocks. Glacio-fluvial, glacio-lacustrine, deltaic, eolian, and interbedded sedimentary deposits, as well as sorted sandy alluvium and unsorted sandy colluvium support the A5 stream type. Valley types that contain A5 stream types are I, III, and VII. The A5 channel has a low width/depth ratio, low sinuosity, and may be situated as a lateral channel to the valley train, or exist as an actively degrading main-stem channel in steep terrain. The channel bed and banks are unstable and very sensitive to induced changes in streamflow regime or in sediment supply. Bedload transport rates are very high, with streambank rejuvenation common. Bank erosion occurs through fluvial erosion, slump/earthflow, surface erosion, dry ravel, freeze/thaw and collapse. The A5a+ types (steeper than 10%) are often associated with mass-wasting erosional processes of debris avalanche and debris torrents.

LEVEL II: THE MORPHOLOGICAL DESCRIPTION

DELINEATIVE CRITERIA (A5)

Landform/soils: Steep slopes in depositional landforms associated with erosional debris or residual soils derived from grussic granite and/or sandstones.

Channel materials: Predominantly sand with lesser amounts of gravel and silt/clay.

Slope Range: .04 - .10 (A5a+ > .10) **Entrenchment Ratio:** < 1.4

Width/depth Ratio: < 12 **Sinuosity:** < 1.2

LEVEL II: THE MORPHOLOGICAL DESCRIPTION

A5a+ - Colorado

A5 - Wyoming

A5 - Nevada

A5a+ - Idaho

LEVEL II: THE MORPHOLOGICAL DESCRIPTION

A5 - Colorado

A5 - Wyoming

A5 - Oregon

LEVEL II: THE MORPHOLOGICAL DESCRIPTION

MORPHOLOGICAL DESCRIPTION AND EXAMPLES OF STREAM TYPES

A6 Stream Type

The A6 stream type is a steep, entrenched and confined channel, incised in cohesive soils. Landforms and soils are observed as highly weathered shales and deposition from lacustrine, glacio-lacustrine, delta deposits, and fine grained alluvial deposits. The A6 stream type is found in valley types I, II, and VII. The channel width/depth ratio is the lowest of the A stream types due to the cohesive nature of the stream bank materials. The A6 stream type has a low sinuosity (< 1.2). Mass-wasting processes characteristic of the A6a channel banks (>10%) are creep, glide, slump/earthflow, and debris slides. The A6a+ stream types may also produce intense debris torrents. In contrast to the high rates of fluvial entrainment due to detachment of bed and bank materials of the A3, A4, and A5 stream types, the bank erosion of A6 streams is often due to liquefaction/saturation/collapse, and negative porewater pressure. The A6 stream type is normally associated with a step/pool profile and for the A6a+ as a cascade channel. Unlike the A3, A4, and A5 stream types, the A6 stream types do not develop a high bedload sediment yield. The A6 stream types are characteristic of a high contribution of wash load to the annual sediment yield.

LEVEL II: THE MORPHOLOGICAL DESCRIPTION

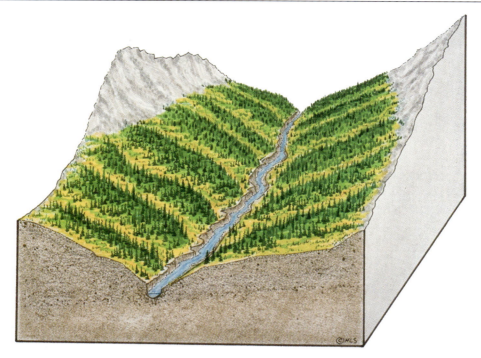

DELINEATIVE CRITERIA (A6)

Landform/soils: Steep slopes or channels dissected in fine alluvial silts and/or clays, or in residual soils derived from siltstone and shales.

Channel materials: Cohesive silt/clay with some sands.

Slope Range: .04 - .10 (A6a+ > .10) **Entrenchment Ratio:** < 1.4

Width/depth Ratio: < 12 **Sinuosity:** < 1.2

LEVEL II: THE MORPHOLOGICAL DESCRIPTION

A6a+ - California

A6 - Colorado

A6 - California

LEVEL II: THE MORPHOLOGICAL DESCRIPTION

A6. - Utah

A6a+. Colorado

A6a+. California

LEVEL II: THE MORPHOLOGICAL DESCRIPTION

MORPHOLOGICAL DESCRIPTION AND EXAMPLES OF STREAM TYPES

B1 Stream Type

The B1 stream type is a moderately entrenched channel associated with bedrock or bedrock controlled drainage ways, faults, folds, and joints, typically seen in valley types II and VI. While channel slopes normally range from 2 to 4%, the B1 type can occur on steeper slopes and still maintain a characteristically moderate (> 12) width/depth ratio. B1 stream types may also have channel slopes less than 2% and are categorized as B1c, to indicate that their morphology (e.g., width/depth ratio, confinement and sinuosity) is still characteristic of the B1 types despite the gentler gradient. Channel materials are dominated by bedrock but can also include boulders, cobble, and sand. The stream banks normally contain finer grained materials than the bed and are typically seen as an unconsolidated mixture of boulders, cobble, and sand. While the channel adjustment of the B1 stream type is primarily the result of lateral extension processes, channel sinuosity is not high (1.1-1.5), but is somewhat greater than the A1 stream type. The B1 stream types are considered as stable channel systems and contribute small amounts of sediment from their beds and banks. The B1 stream type channels are dominated by bed features that produce extensive rapids, with infrequent scour holes for pools. The sequence of the pool-to-pool spacing is irregular and infrequent due to the nature of the bedrock bed.

LEVEL II: THE MORPHOLOGICAL DESCRIPTION

DELINEATIVE CRITERIA (B1)

Landform/soils: Structural controlled narrow valleys with moderate side slopes.

Channel materials: Bedrock bed with streambanks composed of boulders, cobble and gravel.

Slope Range: .02 - .04 (B1c < .02) **Entrenchment Ratio:** 1.4 - 2.2

Width/depth Ratio: > 12 **Sinuosity:** > 1.2

LEVEL II: THE MORPHOLOGICAL DESCRIPTION

B1c - Pennsylvania

B1 - North Carolina

B1c - Wyoming

LEVEL II: THE MORPHOLOGICAL DESCRIPTION

B1 - Oregon

B1 - Maryland

B1 - New Mexico

LEVEL II: THE MORPHOLOGICAL DESCRIPTION

MORPHOLOGICAL DESCRIPTION AND EXAMPLES OF STREAM TYPES

B2 Stream Type

The B2 stream types are seen as moderately entrenched systems, with channel gradients of 2-4%. B2 stream types are typically located in or on coarse alluvial fans, lag deposits from old landslide debris, rockfall and talus areas, coarse colluvial deposits, and structurally controlled drainage-ways. Valley types that normally contain B2 stream channels include types II, III, and VII. The channel bed morphology is dominated by boulder materials, and characterized as series of rapids with irregular spaced scour pools. The B2 stream type has a moderate width/depth ratio and a sinuosity greater than 1.2. Many B2 stream channels have developed in residual materials derived from resistant rock types or from alluvial and/or colluvial deposition. The channel materials are composed primarily of boulders with lesser amounts of cobble, gravel and sand. The boulder materials are often associated with lag deposits originating from both alpine and continental glaciation. The bed and bank materials of the B2 stream types are considered stable and contribute only small quantities of sediment during runoff events.

LEVEL II: THE MORPHOLOGICAL DESCRIPTION

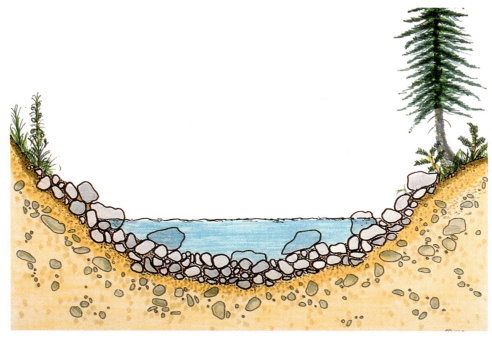

DELINEATIVE CRITERIA (B2)

Landform/soils: Structural controlled narrow valleys associated with colluvium or lag deposits, narrow, moderate to gentle glaciated valleys.

Channel materials: Boulders with smaller amounts of cobble, gravel and sand.

Slope Range: .02 - .04 (B2c < .02)   **Entrenchment Ratio:** 1.4 - 2.2

Width/depth Ratio: > 12   **Sinuosity:** > 1.2

LEVEL II: THE MORPHOLOGICAL DESCRIPTION

STREAM TYPE B2

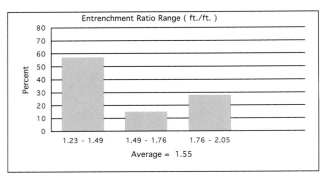
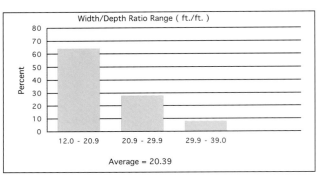
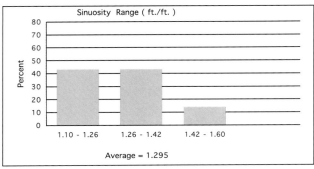
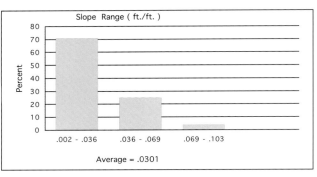
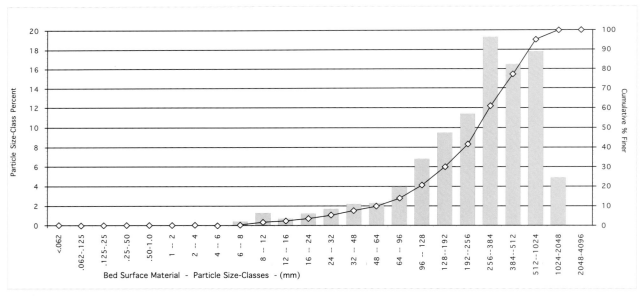

LEVEL II: THE MORPHOLOGICAL DESCRIPTION

B2 - Idaho

B2 - California

B2 - Maryland

LEVEL II: THE MORPHOLOGICAL DESCRIPTION

MORPHOLOGICAL DESCRIPTION AND EXAMPLES OF STREAM TYPES

B3 Stream Type

The B3 stream types are moderately entrenched systems with channel gradients of 2-4%. B3 stream types are typically developed in very coarse alluvial fans, lag deposits from stabilized slide debris, rockfall, talus, and very coarse colluvial deposits and structurally controlled drainage ways. Valley types that contain B3 stream channels include types II, III, and VII. The channel bed morphology is dominated by cobble materials and characterized by a series of rapids with irregular spaced scour pools. The average pool-to-pool spacing for the B3 stream type is 3-4 bankfull channel widths. Pool to pool spacing for the B3c (< 2% slope) is generally 4-5 bankfull channel widths. Pool to pool spacing adjusts inversely to stream gradient. The B3 stream type has a moderate width/depth ratios and sinuosity greater than 1.2. Many B3 stream types are associated with residual materials derived from resistant rock types or from alluvial and/or colluvial deposition. The channel materials are composed primarily of cobble with a few boulders, lesser amounts of gravel and sand. The large cobble materials often have originated from lag deposits that are the result of both alpine and continental glaciation. The bed and bank materials of the B3 stream types are stable and contribute only small quantities of sediment during runoff events. Large woody debris is an important component for fisheries habitat when available.

LEVEL II: THE MORPHOLOGICAL DESCRIPTION

DELINEATIVE CRITERIA (B3)

Landform/soils: Narrow, moderately steep colluvial valleys with gentle side slopes. Soils are colluvium and/or alluvium. Often in fault line valleys or on well vegetated alluvial fans.

Channel materials: Predominantly cobble with lesser amounts of boulders, gravel and sand. Streambanks are stable due to coarse material.

Slope Range: .02 - .04 (B3c, < .02) **Entrenchment Ratio:** 1.4 - 2.2

Width/depth Ratio: > 12 **Sinuosity:** > 1.2

LEVEL II: THE MORPHOLOGICAL DESCRIPTION

STREAM TYPE B3

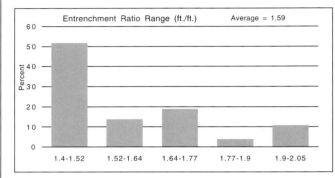
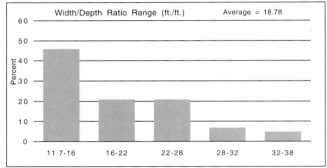
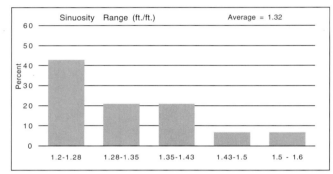
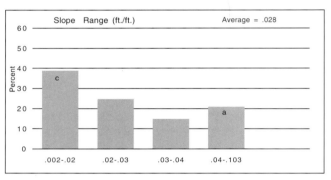
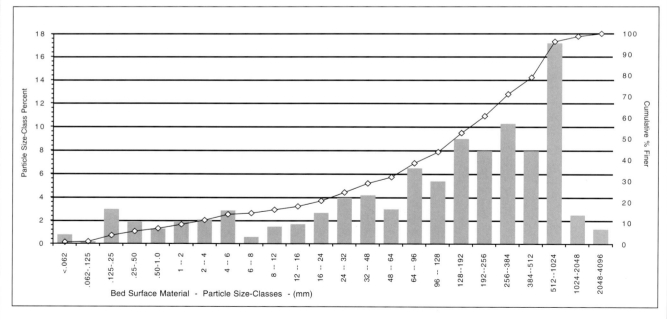

5-70

LEVEL II: THE MORPHOLOGICAL DESCRIPTION

B3 - Wyoming

B3 - Maryland

B3 - Oregon

LEVEL II: THE MORPHOLOGICAL DESCRIPTION

MORPHOLOGICAL DESCRIPTION AND EXAMPLES OF STREAM TYPES

B4 Stream Type

The B4 stream types are moderately entrenched system on gradients of 2-4%. B4 stream types normally develop in stable alluvial fans, colluvial deposits, and structurally controlled drainage ways. Landforms are often gentle to rolling slopes in relatively narrow, colluvial or structurally controlled valleys. Valley types that contain B4 stream channel include are types II, III, and VII. The channel bed morphology is dominated by gravel material and characterized as a series of rapids with irregular spaced scour pools. The average pool-to-pool spacing for the B4 stream type is 3-4 bankfull channel widths, for the B4c (< 2% slope), pool to pool spacing for the B4c (<2%) is generally 4-5 bankfull channel widths. Pool to pool spacing adjusts inversely with stream gradient. The B4 stream type has a moderate width/depth ratio and a sinuosity greater than 1.2. Many B4 stream types are associated with residual materials derived from resistant rock types or from alluvial and/or colluvial deposition. The channel materials are composed predominantly of gravel with lesser amounts of boulders, gravel, and sand. The B4 stream type is considered relatively stable and is not a high sediment supply stream channel. Large, woody, debris is an important component for fisheries habitat when available.

LEVEL II: THE MORPHOLOGICAL DESCRIPTION

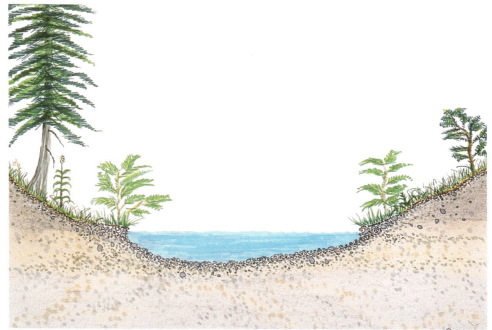

DELINEATIVE CRITERIA (B4)

Landform/soils: Narrow, moderately steep colluvial valleys, occasionally on well vegetated, stable alluvial fan, or in fault line valleys.

Channel materials: Gravel dominated with lesser amounts of boulders, cobble and sand.

Slope Range: .02 - .04 (B4c < .02)   **Entrenchment Ratio:** 1.4 - 2.2

Width/depth Ratio: > 12   **Sinuosity:** > 1.2

LEVEL II: THE MORPHOLOGICAL DESCRIPTION

STREAM TYPE B4

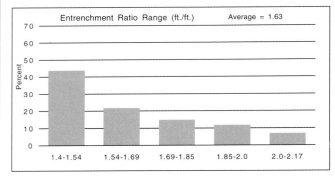
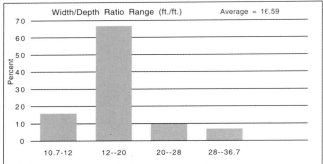
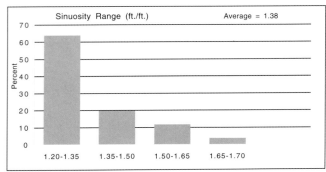
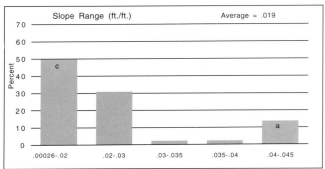
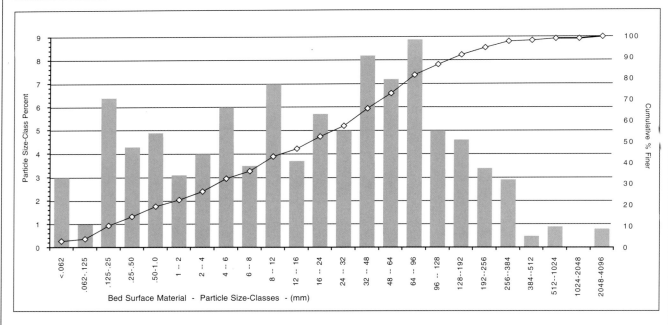

LEVEL II: THE MORPHOLOGICAL DESCRIPTION

B4 - Ontario, Canada

B4 - Alaska

B4 - Maryland

LEVEL II: THE MORPHOLOGICAL DESCRIPTION

MORPHOLOGICAL DESCRIPTION AND EXAMPLES OF STREAM TYPES

B5 Stream Type

The B5 stream types are moderately entrenched systems with channel gradients of 2-4%. B5 stream types are typically established on stable, well vegetated alluvial fans, colluvial deposits and relatively narrow, moderately sloping valleys. Landforms are observed as areas with gentle, rolling slopes in relatively narrow, colluvial valleys, and soils derived from residual materials including grussic granite, eolian sand deposits, and colluvial deposits. Valley types that contain B5 stream channels are types II, III, and VII. The channel bed morphology is dominated by sand-sized materials and characterized as a series of rapids with irregular spaced scour pools.

The average pool-to-pool spacing for the B5 stream type is 3-4 bankfull channel widths. Pool to pool spacing for the B5c (<2% slope) is generally 4-5 bankfull channel widths. Pool to pool spacing adjusts inversely with stream gradient. The B5 stream type has a moderate width/depth ratio and a sinuosity greater than 1.2. The channel materials are composed predominantly of sand and small gravel with occasional amounts of silt/clay. The B5 stream type is relatively stable where the presence of dense riparian vegetation is noted. Large, woody, organic debris is an important component of fisheries habitat where sources are available.

LEVEL II: THE MORPHOLOGICAL DESCRIPTION

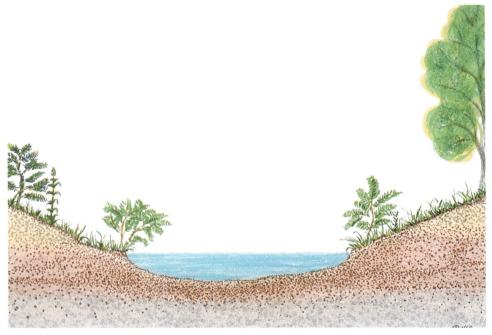

DELINEATIVE CRITERIA (B5)

Landform/soils: Narrow, moderately steep colluvial valleys with gentle sloping side slopes. Can also be located on stable, well vegetated alluvial fans. Soils from residual, eolian, alluvium and/or colluvium.

Channel materials: Predominantly sand with lesser amounts of gravel and silt/clay.

Slope Range: .02 - .04 (B5c, < .02) **Entrenchment Ratio:** 1.4 - 2.2

Width/depth Ratio: > 12 **Sinuosity:** > 1.2

LEVEL II: THE MORPHOLOGICAL DESCRIPTION

B5 - Colorado

B5 - Maryland

B5 - Texas

LEVEL II: THE MORPHOLOGICAL DESCRIPTION

B5 - California

B5 - New Mexico

B5 - Idaho

LEVEL II: THE MORPHOLOGICAL DESCRIPTION

MORPHOLOGICAL DESCRIPTION AND EXAMPLES OF STREAM TYPES

B6 Stream Type

The B6 stream type is a moderately entrenched system, incised in cohesive materials, with channel slopes less than 4%. The B6 stream types are found in narrow valleys containing cohesive residual soils; in depositional landscapes composed of fine, wind deposited (Loess) materials formed as gently sloping terrain; and on well vegetated alluvial fans. Valley types that contain the B6 stream channels include types II, III, and VI. The width/depth ratio of the B6 stream type is generally the lowest of all of the B stream types due to the cohesive nature of the silt/clay streambanks. B6 stream types are generally stable due to the effects of moderate entrenchment and lower width/depth ratios. Additionally, riparian vegetation associated with the B6 type is generally very dense, except in arid environments and plays an important role in maintaining channel stability and lower width/depth ratios. These stream types are "washload" rather than "bedload" streams, and thus, have a characteristically low sediment supply and an infrequent occurrence of sediment deposition.

LEVEL II: THE MORPHOLOGICAL DESCRIPTION

DELINEATIVE CRITERIA (B6)

Landform/soils: Narrow, moderately steep valleys with gentle sloping sideslopes. Soils either residual, alluvial and/or colluvial.

Channel materials: Silt/Clay with lesser amounts of sand.

Slope Range: .02 - .04 (B5c, < .02)   **Entrenchment Ratio:** 1.4 - 2.2

Width/depth Ratio: > 12   **Sinuosity:** > 1.2

LEVEL II: THE MORPHOLOGICAL DESCRIPTION

B6 - Montana

B6 - Texas

LEVEL II: THE MORPHOLOGICAL DESCRIPTION

B6c - Georgia

B6c - Maryland

B6c - Wisconsin

LEVEL II: THE MORPHOLOGICAL DESCRIPTION

MORPHOLOGICAL DESCRIPTION AND EXAMPLES OF STREAM TYPES

C1 Stream Type

Colorado

C1 stream types are slightly entrenched, meandering, alluvial channels with bedrock controlled beds, and occur on gentle gradients in broad valleys. They are characterized by a well developed floodplain constructed of alluvium. The C1 stream type can be found in narrow, structurally controlled valleys as well as in very broad valleys and across a diversity of hydro-physiographic provinces often associated with sedimentary rock. The C1 stream channels are found in valley types IV, VI, and VIII. The C1 stream type adjusts laterally and typically has a high width/depth ratio due to the controlling influence of bedrock materials in the channel bed. The stream gradients are less than 2%. Sinuosity is moderate and meander width ratios are low relative to other alluvial streams. The spacing of pools is related to the nature and resistance of the bedrock, and backwater pools are often created by irregular spacing of large, woody organic debris. Channel materials are predominantly bedrock, although gravels and sand occur in small amounts in depositional sites. The C1 stream type is very stable. Sediment supply for this stream type is very low.

LEVEL II: THE MORPHOLOGICAL DESCRIPTION

DELINEATIVE CRITERIA (C1)

Landform/soils: Broad, gentle gradient structural controlled alluvial valleys.
Channel materials: Bedrock bed with alluvial banks (cobble, gravel and sand).
Slope Range: < .02 (C1c- .001)   **Entrenchment Ratio:** > 2.2
Width/depth Ratio: >12   **Sinuosity:** >1.2

LEVEL II: THE MORPHOLOGICAL DESCRIPTION

C1 - Maryland

C1 - New Mexico

C1 - North Carolina

LEVEL II: THE MORPHOLOGICAL DESCRIPTION

C1 - West Virginia

C1 - Utah

C1 - Texas

LEVEL II: THE MORPHOLOGICAL DESCRIPTION

MORPHOLOGICAL DESCRIPTION AND EXAMPLES OF STREAM TYPES

C2 Stream Type

C2 stream types are boulder-dominated, meandering, high width/depth ratio stream channels with well developed flood plains. They are generally associated with coarse lag deposits originating from extreme past flood events, or from glacial deposition. The C2 stream channel is found in valley types IV, V, VI, and VIII. Modern flows rarely entrain the dominant boulder-sized bed material. The streambanks are composed of coarse, resistant particles and in combination with the boulder materials in the bed, produce a stable stream type. The spacing of pools is related to the nature and resistance of the boulder materials and backwater pools are often created by irregular spacing of large, woody organic debris. These streams have a relatively low bankfull velocity, due to the high relative roughness developing from a high width/depth ratio and large sized channel bed and bank particle. The hydraulic and roughness coefficients of the C2 channel are much different than those for the C1 stream type. Sediment supply in the C2 stream type is very low.

LEVEL II: THE MORPHOLOGICAL DESCRIPTION

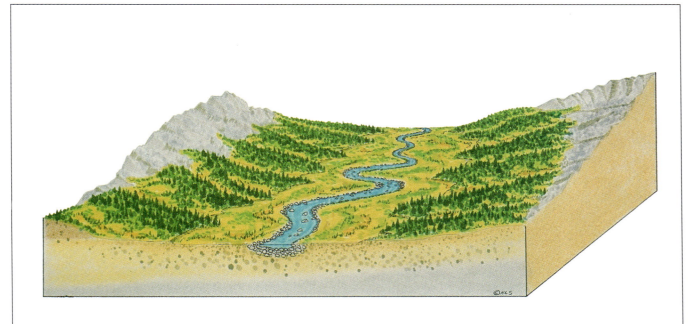

DELINEATIVE CRITERIA (C2)

Landform/soils: Broad, gentle gradient, alluvial valleys associated with lag deposits. Can also be associated with glaciated and/or structural controlled valleys.

Channel materials: Predominantly boulders with lesser amounts of cobble, gravel and sand.

Slope Range: < .02 (C2c- .001) **Entrenchment Ratio:** > 2.2

Width/depth Ratio: >12 **Sinuosity:** >1.2

LEVEL II: THE MORPHOLOGICAL DESCRIPTION

C2 - California

C2 - Nevada

C2 - California

LEVEL II: THE MORPHOLOGICAL DESCRIPTION

C2 - Washington

C2 - Colorado

C2 - California

LEVEL II: THE MORPHOLOGICAL DESCRIPTION

MORPHOLOGICAL DESCRIPTION AND EXAMPLES OF STREAM TYPES

C3 Stream Type

The C3 stream type is a slightly entrenched, meandering, riffle/pool, cobble-dominated channel with a well developed floodplain. The C3 stream type is found in U-shaped glacial valleys; valleys bordered by glacial and Holocene terraces; and in very broad, coarse alluvial valleys typical of the plains areas. The dominant bed material is often originating as a lag deposit from both Pleistocene and Holocene deposition and from extreme, rare floods. The C3 stream channels are found in Valley Types IV, V, VI, and VIII. C3 stream channels have gentle gradients of less than 2%, display a high width/depth ratio, are slightly more sinuous and have a higher meander width ratio than the C1 and C2 stream types. The riffle/pool sequence of the C3 stream type is on average at 5-7 bankfull channel widths. The streambanks are generally composed of unconsolidated, heterogenous, non-cohesive, alluvial materials that are finer than the cobble-dominated bed material. Consequently, the channel is susceptible to accelerated bank erosion. Rates of lateral adjustment are influenced by the presence and condition of riparian vegetation. Sediment supply is low, unless streambanks are in a high erodibility condition. Meander and depositional patterns which modify the condition of this stream type are described in Chapter 6.

LEVEL II: THE MORPHOLOGICAL DESCRIPTION

DELINEATIVE CRITERIA (C3)

Landform/soils: Broad alluvial and glaciated valleys, holocene terraces generally present. Soils are glacial deposition and alluvium.

Channel materials: Predominantly cobble with lesser amounts of gravel and sand. Banks are finer in material size than channel bed.

Slope Range: < .02 (C3c- .001) **Entrenchment Ratio:** > 2.2

Width/depth Ratio: >12 **Sinuosity:** >1.2

LEVEL II: THE MORPHOLOGICAL DESCRIPTION

STREAM TYPE C3

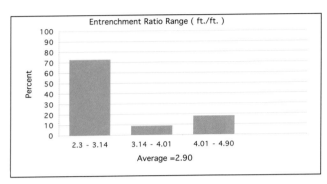
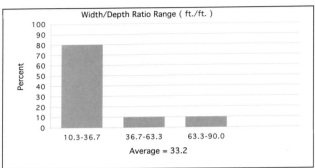
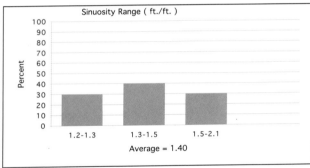
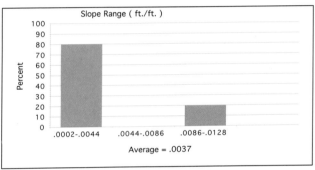
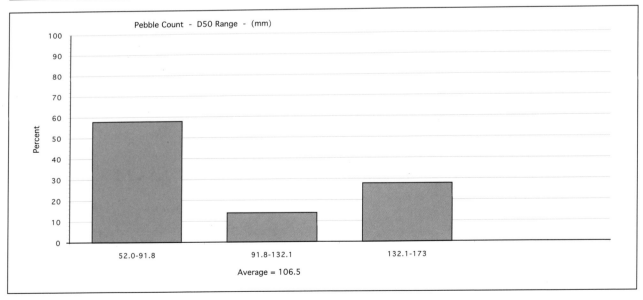

LEVEL II: THE MORPHOLOGICAL DESCRIPTION

C3 - Montana

C3 - Arizona

C3 - California

LEVEL II: THE MORPHOLOGICAL DESCRIPTION

MORPHOLOGICAL DESCRIPTION AND EXAMPLES OF STREAM TYPES

C4 Stream Type

The C4 stream type is a slightly entrenched, meandering, gravel-dominated, riffle/pool channel with a well developed floodplain. The C4 stream type is found in U-shaped glacial valleys; valleys bordered by glacial and Holocene terraces; and in very broad, coarse alluvial valleys typical of the plains areas. Some of the C4 stream types occur in glacial outwash terrain, closer to the lobe where gravel material is present The C4 stream channels are found in Valley Types IV, V, VI, VIII, IX and X. C4 stream channels have gentle gradients of less than 2%, display a high width/depth ratio, are slightly more sinuous and have a higher meander width ratio than the C1, C2 and C3 stream types. The riffle/pool sequence for the C4 stream type average 5-7 bankfull channel widths in length. The streambanks are generally composed of unconsolidated, heterogenous, non-cohesive, alluvial materials that are finer than the gravel-dominated bed material. Consequently, the stream is susceptible to accelerated bank erosion. Rates of lateral adjustment are influenced by the presence and condition of riparian vegetation. Sediment supply is moderate to high, unless streambanks are in a very low erodibility condition. The C4 stream type, characterized by the presence of point bars and other depositional features, is very susceptible to shifts in both lateral and vertical stability caused by direct channel disturbance and changes in the flow and sediment regimes of the contributing watershed. Meander and depositional patterns which modify the condition of this stream type are described in Chapter 6.

LEVEL II: THE MORPHOLOGICAL DESCRIPTION

DELINEATIVE CRITERIA (C4)

Landform/soils: Broad, gentle gradient alluvial valleys and river deltas. Soils are alluvium.

Channel materials: Predominantly gravel, with lesser amounts of cobble, sand and silt/clay.

Slope Range: < .02 (C4c- .001) **Entrenchment Ratio:** > 2.2

Width/depth Ratio: >12 **Sinuosity:** >1.2

LEVEL II: THE MORPHOLOGICAL DESCRIPTION

STREAM TYPE C4

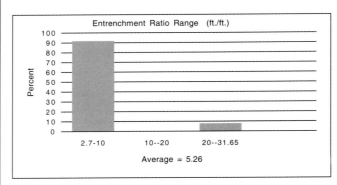

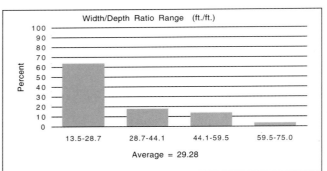

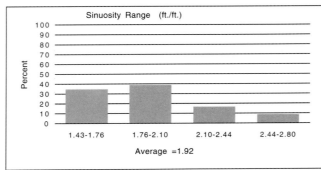

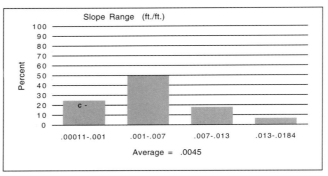

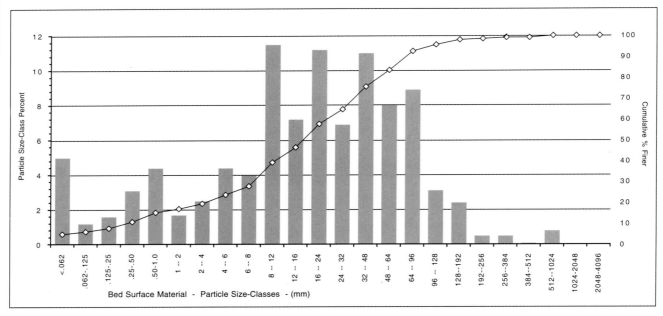

LEVEL II: THE MORPHOLOGICAL DESCRIPTION

C4 - New Mexico

C4 - Colorado

C4 - Wisconsin

LEVEL II: THE MORPHOLOGICAL DESCRIPTION

MORPHOLOGICAL DESCRIPTION AND EXAMPLES OF STREAM TYPES

C5 Stream Type

The C5 stream type is a slightly entrenched, meandering, sand-dominated, riffle/pool channel with a well developed floodplain. The C5 stream type occurs in broad valleys and plains areas with a history of riverine, lacustrine, glacial (outwash and glacio-fluvial), and eolian deposition. The C5 stream type can be found in very low relief basins typical of the interior lowlands, great plains, coastal plains, and in river deltas. Glacial outwash areas can also develop C5 stream types. The C5 stream channels are found in Valley Types IV, V, VI, VIII, IX, X, and XI. It is obvious that the C5 stream type is widely distributed throughout a wide range of physiographic provinces. Generally, C5 stream channels have gentle gradients of less than 2%. Gradients less than 0.001 are denoted as a C5c- to indicate the slope condition of many C5 stream types. The C5 stream channel displays a higher width/depth ratio than the C4 and C3 stream types due to the depositional characteristic of the stream bed and the active lateral migration tendencies. The riffle/pool sequence for the C5 stream type averages 5-7 bankfull channel widths in length. Bed forms of ripples, dunes, and anti-dunes are prevalent. The streambanks are generally composed of sandy material, with stream beds exhibiting little difference in pavement and sub-pavement material composition. Rates of lateral adjustment are influenced by the presence and condition of riparian vegetation. Sediment supply is high to very high, unless streambanks are in a very low erodibility condition. The C5 stream type, characterized by the presence of point bars and other depositional features, is very susceptible to shifts in both lateral and vertical stability caused by direct channel disturbance and changes in the flow and sediment regimes of the contributing watershed. Meander and depositional patterns which modify the condition of this stream type are described in Chapter 6.

LEVEL II: THE MORPHOLOGICAL DESCRIPTION

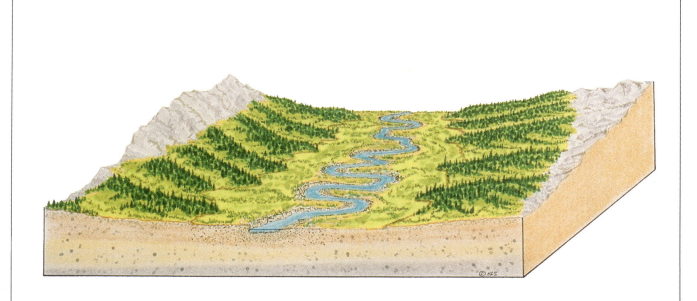

DELINEATIVE CRITERIA (C5)

Landform/soils: Broad, gentle gradient alluvial valleys, river deltas, broad plains. Soils are depositional such as lacustrine, glacial outwash, eolian.

Channel materials: Predominantly sand bed and banks, with occasional gravel and silt/clay. Streambanks may contain finer particles than bed material.

Slope Range: < .02 (C5c- .001) **Entrenchment Ratio:** > 2.2

Width/depth Ratio: >12 **Sinuosity:** >1.2

LEVEL II: THE MORPHOLOGICAL DESCRIPTION

STREAM TYPE C5

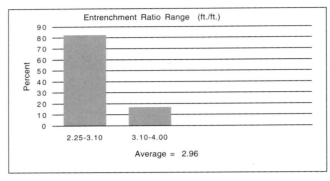
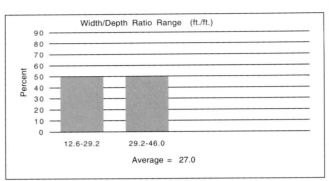
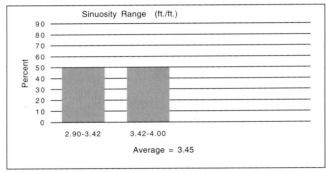
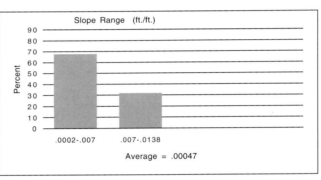
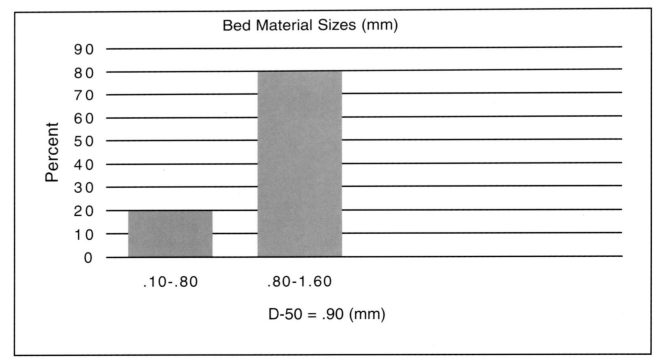

LEVEL II: THE MORPHOLOGICAL DESCRIPTION

C5 - Colorado

C5 - Colorado

C5 - Montana

LEVEL II: THE MORPHOLOGICAL DESCRIPTION

MORPHOLOGICAL DESCRIPTION AND EXAMPLES OF STREAM TYPES

C6 Stream Type

The C6 stream type is a slightly entrenched, meandering, silt-clay dominated, riffle-pool channel with a well developed floodplain. The C6 stream type occurs in broad valleys and plains areas with a history of riverine, lacustrine, and eolian deposition (loess). The C6 stream type can be found in very low relief basins typical of the interior lowlands, great plains, coastal plains, and in river deltas. The lower extremities of glacial outwash areas can also develop C6 stream types. The C6 stream channels are associated with Valley Types IV, V, VI, VIII, IX, X, and XI. It is obvious that the C6 stream type is widely distributed throughout a wide range of physiographic provinces. Generally, C6 stream channels have gentle gradients of less than 2%. Gradients less than 0.001 are denoted as a C6c- to indicate the very low gradients of many C6 stream types. The C6 stream channel displays a lower width/depth ratio than all of the other C stream types due to the cohesive nature of stream bank materials. The riffle/pool sequence for the C6 stream type is, on average, 5-7 bankfull channel widths in length. The streambanks are generally composed of silt-clay and organic materials, with the stream beds exhibiting little difference in pavement and sub-pavement material composition. Rates of lateral adjustment are influenced by the presence and condition of riparian vegetation. Sediment supply is moderate to high, unless streambanks are in a very high erodibility condition. Bedload sediment yields for the stream types are typically low, reflecting the presence of fine bed and bank materials and gentle channel slopes. The C6 stream type is very susceptible to shifts in both lateral and vertical stability caused by direct channel disturbance and changes in the flow and sediment regimes of the contributing watershed. Meander and depositional patterns which modify the condition of this stream type are described in Chapter 6.

LEVEL II: THE MORPHOLOGICAL DESCRIPTION

DELINEATIVE CRITERIA (C6)

Landform/soils: Broad gentle valleys, plains, and deltas. Depositional soils (alluvium), associated with cohesive materials from riverine and lacustrine process. Often associated with tidal influence deltas, marshes and other wetland complexes.

Channel materials: Silt-clay predominates, however many of these C6 stream types are associated with a high organic component including peat.

Slope Range: < .02 (C6c- .001) (.0001 more common) **Entrenchment Ratio:** > 2.2

Width/depth Ratio: >12 (generally lowest of C's) **Sinuosity:** >1.2

LEVEL II: THE MORPHOLOGICAL DESCRIPTION

C6 - Colorado

C6 - Texas

C6 - Georgia

LEVEL II: THE MORPHOLOGICAL DESCRIPTION

C6 - Georgia

C6 - Texas

C6 - Texas

LEVEL II: THE MORPHOLOGICAL DESCRIPTION

MORPHOLOGICAL DESCRIPTION AND EXAMPLES OF STREAM TYPES

D3 Stream Type

The D3 stream types are multiple channel systems, described as braided streams found within broad alluvial valleys and on alluvial fans consisting of coarse depositional materials formed into moderately steep terrain. Primarily, the braided system consists of interconnected distributary channels formed in depositional environments. The D3 stream type occurs in moderately steep, narrow, U-shaped glacial valleys; on alluvial fans; and in gentle gradient alluvial valleys. The D3 stream channels may be found in Valley Types III, V, and VIII. Channel bed materials are predominantly cobble, with a strong bi-modal distribution of sands. Gravel is also present, but often imbedded with sands. The braided channel system is characterized by high bank erosion rates, excessive deposition occurring as both longitudinal and transverse bars, and annual shifts of the bed locations. Bed features are developed from convergence/divergence processes. The channels generally are at the same gradient as their parent valley. A combination of adverse conditions are responsible for channel braiding, including high sediment supply, high bank erodibility, moderately steep gradients, and very flashy runoff conditions which can vary rapidly from a base flow to an over-bank flow on a frequent basis. Characteristic width/depth ratios are very high, exceeding values of 40 to 50 with values of 400 or larger often noted. D3 channel gradients are generally less than 2%; however, D3 types can also develop within alluvial fans which have slopes of 2% to 4% (D3b). Observations have been made of braided streams on alluvial fans with slopes greater than 4% (D3a). The D3 is a very high sediment supply system and typically produces high sediment yields.

LEVEL II: THE MORPHOLOGICAL DESCRIPTION

DELINEATIVE CRITERIA (D3)

Landform/soils: Moderately steep glacial valleys, alluvial fans, narrow fluvial mountain valleys and terraced valleys in coarse alluvium. Bed material can be lag deposit.

Channel materials: Cobble dominated with a mixture of gravel and sand. Bank materials are finer than bed material, and generally actively eroding.

Slope Range: < .04

Entrenchment Ratio: NA (not incised)

Width/depth Ratio: > 40

Sinuosity: Low, channel slope = valley slope

LEVEL II: THE MORPHOLOGICAL DESCRIPTION

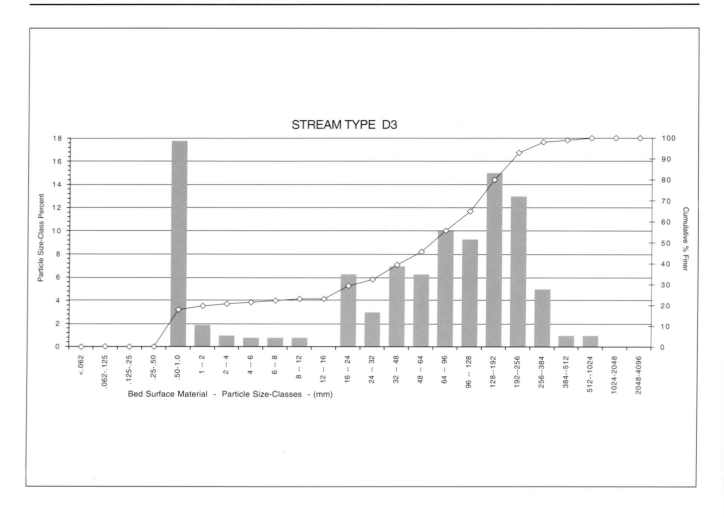

D3 - Montana

LEVEL II: THE MORPHOLOGICAL DESCRIPTION

D3 - Colorado

D3 - Montana

D3 - Colorado

LEVEL II: THE MORPHOLOGICAL DESCRIPTION

MORPHOLOGICAL DESCRIPTION AND EXAMPLES OF STREAM TYPES

D4 Stream Type

The D4 stream types are multiple channel systems, described as braided streams, found within broad alluvial valleys and on alluvial fans consisting of coarse depositional materials formed into moderately steep terrain. Primarily, the braided system consist of interconnected distributary channels formed in depositional environments. The D4 stream type occurs in moderately steep, narrow, U-shaped glacial valleys; on alluvial fans; and in gentle gradient alluvial valleys. This stream type can also occur on low relief river deltas, as well as on the upper lobe of glacial outwash valleys. The D4 stream channels may be found in Valley Types III, V, VIII, IX, X, and XI. Channel bed materials are predominantly gravel, with a strong bi-modal distribution of sands. Cobble may be found in lesser amounts, often imbedded with sands. The braided channel system is characterized by high bank erosion rates, excessive deposition occurring as both longitudinal and transverse bars, and annual shifts of the bed locations. Bed morphology is characterized by a closely spaced series of rapids and scour pools formed by convergence/divergence processes that are very unstable. The channels generally are at the same gradient as their parent valley. A combination of adverse conditions are responsible for channel braiding, including high sediment supply, high bank erodibility, moderately steep gradients, and very flashy runoff conditions which can vary rapidly from a base flow to an over-bank high flow on a frequent basis. Characteristic width/depth ratios are very high, exceeding values of 40 to 50 with values of 400 or larger often noted. D4 channel gradients are generally less than 2%; however, D4 types can also develop within alluvial fans which have slopes of 2% to 4% (D4b). Observations have been made of braided streams on alluvial fans with slopes greater than 4% (D4a). The D4 is a very high sediment supply system, and typically produces high bedload sediment yields.

LEVEL II: THE MORPHOLOGICAL DESCRIPTION

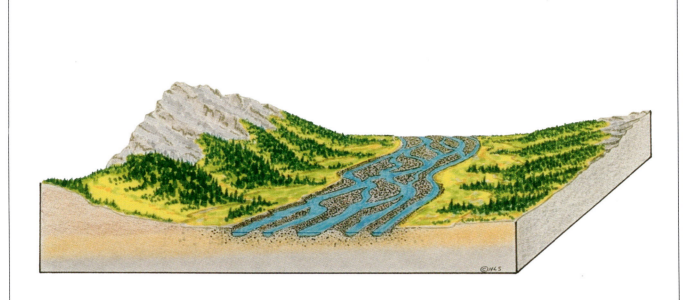

DELINEATIVE CRITERIA (D4)

Landform/soils: Moderately steep glacial valleys, alluvial fans, narrow fluvial mountain valleys and terraced valleys in coarse alluvium. Can occur in gravel splays, and coarse delta deposits.

Channel materials: Gravel bed with smaller quantities of cobble. Typical is a bi-modal distribution of sands. Stream bank materials generally finer than bed, actively eroding.

Slope Range: < .04

Entrenchment Ratio: N/A (not incised)

Width/depth Ratio: > 40

Sinuosity: Low, channel slope = valley slope

LEVEL II: THE MORPHOLOGICAL DESCRIPTION

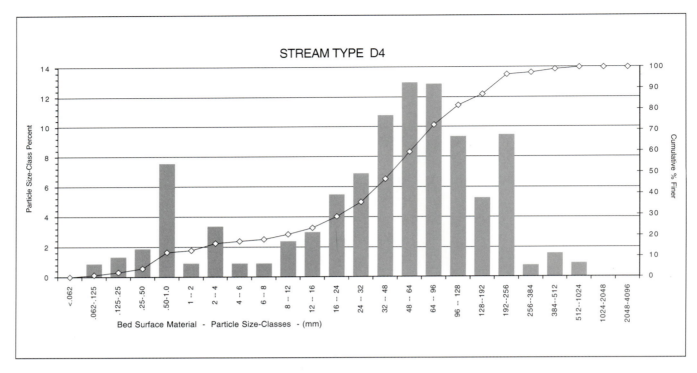

D4 - Colorado

D4 - Colorado

5-114

LEVEL II: THE MORPHOLOGICAL DESCRIPTION

D4 - Montana

D4 - Montana

D4 - Colorado

LEVEL II: THE MORPHOLOGICAL DESCRIPTION

MORPHOLOGICAL DESCRIPTION AND EXAMPLES OF STREAM TYPES

D5 Stream Type

The D5 stream types are multiple channel systems described as braided streams, found within broad alluvial valleys and on alluvial fans consisting of deposited sand-sized materials. The braided system consists of interconnected distributary channels formed in depositional environments. The D5 stream type occurs in gentle gradient, narrow, U-shaped glacial valleys consisting of glacio-lacustrine deposits, sand dunes (eolian); in very low relief alluvial valleys; and in glacial outwash areas and deltas. The D5 stream channels may be found in Valley Types III, V, VIII, IX, X, and XI. Channel bed materials are predominantly sand, with interspersed amounts of silt/clay materials on deltas and in varves of lacustrine depositional areas. The braided channel system is characterized by high bank erosion rates, excessive deposition occurring as both longitudinal and transverse bars, and annual shifts of the bed location. Bed morphology is characterized by a closely spaced series of rapids and scour pools formed by convergence/divergence processes that are very unstable. The channels generally are of the same gradient as their parent valley. A combination of adverse conditions are responsible for channel braiding, including high sediment supply, high bank erodibility, moderately steep gradients, and very flashy runoff conditions which can vary rapidly from a base flow to an over-bank flow on a frequent basis. Characteristic width/depth ratios are very high, exceeding values of 40 to 50 with values of 400 or larger often noted. D5 channel gradients are generally less than 2%; however, D5 types can also develop within alluvial fans which have slopes of 2% to 4% (D5b). Observations have been made of braided streams on alluvial fans with slopes greater than 4% (D5a). The D5 is a very high sediment supply system, and typically produces high bedload sediment yields.

LEVEL II: THE MORPHOLOGICAL DESCRIPTION

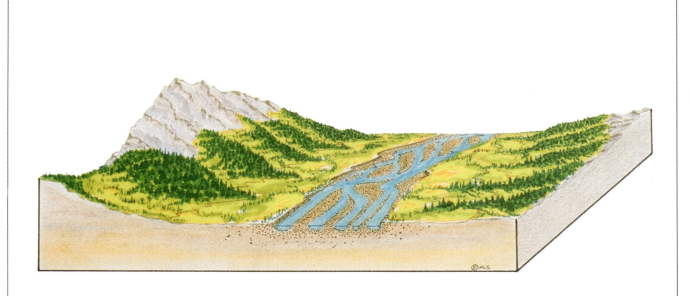

DELINEATIVE CRITERIA (D5)

Landform/soils: Moderately steep to very gentle alluvial valleys, glacial outwash plains, eolian dunes and deltas. Soils are depositional from riverine, lacustrine and eolian processes.

Channel materials: Sand dominated with occasional gravel and silt/clay.

Slope Range: < .04

Entrenchment Ratio: N/A (not incised)

Width/depth Ratio: > 40

Sinuosity: Low, channel slope = valley slope

LEVEL II: THE MORPHOLOGICAL DESCRIPTION

D5 - Wyoming

D5 - Colorado

D5 - Colorado

LEVEL II: THE MORPHOLOGICAL DESCRIPTION

D5 - Colorado

D5 - Texas

D5 - Montana

LEVEL II: THE MORPHOLOGICAL DESCRIPTION

MORPHOLOGICAL DESCRIPTION AND EXAMPLES OF STREAM TYPES

D6 Stream Type

Alaska

The D6 stream types are multiple channel systems described as braided streams, found within broad alluvial valleys and in river deltas consisting of cohesive silt-clay depositional materials. The braided system consists of interconnected distributary channels formed in depositional environments. The D6 stream type occurs typically in wide valleys with very gentle gradients. Lacustrine deposits and river deltas are the most typical landform/materials. The D6 stream channels are found in Valley Types VIII, X and XI. Channel bed materials are predominantly silt-clay with a contribution of organic material such as peat. The braided channel system is characterized by high bank erosion rates, excessive deposition occurring as both longitudinal and transverse bars, and annual shifts of the bed location. Bed morphology is characterized by a closely spaced series of rapids and scour pools formed by convergence/divergence processes that are very unstable. The channels generally are of the same gradient as their parent valley. A combination of adverse conditions are associated with braiding, including high sediment supply, high bank erodibility, moderately steep gradients, and very flashy runoff conditions which can vary rapidly from a base flow to an over-bank flow on a frequent basis. Characteristic width/depth ratios are very high, exceeding values 40 to 50, with values of 400 or larger often noted. D6 channel gradients are generally less than .001% slope. The D6 is a high sediment supply system that typically produces low bedload sediment yields and high suspended or washload sediment yields.

LEVEL II: THE MORPHOLOGICAL DESCRIPTION

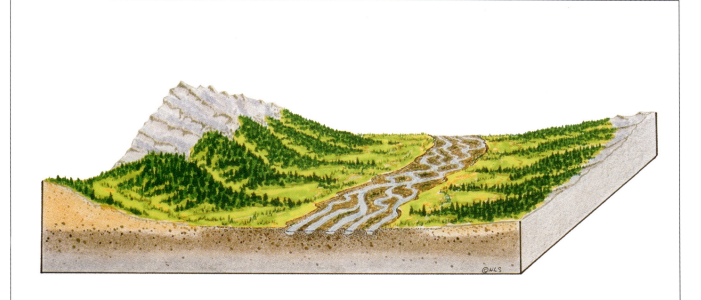

DELINEATIVE CRITERIA (D6)

Landform/soils: Gentle, wide alluvial valleys, Lacustrine deposits, and deltas. Cohesive depositional soils. Organic lenses and peat common in deltas and some lacustrine features.

Channel materials: Silts and/or clays dominate the D6 stream type. Organic contributions are common such as peat.

Slope Range: < .02 (generally less than .0001) **Entrenchment Ratio:** N/A (not incised)

Width/depth Ratio: > 40 **Sinuosity:** Low, channel slope = valley slope

LEVEL II: THE MORPHOLOGICAL DESCRIPTION

MORPHOLOGICAL DESCRIPTION AND EXAMPLES OF STREAM TYPES

DA Stream Type

British Columbia — Photo by D. Smith

The DA stream types are highly interconnected channel systems developing in gentle relief terrain areas consisting of cohesive soil materials and exhibiting wetland environments with stable channel conditions. Landforms are seen as unconfined, broad valleys with well developed floodplains, or delta areas that are more typical of marshes with stable channels. The DA stream type is characteristic of a balance between the rate of basin filling and basin subsidence such that an equilibrium condition is maintained over time. Valley types supporting the DA stream systems are types X and XI. The delta type XIe is considered as a stable delta system, supporting wetlands such as fresh water marshes and tidal influenced salt marshes, with system base levels controlled by lakes or sea level, to maintain stable elevations. In broad alluvial valleys, the anastomosed channels are generally stable due to presence of cohesive bank materials and extensively developed riparian vegetation. The stream beds are often vertically accreting, but kept in balance due to the subsidence effects of tectonically active basins (Smith and Putnam, 1980). The anastomosed channel patterns typically display a range of low to high width/depth ratios and a similar range of sinuosities, with high meander belt widths. Channel slopes are less than 0.5% but typically in the .01% range. Dominant channel bed particle sizes for this stream type vary from gravel (DA4), to sand (DA5), to silt-clay (DA6). Most of the channel banks, however, contain a highly cohesive material component, intermixed with a dense root mass. Peat is commonly found. Sediment supply and bedload contributions are low, as sediment transport is dominated by wash load or suspended sediment.

LEVEL II: THE MORPHOLOGICAL DESCRIPTION

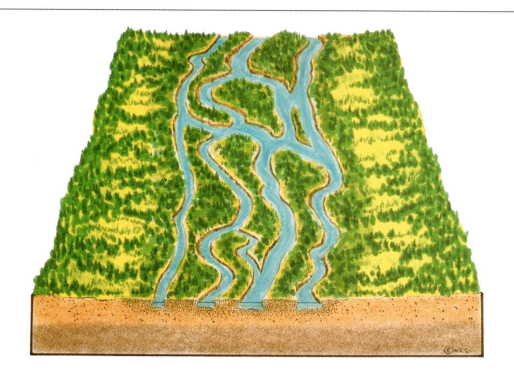

DELINEATIVE CRITERIA (DA4-DA6))

Landform/soils: Broad gentle valleys and deltas. Wetland environments with stable islands, often cohesive banks mixed with organic material. Depositional soils.

Channel materials: The materials vary from gravel (DA4), sand (DA5), to silt/clay (DA6). Peat and other organic materials are very common with these streams.

Slope Range: < .005 (average closer to .0001)   **Entrenchment Ratio:** N/A (not incised)

Width/depth Ratio: Highly variable   **Sinuosity: Highly variable**

LEVEL II: THE MORPHOLOGICAL DESCRIPTION

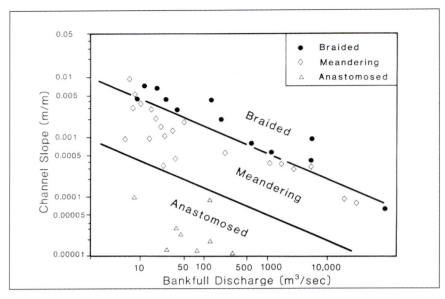

Relations of channel patterns versus slope and bankfull discharge. (Smith and Putnam, 1980).

DA - Anastomosed River.

DA - British Columbia. Anastomosed river channels.

LEVEL II: THE MORPHOLOGICAL DESCRIPTION

DA - Anastomosed River.

DA - Anastomosed River.

DA - Anastomosed River.

LEVEL II: THE MORPHOLOGICAL DESCRIPTION

MORPHOLOGICAL DESCRIPTION AND EXAMPLES OF STREAM TYPES

E3 Stream Type

Colorado

The E3 stream types are seen as systems with moderate sinuosity, with gentle to moderately steep channel gradients, and with very low width/depth ratios. The E stream types are located in a variety of land forms including high mountain meadows, alpine tundra, and broad alluvial valleys with well developed floodplains. The E3 stream channels are found in valley types VIII and X. The E3 stream type exhibits predominantly cobble-sized bed materials, with channel slopes less than 2 %; however, they can also develop with slopes of 2-4% (E3b) and in some cases with slopes greater than 4% (E3a). The E3 stream type can develop with a wide range of channel slopes due to the nature of the inherently stable bed and banks. Sinuosities and meander width ratios decrease, however, with an increase in channel slope. Streambanks are composed of materials finer than that of the dominant channel bed materials, and are typically stabilized with dense riparian or wetland vegetation that forms densely rooted sod mats from grasses and grass like plants as well as woody species. Typically the E3 stream channels have high meander width ratios, high sinuosities, and low width/depth ratios. The E3 stream types are hydraulically efficient channel forms, and they maintain a high sediment transport capacity. The narrow and relatively deep channels maintain a high resistance to plan form adjustment, which results in channel stability without significant downcutting. The E3 stream channels are very stable unless the streams are disturbed, and significant changes in sediment supply and/or streamflow occur.

LEVEL II: THE MORPHOLOGICAL DESCRIPTION

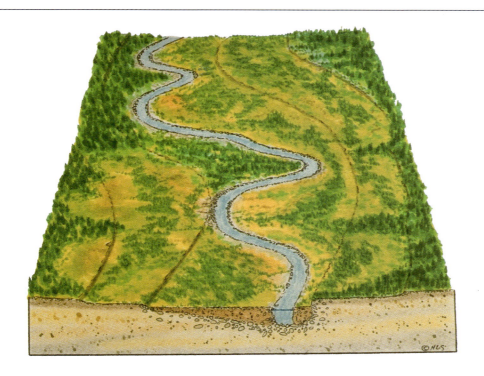

DELINEATIVE CRITERIA (E3)

Landform/soils: Broad, gentle to moderately steep alluvial valleys.

Channel materials: Cobble dominated with fewer accumulations of gravel and sand. Stream banks have gravel/sand matrix mixed with dense root mats/organic material. Very stable.

Slope Range: < .02 ($E3_b$, .02-.04)

Width/depth Ratio: < 12

Entrenchment Ratio: > 2.2

Sinuosity: > 1.5 (less if $E3_b$)

LEVEL II: THE MORPHOLOGICAL DESCRIPTION

STREAM TYPE E3

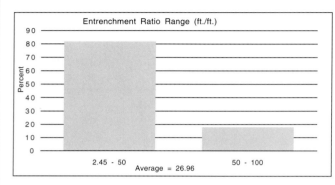
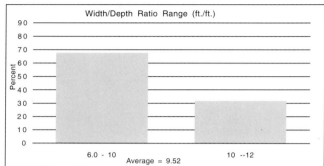
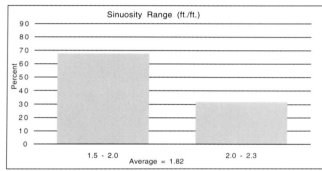
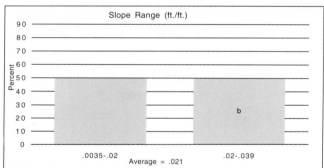
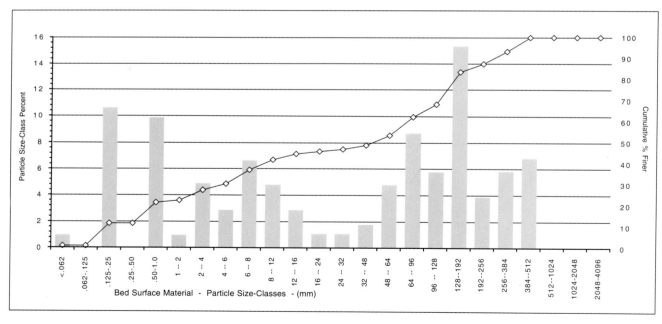

LEVEL II: THE MORPHOLOGICAL DESCRIPTION

E3 - Wyoming

E3 - Colorado

E3 - Colorado

LEVEL II: THE MORPHOLOGICAL DESCRIPTION

MORPHOLOGICAL DESCRIPTION AND EXAMPLES OF STREAM TYPES

E4 Stream Type

The E4 stream types are channel systems with low to moderately sinuosity, gentle to moderately steep channel gradients, with very low channel width/depth ratios. The E4 type is a riffle/pool stream found in a variety of land forms including high mountain meadows, alpine tundra, deltas, and broad alluvial valleys with well developed floodplains. The E4 stream channels are found in valley types VIII, X, and XI. The E4 stream type exhibits predominantly gravel sized bed materials, with channel slopes less than 2%; however, they can also develop with slopes of 2-4% (E4b) . Due to the inherently stable nature of the bed and banks, this stream type can develop with a wide range of channel slopes. Sinuosities and meander width ratios decrease, however, with an increase in slope. Streambanks are composed of materials finer than that of the dominant channel bed materials, and are typically stabilized with extensive riparian or wetland vegetation that forms densely rooted sod mats from grasses and grass like plants, as well as woody species. Typically the E4 stream channels have high meander width ratios, high sinuosities, and low width/depth ratios. The E4 stream types are hydraulically efficient channel forms and they maintain a high sediment transport capacity. The narrow and relatively deep channels maintain a high resistance to plan form adjustment which results in channel stability without significant downcutting. The E4 stream channels are very stable unless the stream banks are disturbed, and significant changes in sediment supply and/or streamflow occur.

LEVEL II: THE MORPHOLOGICAL DESCRIPTION

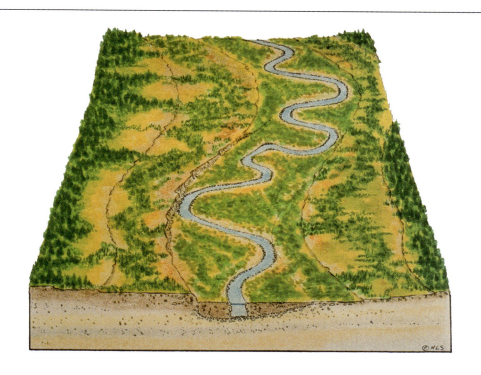

DELINEATIVE CRITERIA (E4)

Landform/soils: Gentle slopes in broad riverine or lacustrine valleys and river deltas.

Channel materials: Gravel dominated bed with smaller accumulations of sand and occasional cobble. Streambanks composed of sandy/gravel mixture with dense root mat.

Slope Range: < 0.02   **Entrenchment Ratio:** > 2.2

Width/depth Ratio: < 12   **Sinuosity:** > 1.5

LEVEL II: THE MORPHOLOGICAL DESCRIPTION

STREAM TYPE E4

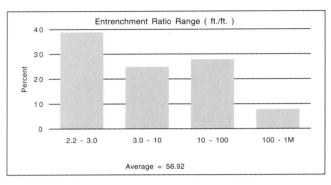
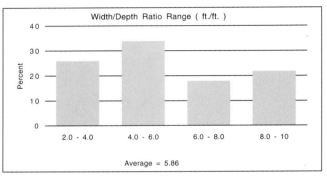
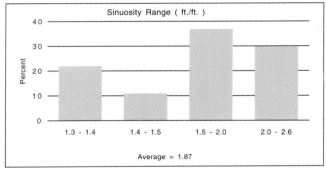
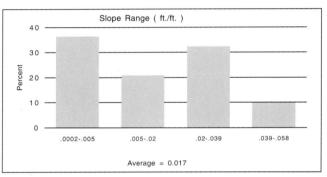
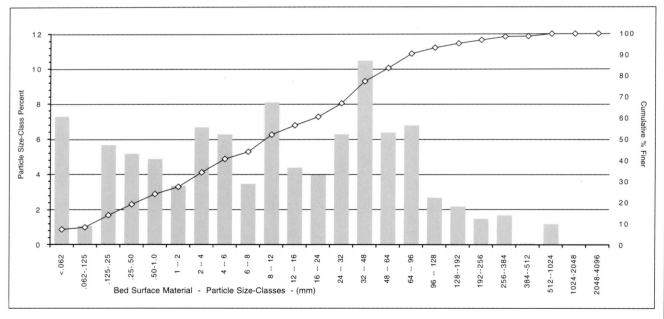

LEVEL II: THE MORPHOLOGICAL DESCRIPTION

E4 - Wyoming

E4 - Colorado

E4 - Wyoming

LEVEL II: THE MORPHOLOGICAL DESCRIPTION

MORPHOLOGICAL DESCRIPTION AND EXAMPLES OF STREAM TYPES

E5 Stream Type

Colorado

The E5 stream types are channel systems with low to moderately sinuosity, gentle to moderately steep channel gradients, and very low channel width/depth ratios. The E5 stream type is found in a variety of land forms including high mountain meadows, alpine tundra, deltas, lacustrine valleys, and broad alluvial valleys with well developed floodplains. The E5 stream channels are found in valley types VIII, X, and XI. The E5 stream type is typically seen as a riffle/pool system with channel slopes less than 2%. Due to the inherently stable nature of the bed and banks, this stream type can develop with a wide range of channel slopes. Sinuosities and meander width ratios decrease, however, with an increase in slope. Streambanks are composed of materials finer than that of the dominant channel bed materials, and are typically stabilized with extensive riparian or wetland vegetation that forms densely rooted sod mats from grasses and grass like plants, as well as woody species. Typically the E5 stream channel has high meander width ratios, high sinuosities, and low width/depth ratios. The E5 stream types are hydraulically efficient channel forms and they maintain a high sediment transport capacity. The narrow and relatively deep channels maintain a high resistance to plan form adjustment which results in channel stability without significant downcutting. The E5 stream channels are very stable unless the streambanks are disturbed, and significant changes in sediment supply and/or streamflow occur.

LEVEL II: THE MORPHOLOGICAL DESCRIPTION

DELINEATIVE CRITERIA (E5)

Landform/soils: Gentle slopes in broad riverine or lacustrine valleys and river deltas. Can be laterally contained in entrenched valley, evolving to a channel inside a previous channel.

Channel materials: Sand dominated bed with smaller accumulations of gravel and occasional silt/clay. Streambanks composed of sandy/silt/clay mixture with dense root mat.

Slope Range: < 0.02 **Entrenchment Ratio:** > 2.2

Width/depth Ratio: < 12 **Sinuosity:** > 1.5

LEVEL II: THE MORPHOLOGICAL DESCRIPTION

STREAM TYPE E5

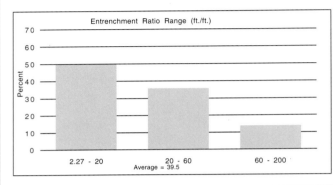
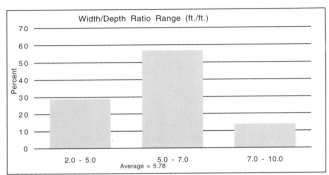
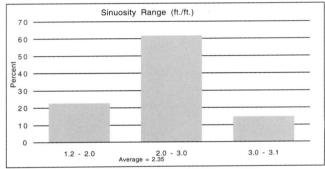
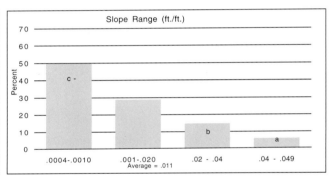
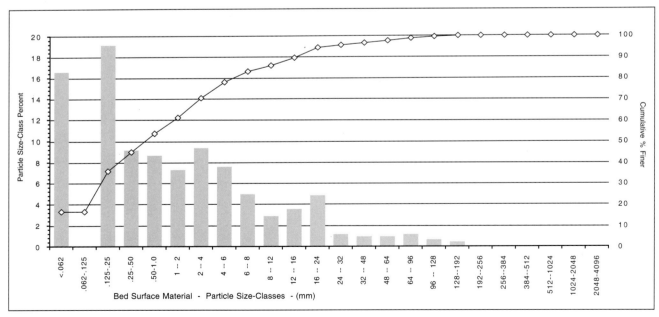

LEVEL II: THE MORPHOLOGICAL DESCRIPTION

E5 - Colorado

E5 - Wisconsin

E5 - Colorado

LEVEL II: THE MORPHOLOGICAL DESCRIPTION

MORPHOLOGICAL DESCRIPTION AND EXAMPLES OF STREAM TYPES

E6 Stream Type

Maryland

The E6 stream types are channel systems with moderate to high sinuosity, gentle to moderately steep channel gradients, and very low channel width/depth ratios. The E6 stream types are found in a variety of land forms including high mountain meadows, alpine tundra, deltas, lacustrine valleys, and broad alluvial valleys with well developed floodplains. The E6 stream channels are found in valley types VIII, X, and XI. The E6 stream type is typically seen as a riffle/pool system with the dominant channel materials composed of silt-clay, interspersed with organic materials. Channel slopes are less than 2%, with a high number having slopes of less than .01%. Due to the inherently stable nature of the bed and banks, this stream type can exist on a wide range of slopes. Sinuosities and meander width ratios decrease, however, with an increase in slope. Streambanks are composed of materials similar to those of the dominant bed materials and are typically stabilized with riparian or wetland vegetation that forms densely rooted sod mats from grasses and grass like plants as well as woody species. Typically the E6 stream channel has high meander width ratios, high sinuosities, and low width/depth ratios. The E6 stream types are hydraulically efficient forms as they require the least cross-sectional area per unit of discharge. The narrow and relatively deep channels maintain a high resistance to plan form adjustment which results in channel stability without significant downcutting. The E6 stream channels are very stable unless the streambanks are disturbed and significant changes in sediment supply and/or streamflow occur.

LEVEL II: THE MORPHOLOGICAL DESCRIPTION

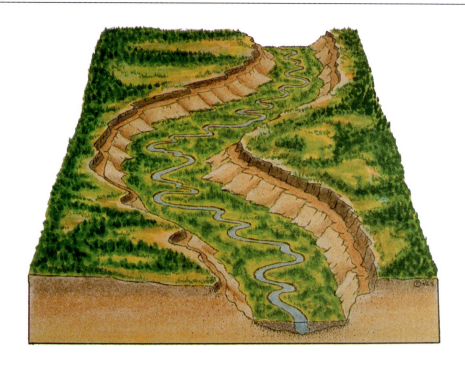

DELINEATIVE CRITERIA (E6)

Landform/soils: Gentle slopes in broad riverine or lacustrine valleys and river deltas. Can be laterally contained in entrenched valley, evolving to a channel inside a previous channel.

Channel materials: Silt/clay dominated cohesive channel materials with accumulations of organic material including peat. Dense root mat on streambanks.

Slope Range: < .02 (often < .0001)   **Entrenchment Ratio:** > 2.2

Width/depth Ratio: < 12   **Sinuosity:** > 1.5

LEVEL II: THE MORPHOLOGICAL DESCRIPTION

E6 - Colorado

E6 - Nevada

E6 - Colorado

LEVEL II: THE MORPHOLOGICAL DESCRIPTION

E6 - Texas

E6 - Utah

E6 - Nevada

LEVEL II: THE MORPHOLOGICAL DESCRIPTION

MORPHOLOGICAL DESCRIPTION AND EXAMPLES OF STREAM TYPES

F1 Stream Type

Ontario, Canada

The F1 stream type is an entrenched, meandering, high width/depth ratio channel that is deeply incised in valleys that are structurally controlled with bedrock. The F1 stream channels are often entrenched in highly weathered rock formations configured as low relief landforms with low valley gradients. Side slopes of the F1 stream types are often vertical and confine the river laterally for great distances. The F1 stream channel are located in Valley Type IV and VI. The F1 stream type does not have developed floodplains, with all of the natural range of flows contained in a similar width channel. The dominant channel materials are principally bedrock, with boulders, cobble, and gravel present in fewer quantities. The F1 stream type has a relatively low to moderate sinuosity and low meander width ratios due to the degree of natural entrenchment and lateral containment. The "top of banks" of this stream type cannot be reached by floods that may be developed with the modern-day climate. The F1 stream type typically exhibits low sediment deposition, due to the low sediment supply from the relatively stable bed and banks. These systems are considered very stable stream types due to the resistant nature of their channel materials, and basically have not changed or significantly adjusted in modern times.

LEVEL II: THE MORPHOLOGICAL DESCRIPTION

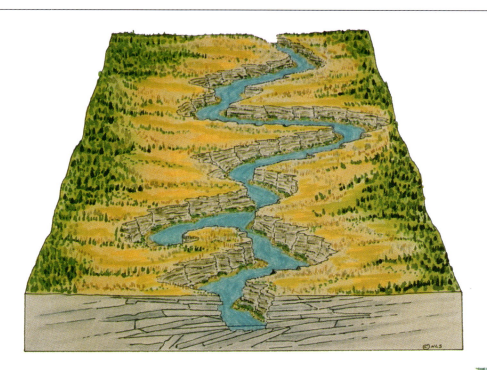

DELINEATIVE CRITERIA (F1)

Landform/soils: The F1 stream type is associated with deeply entrenched, structurally controlled, gentle gradient valleys and gorges. The F1 stream type is associated with highly weathered bedrock in a combination of river downcutting and uplift of valley walls.

Channel materials: Bedrock dominated channel with accumulations of boulders, cobble and gravel. Some sand deposits in pools and backwater eddies.

Slope Range: < .02

Width/depth Ratio: > 12

Entrenchment Ratio: < 1.4

Sinuosity: > 1.2

LEVEL II: THE MORPHOLOGICAL DESCRIPTION

F1 - Texas

F1 - Colorado

F1 - Texas

LEVEL II: THE MORPHOLOGICAL DESCRIPTION

F1 - New Mexico

F1 - Texas

LEVEL II: THE MORPHOLOGICAL DESCRIPTION

MORPHOLOGICAL DESCRIPTION AND EXAMPLES OF STREAM TYPES

F2 Stream Type

The F2 stream type is an entrenched, meandering, high width/depth ratio channel that is deeply incised in valleys that are structurally controlled with boulder materials. The F1 stream channels are often entrenched in highly weathered rock formations configured as in low relief landforms with low valley gradients. Side slopes of the F1 stream types are often vertical and confine the river laterally for great distances. The F2 stream channels are located in Valley Type IV and VI. The F2 stream type does not have developed floodplains, with all of the natural range of flows contained in a similar width channel.. The dominant channel materials are boulders, with cobble and gravel present in fewer quantities. The F2 stream type has a relatively low to moderate sinuosity and low meander width ratios due to the degree of natural entrenchment and lateral containment. The "top of banks" of this stream type cannot be reached by floods that may be developed with the modern-day climate. The F2 stream type typically exhibits low sediment deposition due to the low sediment supply from the relatively stable bed and banks. These systems are considered very stable stream types due to the resistant nature of their channel materials, and basically have not changed or significantly adjusted in modern times.

LEVEL II: THE MORPHOLOGICAL DESCRIPTION

DELINEATIVE CRITERIA (F2)

Landform/soils: The F2 stream type is associated with deeply entrenched, structurally controlled, gentle gradient valleys and gorges. The F2 stream type is associated with highly weathered bedrock in a combination of river downcutting and uplift of valley walls.

Channel materials: Boulder dominated channel with accumulations of cobble and gravel. Some sand deposits in pools and backwater eddies.

Slope Range: < .02

Width/depth Ratio: > 12

Entrenchment Ratio: < 1.4

Sinuosity: > 1.2

LEVEL II: THE MORPHOLOGICAL DESCRIPTION

F2 - Utah

F2 - Texas

F2 - California

LEVEL II: THE MORPHOLOGICAL DESCRIPTION

F2 - Utah

F2 - Montana

LEVEL II: THE MORPHOLOGICAL DESCRIPTION

MORPHOLOGICAL DESCRIPTION AND EXAMPLES OF STREAM TYPES

F3 Stream Type

The F3 stream type is a cobble dominated, entrenched, meandering channel, deeply incised in gentle terrain. The "top of banks" elevation for this stream type is much greater than the bankfull stage, which is indicative of the deep entrenchment. The F3 stream type can be incised in alluvial valleys, resulting in the abandonment of former floodplains. The F3 stream channels are found in Valley Types IV, VI, VIII, and X. The F3 channels have slopes that are generally less than 2%, exhibit riffle/pool bed features, and have width/depth ratios that are high to very high. The dominant channel materials are cobble, with lesser accumulations of gravel and sand. Often the sands will be imbedded with the large cobble sizes. Sediment supply in the F3 stream types is moderate to high, depending on bank erodibility conditions. Depositional features (central and transverse bars) are common, and related to the high sediment supply from streambanks and the high width/depth ratio. Riparian vegetation plays a marginal role in streambank stability due to the typically very high bank heights which extend beyond the rooting depth of riparian plants. Exceptions to this are the F3 stream types in the Northeast, Northwest, and Southeast United States where the relatively longer growing seasons and ample precipitation results in the establishment of riparian vegetation that tends to cover the entire slope facet of channel banks.

LEVEL II: THE MORPHOLOGICAL DESCRIPTION

DELINEATIVE CRITERIA (F3)

Landform/soils: The F3 stream type is associated with deeply entrenched, structurally controlled, gentle gradient valleys and gorges. The F3 stream type is associated with highly weathered bedrock or depositional soils involving a combination of river downcutting and uplift of valley walls.

Channel materials: Cobble dominated channel with accumulations of gravel and sand. Streambanks generally gravel/sand matrix, unstable, unless well vegetated.

Slope Range: < .02

Width/depth Ratio: > 12

Entrenchment Ratio: < 1.4

Sinuosity: > 1.2

5-151

LEVEL II: THE MORPHOLOGICAL DESCRIPTION

STREAM TYPE F3

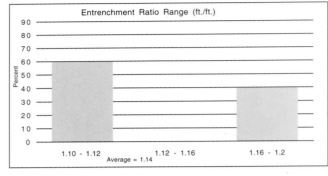
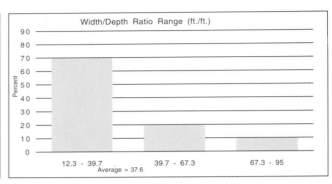
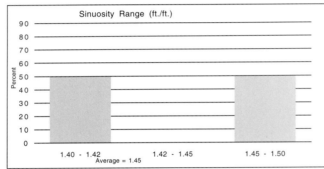
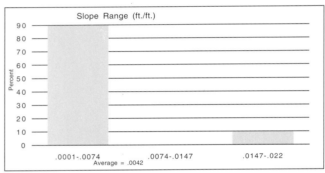
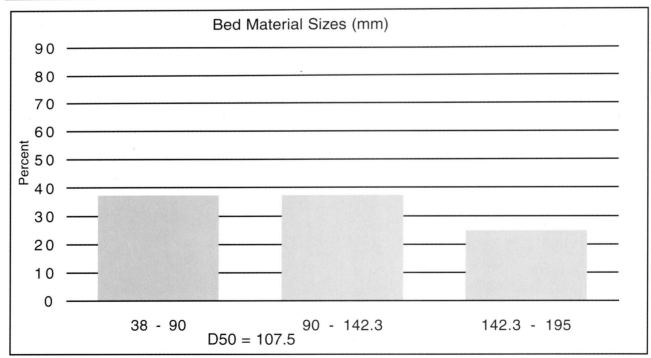

LEVEL II: THE MORPHOLOGICAL DESCRIPTION

F3 - North Carolina

F3 - Arizona

F3 - Colorado

LEVEL II: THE MORPHOLOGICAL DESCRIPTION

MORPHOLOGICAL DESCRIPTION AND EXAMPLES OF STREAM TYPES

F4 Stream Type

New Mexico

The F4 stream type is a gravel dominated, entrenched, meandering channel, deeply incised in gentle terrain. The "top of banks" elevation for this stream type is much greater than the bankfull stage, which is indicative of the deep entrenchment. The F4 stream type can be incised in alluvial valleys, resulting in the abandonment of former floodplains. The F4 stream channels are found in Valley Types IV, VI, VIII, X, and XI. The F4 channels have slopes that are generally less than 2%, exhibit riffle/pool bed features, and have width/depth ratios that are high to very high. The dominant channel materials are gravel, with lesser accumulations of cobble and sands. Often the sand will be imbedded with the cobble and gravel. Sediment supply in the F4 stream types is moderate to high, depending on stream bank erodibility conditions. Depositional features are common in this stream type, and over time, tend to promote development a flood plain inside of the bankfull channel (see Chapter 6). Central and transverse bars are common, and related to the high sediment supply from streambanks and the high width/depth ratio. Stream bank erosion rates are very high due to side slope rejuvenation and mass-wasting processes which enhance the fluvial entrainment. Riparian vegetation plays a marginal role in streambank stability due to the typically very high bank heights, which extend beyond the rooting depth of riparian plants. Exceptions to this are the F4 stream types in the Northeast, Northwest, and Southeast United States where the relatively longer growing seasons and ample precipitation results in the establishment of riparian vegetation that tends to cover the entire slope face of channel banks.

LEVEL II: THE MORPHOLOGICAL DESCRIPTION

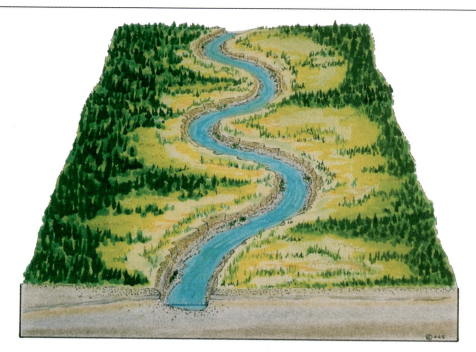

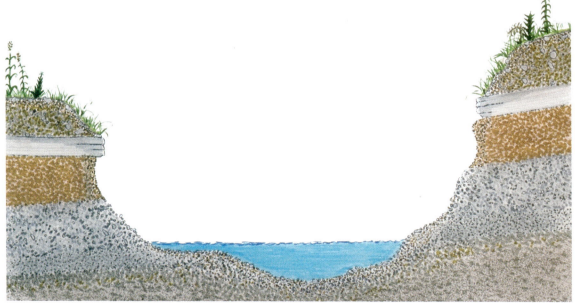

DELINEATIVE CRITERIA (F4)

Landform/soils: The F4 stream type is associated with deeply entrenched, structurally controlled, gentle gradient valleys and gorges. The F4 stream type is associated with highly weathered bedrock or depositional soils involving a combination of river downcutting and uplift of valley walls.

Channel materials: Gravel dominated channel with some cobble and sand accumulations. Streambanks are generally eroding unless stabilized with massive riparian vegetation.

Slope Range: < .02   **Entrenchment Ratio:** < 1.4

Width/depth Ratio: > 12   **Sinuosity:** > 1.2

LEVEL II: THE MORPHOLOGICAL DESCRIPTION

STREAM TYPE F4

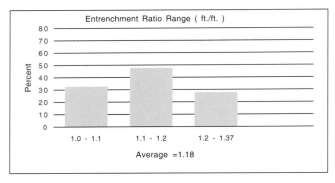
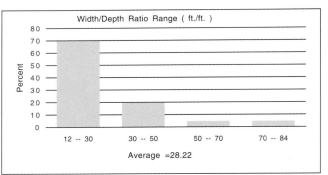
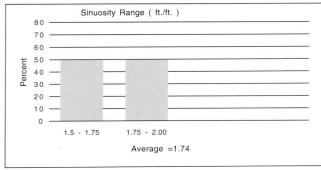
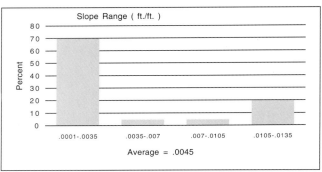
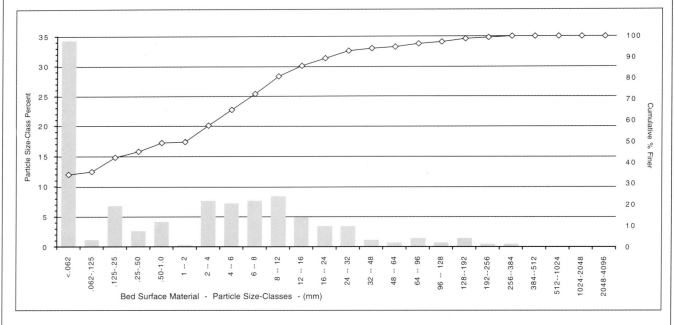

LEVEL II: THE MORPHOLOGICAL DESCRIPTION

F4 - Colorado

F4 - California

F4 - Texas

LEVEL II: THE MORPHOLOGICAL DESCRIPTION

MORPHOLOGICAL DESCRIPTION AND EXAMPLES OF STREAM TYPES

F5 Stream Type

The F5 stream type is a sand dominated, entrenched, meandering channel, deeply incised in gentle terrain. The "top of banks" elevation for this stream type is much greater than the bankfull stage which is indicative of the deep entrenchment. The F5 stream type can be deeply incised in alluvial valleys or in lacustrine deposits, resulting in the abandonment of former floodplains. The F5 stream channels are found in Valley Types IV, VI, VIII, X, and XI. The F5 channels have slopes that are generally less than 2%, exhibit riffle/pool bed features, and have width/depth ratios that are high to very high. The dominant channel materials are sand with lesser accumulations of gravel and some silt-clay. Sediment supply in the F5 stream types is moderate to high, depending on stream bank erodibility conditions. Depositional features are common in this stream type, and over time, tend to promote development a flood plain inside of the bankfull channel (see Chapter 6). Central and transverse bars are common, and related to the high sediment supply from streambanks and the high width/depth ratio. Stream bank erosion rates are very high due to side slope rejuvenation and mass-wasting processes which enhance the fluvial entrainment of eroded bank materials. Riparian vegetation plays a marginal role in streambank stability due to the typically very high bank heights which extend beyond the rooting depth of riparian plants. Exceptions to this are the F5 stream types in the Northeast, Northwest, and Southeast United States where the relatively longer growing seasons and ample precipitation results in the establishment of riparian vegetation that tends to cover the entire slope face of channel banks.

LEVEL II: THE MORPHOLOGICAL DESCRIPTION

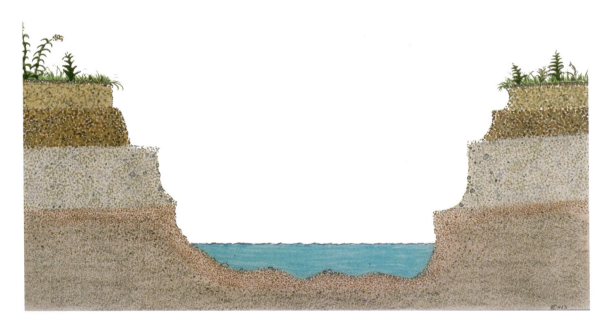

DELINEATIVE CRITERIA (F5)

Landform/soils: The F5 stream type is associated with deeply entrenched channels in alluvium or in structurally controlled, gentle gradient valleys and gorges. The F5 stream type is associated with highly weathered rock or depositional soils involving a combination of river downcutting and/or uplift of the valley walls.

Channel materials: This is a sand dominated channel, both bed and streambanks.

Slope Range: < .02 **Entrenchment Ratio:** < 1.4

Width/depth Ratio: > 12 **Sinuosity:** > 1.2

LEVEL II: THE MORPHOLOGICAL DESCRIPTION

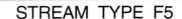

STREAM TYPE F5

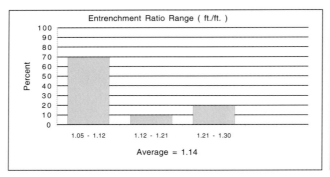

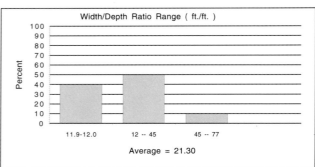

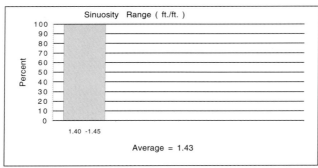

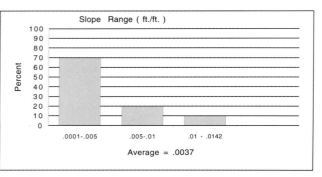

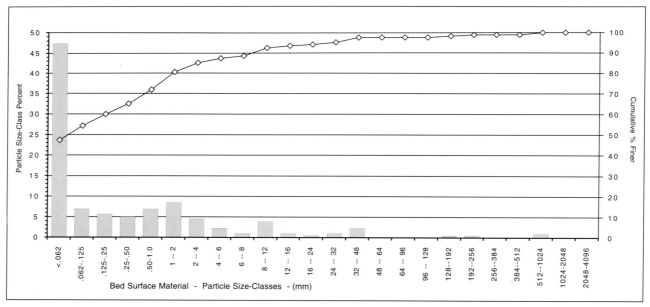

5-160

LEVEL II: THE MORPHOLOGICAL DESCRIPTION

F5 - Colorado

F5 - Texas

F5 - Maryland

LEVEL II: THE MORPHOLOGICAL DESCRIPTION

MORPHOLOGICAL DESCRIPTION AND EXAMPLES OF STREAM TYPES

F6 Stream Type

The F6 stream types are entrenched, meandering, gentle gradient streams deeply incised in cohesive sediments of silt and clay. The F6 stream channels have very high width/depth ratios, moderate sinuosities, and low to moderate meander width ratios. The "top of bank" elevation for this stream type is much greater than the bankfull stage which is indicative of the deep entrenchment. The related landforms are often seen as terrace bounded alluvial valleys, deltas, and coastal plains. The F6 stream channels are found in Valley Types IV, VIII, X and XI. Depositional soils in these valleys often originate from fine alluvial, eolian (loess), and lacustrine deposits. The F6 stream banks are relatively more stable than the F3, F4, or F5 stream banks, due to their inherent cohesive nature and ability to "stand" much steeper. Deep rooted riparian vegetation is much more effective at maintaining stability in the cohesive bank materials. However, mass wasting due to bank saturation/liquefaction and collapse is still a prevalent process in hydro-physiographic provinces where the composition of riparian vegetation is poor and natural densities have been reduced. The F6 stream systems produce relatively low bedload sediment yields due to the lack of coarse material in the channels, thus, excessive bar deposition is not generally observed with the F6 stream type. These stream types are very sensitive to disturbance and adjust rapidly to changes in flow regime and sediment supply from the watershed.

LEVEL II: THE MORPHOLOGICAL DESCRIPTION

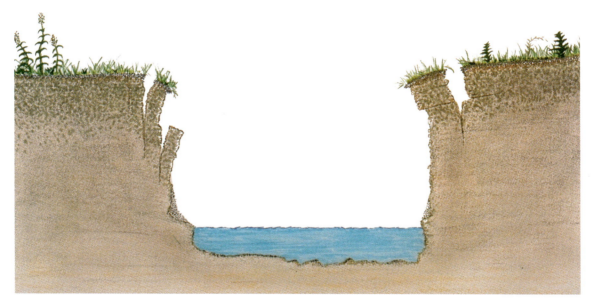

DELINEATIVE CRITERIA (F6)

Landform/soils: The F6 stream type is associated with deeply entrenched channels in alluvium or in structurally controlled, gentle gradient valleys and gorges. The F6 stream type is associated with highly weathered rock or depositional soils involving a combination of river downcutting and/or uplift of the valley walls. Cohesive soils with occasional mass-wasting slump blocks.

Channel materials: Silt and or clay

Slope Range: < .02   **Entrenchment Ratio:** < 1.4

Width/depth Ratio: > 12   **Sinuosity:** > 1.2

5-163

LEVEL II: THE MORPHOLOGICAL DESCRIPTION

STREAM TYPE F6

Entrenchment Ratio Range (ft./ft.)
- 1.10 - 1.14
- 1.14 - 1.20
- Average = 1.15

Width/Depth Ratio Range (ft./ft.)
- 14.0 - 27.9
- 27.9 - 42.0
- Average = 23.1

Sinuosity Range (ft./ft.)
- 1.58 - 1.60
- Average = 1.60

Slope Range (ft./ft.)
- .0005 - .012
- .012 - .024
- Average = .0056

Bed Material Size = less than .06 mm.

F6 - Colorado

LEVEL II: THE MORPHOLOGICAL DESCRIPTION

F6 - Virginia

F6 - Texas

F6 - Maryland

LEVEL II: THE MORPHOLOGICAL DESCRIPTION

MORPHOLOGICAL DESCRIPTION AND EXAMPLES OF STREAM TYPES

G1 Stream Type

Washington

G1 stream channels are deeply entrenched into bedrock and have moderate channel gradients, low width/depth ratios, and randomly spaced steps and plunge pools. The G1 stream type patterns, profiles, and dimensions are structurally controlled and related to the presence of faults and joints or erosion into highly weathered bedrock. The stream type is very stable, with limited rates of lateral or vertical adjustment. The G1 stream type is a step/pool system with low sediment storage capacities and a low sediment supply, due to the stable nature of the channel bed and bank materials. The G1 is similar to the A1 stream type with the exception that the G1 occurs primarily on moderate slopes and has a slightly higher channel sinuosity. G1 channels can also occur as narrow, deep gorges on larger rivers, where the reach has a gradient of 2-4 per cent and produces the more difficult class 4 and 5 rapids, which are often used for recreational boating.

LEVEL II: THE MORPHOLOGICAL DESCRIPTION

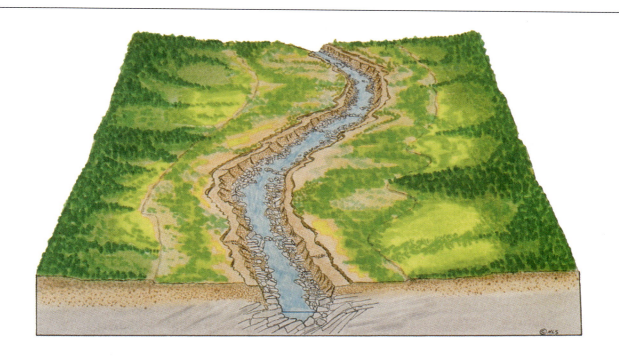

DELINEATIVE CRITERIA (G1)

Landform/soils: The G1 stream type is associated with moderately steep, structural controlled, narrow valleys. They are similar to the A1, but not as steep.

Channel materials: Bedrock with a mixture of boulders and cobble with some minor amounts of gravel.

Slope Range: < .04   **Entrenchment Ratio:** < 1.4

Width/depth Ratio: < 12   **Sinuosity:** > 1.2

LEVEL II: THE MORPHOLOGICAL DESCRIPTION

G1 - Arizona

G1 - Colorado

G1 - Idaho

LEVEL II: THE MORPHOLOGICAL DESCRIPTION

G1 - Colorado

G1 - Oregon

G1 - New Mexico

LEVEL II: THE MORPHOLOGICAL DESCRIPTION

MORPHOLOGICAL DESCRIPTION AND EXAMPLES OF STREAM TYPES

G2 Stream Type

Colorado

The G2 stream channels are deeply entrenched, slightly meandering, step/pool systems, with the dominant channel bank and bed materials appearing as boulders. The G2 is very stable, with moderate channel gradients of 2 to 4 per cent and a low width/depth ratio. The "slope continuum" concept is applied to the stream type description if the observed reach exhibits slopes of less than 2 per cent. Such a reach would be designated as a G2c, while maintaining a similar morphology, dimension, and pattern. For those channels with flatter slopes, the width/depth ratio and sinuosity tends to slightly increase above values observed for G2 channels with steeper slopes. The G2 stream type exhibits a channel bed dominated by boulder materials, while the channel banks generally have a higher percentage of cobble, gravel, and some sands mixed with scattered boulders. The G2 stream type is associated with very coarse alluvial fans, boulder debris from landslides, wedges below talus fields, colluvial deposits from up-slope gravitational erosion, and structurally controlled slopes. The G2 stream types can also occur as narrow, deep gorges on larger rivers, where the reach has a gradient of 2-4 percent, and produces the more difficult class 4 and 5 rapids which are often used for recreational boating.

LEVEL II: THE MORPHOLOGICAL DESCRIPTION

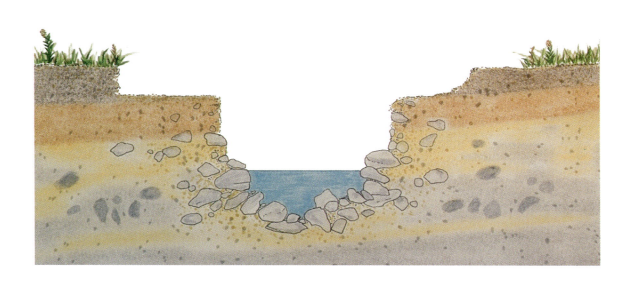

DELINEATIVE CRITERIA (G2)

Landform/soils: The G2 stream type is associated with structural controlled, narrow, moderately steep valleys. They are similar to the A2 stream types but not as steep.

Channel materials: Boulder dominated channel with cobble and gravel.

Slope Range: < .04 **Entrenchment Ratio:** < 1.4

Width/depth Ratio: < 12 **Sinuosity:** > 1.2

LEVEL II: THE MORPHOLOGICAL DESCRIPTION

G2 - Colorado

G2 - Arizona

LEVEL II: THE MORPHOLOGICAL DESCRIPTION

G2 - Colorado

G2 - Colorado

LEVEL II: THE MORPHOLOGICAL DESCRIPTION

MORPHOLOGICAL DESCRIPTION AND EXAMPLES OF STREAM TYPES

G3 Stream Type

Nevada

The G3 stream type is deeply incised in depositional material primarily comprised of an unconsolidated, heterogenous mixture of cobble, gravel, and sand. The G3 stream type is highly unstable due to the very high sediment supply available from both upslope and channel derived sources. The G3 channels have a moderate gradient, a low width/depth ratio, a characteristic step/pool morphology, and low sinuosities except when deeply incised in a previously sinuous channel. Bank erosion and bedload transport is typically very high in the G3 stream channels due to the combined efforts of low width/depth ratios, moderate channel gradients, and the high sediment supply. The ratio of bedload to total sediment load often exceeds 50%. The observed effects of vertical and lateral instability processes are primarily due to the combination of high streamflow energy and high available sediment supply. The G3 stream types are usually found in landform features such as alluvial fans, and landslide debris, and often seen as headcut gullies deeply incised in meadows, fluvial terraces, and in the bottom of previous channels. These stream types are very sensitive to disturbance and tend to make significant adverse channel adjustments to changes in flow regime and sediment supply from the watershed.

LEVEL II: THE MORPHOLOGICAL DESCRIPTION

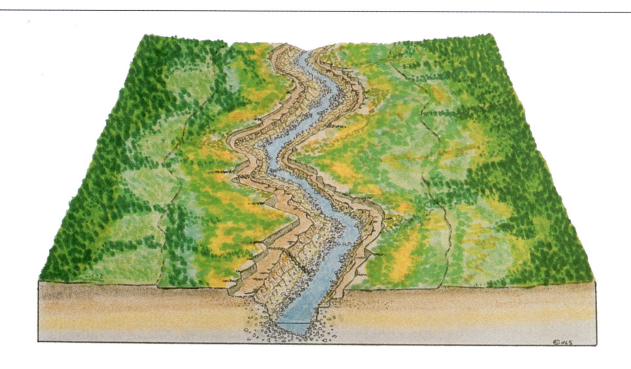

DELINEATIVE CRITERIA (G3)

Landform/soils: The G3 stream type is associated with moderately steep, fluvial dissected landforms, alluvial fans or down cut in alluvial or colluvial valleys. Soils are a heterogeneous mixture of unconsolidated non-cohesive material generally in alluvium and colluvium.

Channel materials: Cobble dominated channel with a mixture of gravel and sand.

Slope Range: < .04 **Entrenchment Ratio:** < 1.4

Width/depth Ratio: < 12 **Sinuosity:** > 1.2

LEVEL II: THE MORPHOLOGICAL DESCRIPTION

STREAM TYPE G3

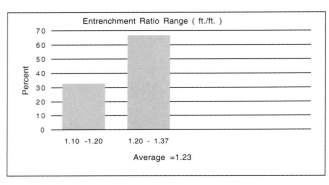
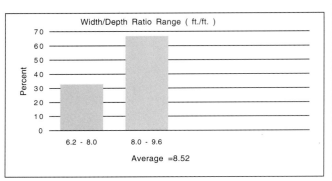
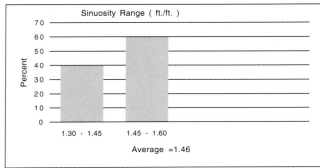
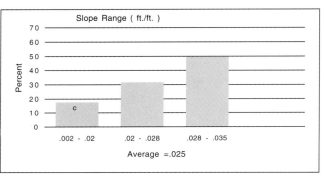
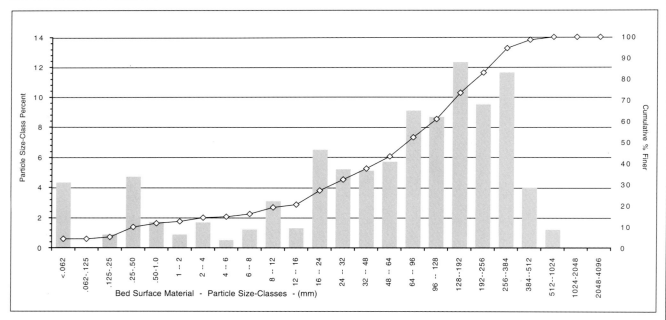

LEVEL II: THE MORPHOLOGICAL DESCRIPTION

G3 - Colorado

G3 - California

LEVEL II: THE MORPHOLOGICAL DESCRIPTION

MORPHOLOGICAL DESCRIPTION AND EXAMPLES OF STREAM TYPES

G4 Stream Type

The G4 stream type is deeply incised in depositional material primarily comprised of an unconsolidated, heterogenous mixture of gravel, some small cobble, and sand. The G4 stream type is very unstable due to the very high sediment supply available from both upslope and channel derived sources. The G4 stream channels have a moderate gradient, a low width/depth ratio, a characteristic step/pool morphology and low sinuosities; except when deeply incised in a previously sinuous channel. Pools are often filling with bedload, as the potential for sediment storage is high. Bank erosion and bedload transport rates are typically high in the G4 stream channel due to the combined effects of low width/depth ratios, moderate channel gradients, and the high sediment supply. The ratio of bedload to total sediment load often exceeds 50%. The observed effects of vertical and lateral instability processes are primarily due to the combination of high streamflow energy and high available sediment supply. The G4 stream types are usually observed in landform features such as alluvial fans, landslide debris, and are often seen as deeply incised headcut gullies in meadows, fluvial terraces, and in the bottom of previous channels. These stream types are very sensitive to disturbance and tend to make significant adverse channel adjustments to changes in flow regime and sediment supply from the watershed.

LEVEL II: THE MORPHOLOGICAL DESCRIPTION

DELINEATIVE CRITERIA (G4)

Landform/soils: The G4 stream type is associated with moderately steep, fluvial dissected landforms, alluvial fans or down cut in alluvial or colluvial valleys. Soils are a heterogeneous mixture of unconsolidated non-cohesive material generally in alluvium and colluvium.

Channel materials: Gravel dominated channel with mixtures of sand and some cobble.

Slope Range: < .04   **Entrenchment Ratio:** < 1.4

Width/depth Ratio: < 12   **Sinuosity:** > 1.2

LEVEL II: THE MORPHOLOGICAL DESCRIPTION

STREAM TYPE G4

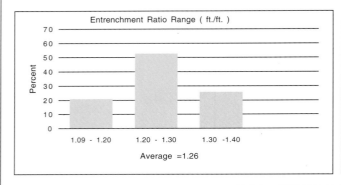
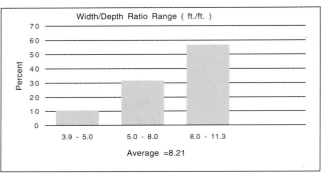
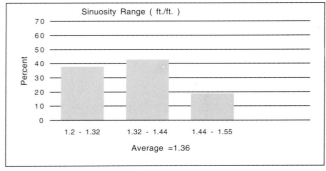
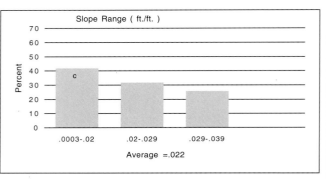
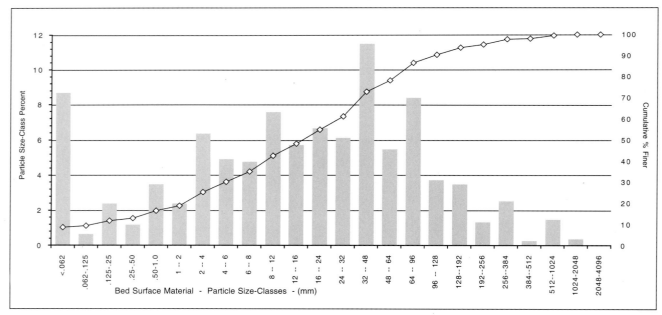

5-180

LEVEL II: THE MORPHOLOGICAL DESCRIPTION

G4 - Nevada

G4 - California

G4 - Maryland

LEVEL II: THE MORPHOLOGICAL DESCRIPTION

MORPHOLOGICAL DESCRIPTION AND EXAMPLES OF STREAM TYPES

G5 Stream Type

New Mexico

The G5 stream type is an entrenched, moderately steep, step/pool channel deeply incised in sandy materials. Channel sinuosities are relatively low, as are width/depth ratios. These "sandy gully" stream types transport great amounts of sediment due to the ease of particle detachment and fluvial entrainment. The G5 stream channels are generally in a degradation mode derived from near continuous channel adjustments, due to excessive bank erosion. Bedload transport rates can easily exceed 50 % of total load; with active, extensive, consistent channel erosion more typical than not. Exceptions may occur where very dense woody vegetation helps stabilize the toe of the stream bank slopes. The G5 stream type is similar in character to A5 channels, except G5 channel gradients are less than 4% and, tend to be more sinuous with somewhat higher width/depth ratios, due to the gentler channel slopes. The "slope continuum" concept is applied for the "gully" stream types if the observed reach exhibits slopes less than 2 %. Such a reach is given the designation of G5c. The lower gradient gully reaches are generally observed developing within a previously meandering, low gradient system with floodplains such as a C5 situated in wide alluvial valleys. These stream types are very sensitive to disturbance and tend to make significant adverse channel adjustments to changes in flow regime and sediment supply from the watershed.

LEVEL II: THE MORPHOLOGICAL DESCRIPTION

DELINEATIVE CRITERIA (G5)

Landform/soils: The G5 stream type is associated with moderately steep, fluvial dissected landforms, alluvial fans or down cut in alluvial or colluvial valleys. Soils are a heterogeneous mixture of unconsolidated non-cohesive material generally in alluvium, and colluvium, eolian (sand) deposition and residual soils such as those derived from grussic granite.

Channel materials: Sand dominated channel with mixtures of Gravel and some silt/clay.

Slope Range: < .04

Entrenchment Ratio: < 1.4

Width/depth Ratio: < 12

Sinuosity: > 1.2

LEVEL II: THE MORPHOLOGICAL DESCRIPTION

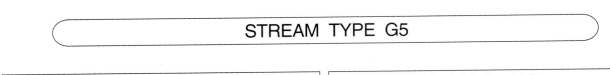

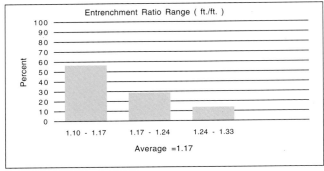
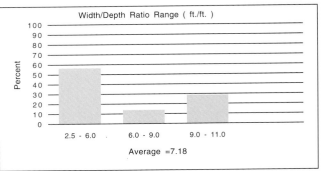
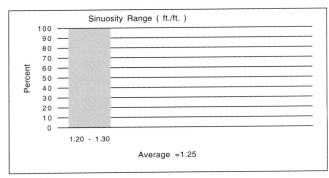
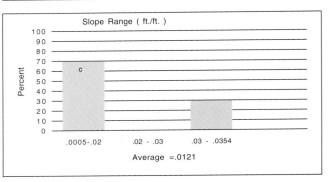
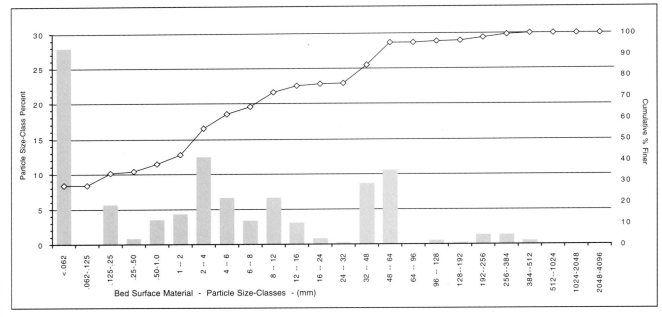

LEVEL II: THE MORPHOLOGICAL DESCRIPTION

G5 - Colorado

G5 - Arizona

G3 - Nevada

LEVEL II: THE MORPHOLOGICAL DESCRIPTION

MORPHOLOGICAL DESCRIPTION AND EXAMPLES OF STREAM TYPES

G6 Stream Type

The G6 stream type is an entrenched gully system with gentle to moderately steep channel gradients; that is deeply incised in cohesive materials of silts and clays. Bedload sediment transport rates are relatively low, and replaced by high washload and suspended sediment yields that commonly occur within the stream type. The bed features are generally observed as an unstable, degrading step/pool morphology. The dominate lithology for the G6 types include shales and depositional environments such as fans, deltas, lacustrine landforms, and other features that have cohesive, silt/clay deposits. Streambank erosion processes acting on the typically steep banks produce very high amounts of erodible material, especially within delta and lacustrine landforms. Woody riparian vegetation can have a bank stabilizing tendency if the vegetation densities are very high. The G6 stream types are very sensitive to disturbance and tend to make significant adverse channel adjustments to changes in flow regime and sediment supply from the watershed. The G6 stream type is generally considered to be experiencing near continuous degradational processes. It is not unusual to observe channel gradients of less than 2% (G5c), or even channel slopes less than .1% (G5c-).

LEVEL II: THE MORPHOLOGICAL DESCRIPTION

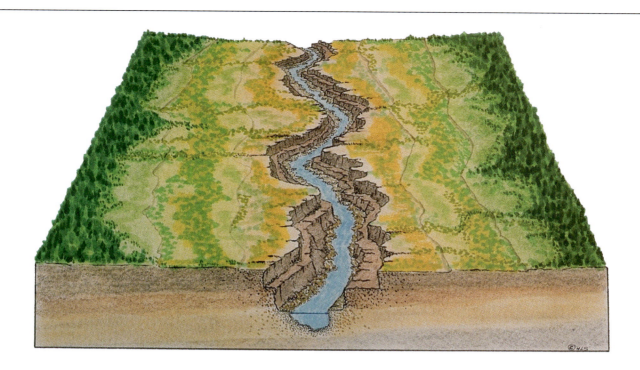

DELINEATIVE CRITERIA (G6)

Landform/soils: The G6 stream type is associated with moderately steep, fluvial dissected landforms, alluvial fans or down cut in alluvial or colluvial valleys. Soils are cohesive materials generally in alluvium, colluvium, eolian deposits (loess), and residual soils such as those derived from shales.

Channel materials: Silt/clay dominated channel with mixtures of gravel and some silt/clay.

Slope Range: < .04 **Entrenchment Ratio:** < 1.4

Width/depth Ratio: < 12 **Sinuosity:** > 1.2

LEVEL II: THE MORPHOLOGICAL DESCRIPTION

G6 - Colorado

G6 - Utah

G6 - California

LEVEL II: THE MORPHOLOGICAL DESCRIPTION

G6 - Utah

G6 - Colorado

G6 - Nevada

CHAPTER 6

LEVEL III: ASSESSMENT OF STREAM CONDITION AND DEPARTURE FROM ITS POTENTIAL

"A science of land health needs, first of all, a base datum of normality, a picture of how healthy land maintains itself as an organism."
ALDO LEOPOLD, <u>A SAND COUNTY ALMANAC</u>

Stream morphology, as defined by Level II criteria, serves as the basic physical stream template. A number of hydrologic, biological, ecological, and human factors, in turn, influence the state of a stream having a given morphology. Level III analyses incorporate these additional factors as an overlay to the morphological template in order to further describe the existing stream condition or "state" (***Figure 6-1a***).

The stability of a stream is a major determinant of its condition and a prerequisite for its optimum functioning. Stream stability is morphologically defined as the ability of the stream to maintain, over time, its dimension, pattern, and profile in such a manner that it is neither aggrading nor degrading

LEVEL III: ASSESSMENT OF STREAM CONDITION AND DEPARTURE

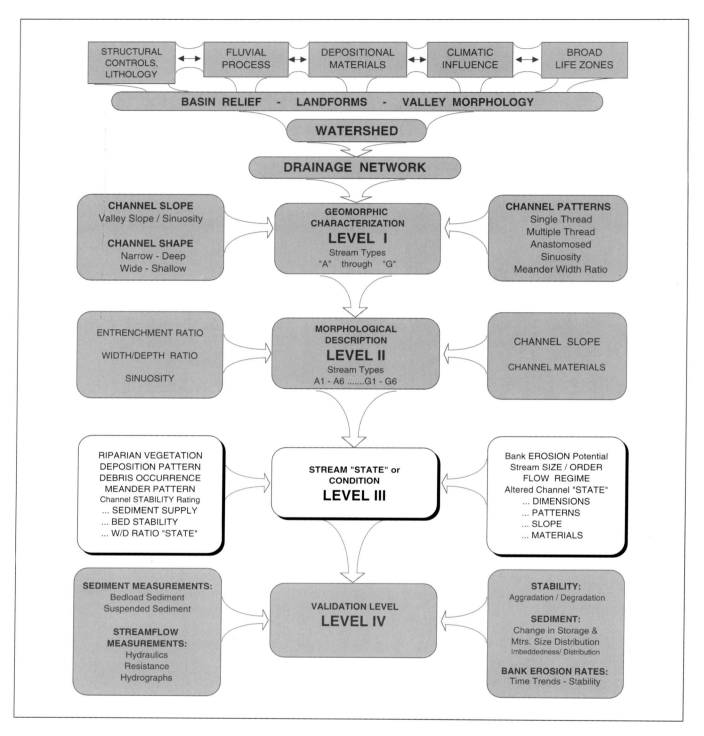

FIGURE 6-1a. Stream state or condition (Level III) in relation to the hierarchical river inventory.

and is able to transport without adverse consequence the flows and detritus of its watershed. Morphologic stability permits the full expression of natural stream characteristics. Self-formed streams that are stable, and self-maintained, and whose physical and biological function is at an optimum are said to be operating at their full potential. A major impetus for studying streams is to determine how well their current condition matches their operational potential. Stream potential has been described in terms of "Desired Future Condition" (USDA Forest Service, 1992 and Bauer and Burton 1993), and "Proper Functioning Condition" (USDI

LEVEL III: ASSESSMENT OF STREAM CONDITION AND DEPARTURE

Bureau of Land Management, 1993). These definitions concentrate on comparing an existing condition to a functioning and stable operational potential for a given stream reach. The concepts of a future or proper functioning condition have been initially set as water resource management goals to compare and evaluate the consequences of various resource development activities such as timber harvest, road construction, and riparian grazing.

A number of factors can change the stability and function of streams including; changes in streamflow, sediment regime, land use within the watershed, and direct disturbances (e.g., channelization, culverts, bridges). The response of a stream to a natural or imposed disturbance varies by stream type. The ability to characterize these responses and the associated physical effects by stream type is important to: (a) assess past impacts, (b) anticipate future consequences of alternative management strategies, (c) evaluate the potential for natural recovery, (d) determine the evolutionary stages of channel adjustment, (e) determine the feasibility of restoration, and (f) develop restoration designs that match or accommodate the functioning of a system's natural stable tendencies.

The Level III analysis process results in a description of stream condition as it relates to stream stability, potential, and function. The objectives of Level III analyses are to:

1. Develop a quantitative basis for comparing streams having similar morphologies, but which are in different states or condition.
2. Describe the potential natural stability of a stream, as contrasted with its existing condition.
3. Determine the departure of a stream's existing condition from a reference baseline.
4. Provide guidelines for documenting and evaluating additional field parameters that influence stream state (e.g., flow regime, stream size, sediment supply, channel stability, bank erodibility, and direct channel disturbances).
5. Provide a framework for integrating companion studies (e.g., fish habitat indices, and composition and density of riparian vegetation).
6. Develop and/or refine channel stability prediction methods.
7. Provide the basis for efficient Level IV validation sampling and data analyses.

The importance of the last objective should not be underestimated. As will be described in this chapter, Level III analyses infer condition from patterns that are observed in the field. Careful documentation of pattern and process observations permits the development of informed conclusions about channel condition. Level IV monitoring activities are required to verify the extent and magnitude of stream channel adjustment processes that may be indicated from collected dimension, profile, and pattern data. Such verifications based on collected data and related analyses are extremely valuable because the process of verification permits quantitative extrapolation of stream condition to other areas having similar morphologic and physiographic characteristics as determined from a Level III inventory. The integration of companion inventories with the additional variables that influence stream state is shown in *Figure 6-1b*. This holistic approach integrates both physical and biological function within a watershed context, providing a wide range of interpretation's for management applications.

This chapter is divided into three parts: (1) a determination of stream condition and departure from potential state, (2) a description of each of the Level III field parameters and their relationship to companion stream system and watershed analyses, and (3) a summary of a Level III assessment of stream condition.

STREAM CONDITION AND STREAM DEPARTURE ANALYSIS

Stream Potential

Stream potential is defined as the best channel condition, based on quantifiable morphological characteristics, for each stream type. *Figure 6-2a* and *Figure 6-2b* visually illustrates this concept. The C4 stream types shown in each photo have the same slope, depositional history, channel material,

LEVEL III: ASSESSMENT OF STREAM CONDITION AND DEPARTURE

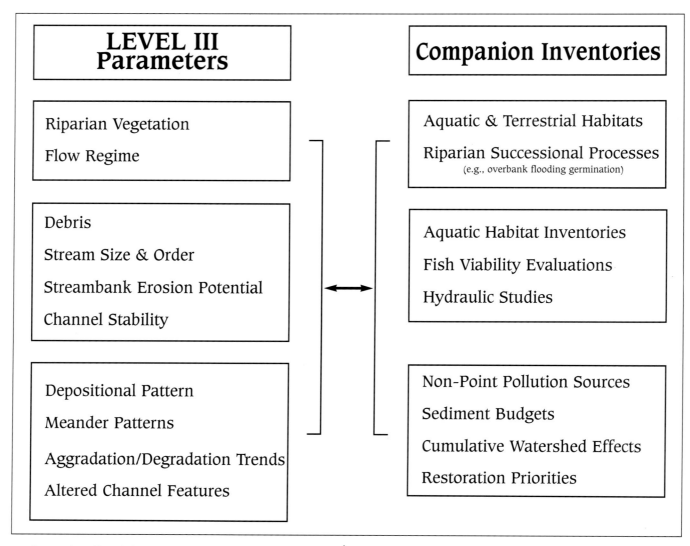

FIGURE 6-1b. Primary relationships between Level III parameters and companion inventories.

and similar floodplain and terrace characteristics. The depicted streams differ primarily in composition and density of riparian vegetation. The stream shown in *Figure 6-2b* is considered at full operating potential, with a low width/depth ratio, and the right bank terrace stabilized by a dense willow grass stand. The channel is competent to transport sediment without excessive sediment being contributed from the banks. In contrast, the condition of the stream shown in *Figure 6-2a* has departed significantly from its potential. The "unstable" C4 stream reach is being further destabilized laterally by accelerated bank erosion and vertically by deposition, resulting in decreased competence to transport sediment.

The preceding photographs were selected to illustrate stream potential. Visual characteristics, while compelling, are frequently interpreted differently by different people or even by the same person at different times. Moreover, many streams have been degraded for so long that degraded conditions have become associated with "natural" or baseline conditions. Stream classification as an analysis process permits a quantitative assessment of the degree to which existing conditions differ from an accepted range of morphological values documented for different stream types. For example, streams functioning at full potential exhibit a set of stability or condition characteristics that may be quantitatively described in terms of channel size and shape, low erodibility factors, low lateral migration rates,

LEVEL III: ASSESSMENT OF STREAM CONDITION AND DEPARTURE

FIGURE 6-2a. "Disturbed state" of a C4 stream type (note high bank on right with grass/forb community).

FIGURE 6-2b. "Stable" C4 stream type functioning at its potential (note high bank on right with mixed willow and grass).

ing stream condition can be compared to geomorphological data base for similar stream types to see if one or more key stability or condition criteria are close to or beyond the defined range of desired characteristic values. Alternatively, the same stream reach can be compared at different points in time through the use of historical photography. Where photography and ancillary data are available, a "before/after" comparisons may help identify the factors that caused the change in river condition. Finally, departures from potential or desired condition can be determined by comparing river condition at different points in space, i.e., upstream and downstream of human - or naturally - induced changes to the stream system. The streams shown in *Figure 6-3* are upstream/downstream comparisons of stream channel condition or "state" brought about primarily by changes in riparian vegetation and bank stability due to the impacts of grazing practices within the riparian area. The comparisons, of course, must be made between similar stream types as quantitatively determined using Level II classification criteria. Additional field monitoring methods to quantitatively validate channel process interpretations and stream response predictions are described in Chapter 7.

and comparatively low rates of sediment supply. Additional examples of morphologic variables that indicate stream condition and methods of assessment are presented later in this chapter.

The degree of departure for an existing stream condition from its full operating potential can be determined in several ways. Data describing exist-

LEVEL III: ASSESSMENT OF STREAM CONDITION AND DEPARTURE

FIGURE 6-3a. Reach "A," depicting a stable C4 stream type existing at its potential.

FIGURE 6-3b. Reach "B," located 100 yards downstream of Reach "A," is in an unstable condition due to the effects of over-grazing with a riparian vegetation conversion from willows to grass.

FIGURE 6-3c. Note the "fence-line" contrast due to different grazing practices and resultant changes in stream stability.

FIGURE 6-3. An example of riparian vegetation changes which have influenced the channel "state" or condition. (Colorado)

LEVEL III: ASSESSMENT OF STREAM CONDITION AND DEPARTURE

EVOLUTIONARY STAGES OF CHANNEL ADJUSTMENT

To understand the evolutionary tendencies of rivers, following either a natural or an imposed change is to improve one's ability to "read the river" in its current state. A proper interpretation for the current condition of a river reach in comparison to its potential can provide a valuable guide for management direction and/or restoration. A given classification of a river reach does not indicate that the stream is necessarily in a stable pattern or is functioning close to its "potential." Rather, the stream type classification developed with a Level II analysis describes only the existing morphologic conditions. The desire to make a stream into what is conceived to be "good," has to be balanced with an individual's understanding of the morphological features of the natural stable stream system. The self-stabilization tendencies of a stream system and the natural tendency to evolve into a particular morphological form needs to be understood to provide an individual with a "blueprint" for the river's future. Watershed management and stream restoration can be effective when such activities and practices are designed to be compatible with the "most probable stable form" of rivers. The most probable state of rivers is best described in the book, "A View of the River" by L.B. Leopold (1994). The intricacies of the multiple interacting variables which form and maintain the river are well described by Leopold (1994), where the concept of entropy (energy distribution), longitudinal profiles, and principles of minimum variance are used to describe the progression toward the most probable form. These fluvial process tendencies toward a uniformity of energy expenditure in open, steady state systems are complex, which often makes communication of the related principles difficult in terms of description and understanding. Field evidence collected over time can help provide insight into observed changes in river morphology, in the presence of changing flow and sediment regimes that may be influenced by changes in watershed condition.

Adjustment Examples

Rivers generally do not change instantaneously, under a geomorphic exceedance or "threshold". Rather, they undergo a more consistent series of channel adjustments over time to accommodate changes or alterations in the "driving" variables. Many of the individual adjustments can occur quite rapidly, however. The dimensions, pattern, and profile of the river reflect the combined processes of adjustment which are presently responsible for the form and function of the river. The rate and direction of channel adjustment is a function of the nature and magnitude of perhaps the change in climate or land use and the stream type involved. Some stream types can change or evolve rapidly, while others are comparatively slow in their response.

In reviewing historical aerial photos, observations can be made of progressive stages in channel adjustment. Adjustments occur partially as a result of a change in the streamflow magnitude and/or timing, sediment supply and/or size, direct channel disturbance, and riparian vegetation changes. Observed changes in channel morphology over time can be quantified and communicated in terms of stream type changes. For example, within a given stream reach and perhaps due to streambank instability, with a resultant increase in bank erosion rate, field observations would normally indicate: an increasing width/depth ratio; decreased sinuosity; increased slope; establishment of a bi-modal particle size distribution; increased bar deposition; accelerated bank erosion; increased sediment supply; decreased sediment transport capacity; a decreased meander width ratio; and channel aggradation. The extended changes in process and condition can be described more simply as a series of progressive physical responses or channel adjustments resulting in the evolution of a stream type from an E4 to C4 to C4 (bar-braided) to D4 (*Figure 6-4*), (Rosgen, 1994).

Corresponding changes in channel dimension, pattern, and profile that would progressively evolve from an E4 to C4 to G4 to F4 to E4 stream type is shown in *Figure 6-5* (Rosgen 1994). As the local reach slope steepens in conjunction with a higher

LEVEL III: ASSESSMENT OF STREAM CONDITION AND DEPARTURE

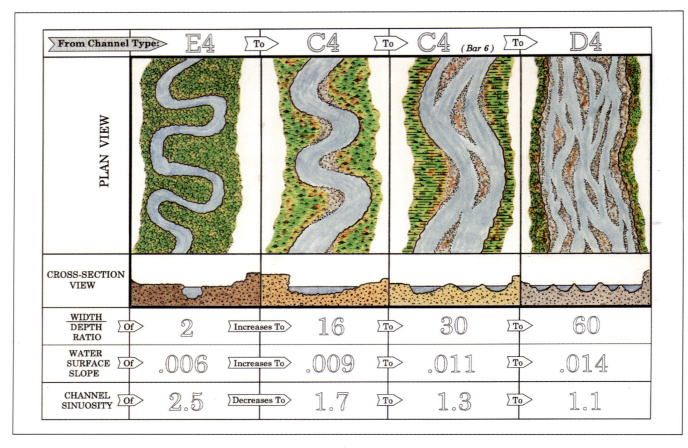

FIGURE 6-4. Example of progressive stages of channel adjustment due to an imposed change in stream bank stability.

width/depth ratio (conversion of E4 to C4 stream type), chute cutoffs develop across large point bars that begin down cutting, eventually into a steeper, entrenched gully (G4 stream type). The stream then abandons its floodplain, followed by a decrease in width/depth ratio. The degradation process that is initiated results in a lowering of the base level, thereby rejuvenating (over-steepening) all the tributaries to the main-stem river. The resultant headward advancement of the drainage network adds an accelerated excess sediment supply originating from both channel degradation and bank erosion processes. As the banks continue to erode, the belt width and width/depth ratio increase, which eventually leads to an F4 stream type. The natural tendency for a river is to balance its slope with that of its valley and rebuild a new floodplain. In order to decrease the stream slope and rebuild a new floodplain, the stream must progressively increase sinuosity and belt width. An increase in the dimensions of belt width and sinuosity can only happen through a process of lateral extension. The tendency for lateral extension of natural channels is predictable and the accelerated rate of bank erosion can be readily observed in the F4 stream type. The next series of progressive adjustments lead to a C4 stream type which eventually evolves back to the original E4 stream type. The resulting E4 morphology is a lower width/depth ratio, a reduction in channel slope, and an increase in sinuosity and meander width ratio. The previously over-widened bed of the F4 stream type is now the elevation of the new floodplain for the new C4 stream type, which gradually incises, reducing the width/depth ratio and increasing the entrenchment ratio. These channel adjustments as described above, signal the start of a new E4 stream type. The stream can eventually evolve, under a changed sediment and flow regime, into a sinuous, low gradient, low width/depth ratio channel with a well developed floodplain which matched the original, or pre-distur-

LEVEL III: ASSESSMENT OF STREAM CONDITION AND DEPARTURE

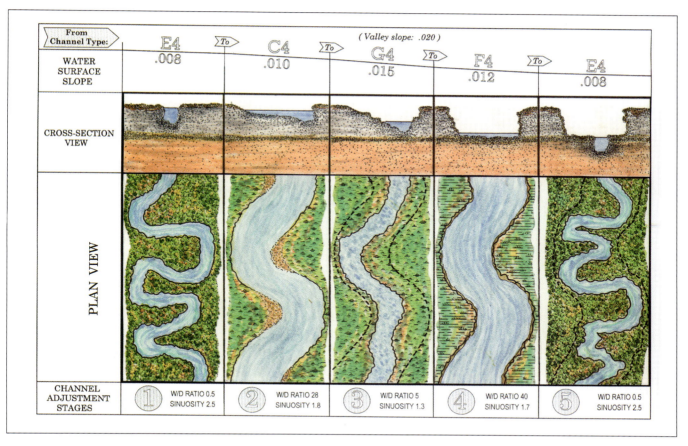

FIGURE 6-5. Adjustment of stream types in five progressive stages.

bance stream type morphology. The original morphology can be self-replicated; however, the new local base level will now exist at a lower elevation in the valley, which will continue to advance the drainage network of the tributaries. The recurring evolutionary sequence takes place in the presence of "good" riparian conditions, where vegetation provides the necessary resistance to flow forces, illustrating the stream's ability to reach a condition or state described as natural stability. As you will recall from the previous definition of stability, the stream has to be able to properly distribute the flow and sediment produced by its watershed in order to maintain the dimension, pattern and profile without either aggrading or degrading. The evolution from a high width/depth ratio (F4) to the lower width/depth ratio stream types (C4 and E4) improves the sediment transport capacity. The increase in sediment transport capacity is due to the change in boundary stress distribution and an increase in stream power (due to increased velocity and depth). Even though the C4 and E4 stream types have a more gentle slope than the F4, these stream types are more efficient at moving larger sizes and volumes of sediment since they require less cross-sectional area for the same discharge resulting in a higher mean velocity.

The above process of stream type development can be more simply described as an adjustment from stream type E4 to C4 to G4 to F4 and eventually back to E4 (*Figure 6-5*). Another series of illustrations, depicting changes in cross-section and plan-view corresponding to the adjustments shown in *Figure 6-5* are shown in *Figure 6-6*. Commonly such land-use activities as livestock grazing under saturated soil conditions that can lead to streambank trampling, along with heavy utilization of riparian vegetation will result in a corresponding decrease in streambank stability sufficient to initiate a shift in stream type. The stream type or stability shift brought on by a natural or imposed change

LEVEL III: ASSESSMENT OF STREAM CONDITION AND DEPARTURE

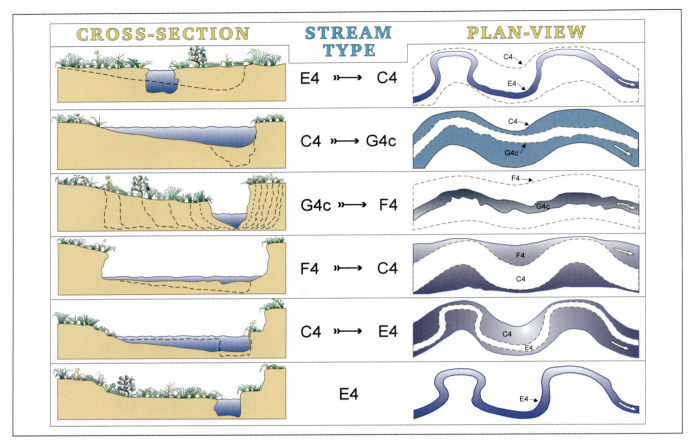

FIGURE 6-6. Adjustments of channel cross-section and plan-view patterns, as stream types change or shift through an evolutionary cycle.

exceedance of a "geomorphic threshold" leads to changes in the morphological variables of width/depth ratio, slope, sinuosity, and meander width ratio. Examples of some typical stream type evolution sequences depicting a wide range of channel adjustments are shown in the photographs of *Figures 6-7a* to *6-7f* and *Figures 6-8a* to *6-8f*.

The evolutionary sequences shown in *Figures 6-4* through *6-6* and the photographic examples in *Figure 6-7* and *Figure 6-8* are only a few of many potential scenarios of stream type shifts. Often "B" stream types evolve from "G" stream types located on alluvial fans when erosional sequences create a moderate entrenchment, an increase in width/depth ratio, and emerging riparian vegetation gradually stabilizing the stream banks.

Stream channel adjustments resulting from the influence of various physiographic processes, including climate change, adverse watershed impacts, vegetative composition changes, reservoir construction, and direct channel disturbances have been well documented throughout western North America. The knowledge provided by observing and documenting historical adjustments, studying time trends from relict photographs, and understanding the natural tendency of rivers to regain a condition or state of stability can assist in managing and restoring disturbed river systems. Additional information concerning channel shifts can be obtained from data related to long-term U.S. Geological Survey streamflow stations. Since most stream systems are always seeking a stable condition, by the time all the research and development for restoration plans and permits are obtained, we often cannot get there fast enough to "fix" streams before they have already fixed themselves. An important challenge for us all is to learn how to recognize the tendency for stream systems to develop a natural stability and to understand the time-trends involved. Often the directions of management that are designed to "restore" streams into a state or con-

LEVEL III: ASSESSMENT OF STREAM CONDITION AND DEPARTURE

FIGURE 6-7a. C5 stream type shifting to D5 (note high width/depth ratio, aggradation, and deposition on flood plain).

FIGURE 6-7b. Change from a C4 to a G4 stream type.

FIGURE 6-7c. Evolution from C4 to G4 to F4 to C4. Bed of a previous F4 is flood plain of new C4.

FIGURE 6-7d. Same evolution sequence as in Figure 6-7c except located in the maritime climate mountainous region of northern California.

FIGURE 6-7e. C4 stream type East Fork, San Juan River, Colorado.

FIGURE 6-7f. East Fork, San Juan River, Colorado 2 miles upstream from C4 stream type (e) converted to D4 stream type due to vegetation conversion.

LEVEL III: ASSESSMENT OF STREAM CONDITION AND DEPARTURE

FIGURE 6-8a. Change from an E4-C4-G4- currently entrenched F4 stream type due to a major flood, 1979.

FIGURE 6-8b. Same reach as (a) but 10 years later - evolved from a F4 to an E4 stream type. Bed of old F4 is now the flood plain of the new E4.

FIGURE 6-8c. Example of evolution to a stable E5 stream type from a high width/depth ratio entrenched meandering F5.

FIGURE 6-8d. Example of conversion from G3-F3b-B3 stream type.

FIGURE 6-8e. Example of a E5 streamtype evolved from a G5-F5-C5 type conversion.

FIGURE 6-8f. Example of G5 stream type 150 feet downstream of (6-8e) due to culvert crossing. An important comparison of adjustment, reverting back to a G5 from an E5 streamtype.

LEVEL III: ASSESSMENT OF STREAM CONDITION AND DEPARTURE

dition that does not match the dimension, pattern and slope of the original natural, stable form only serve to provide a trial and error method of learning. If one is assigned to "restore" a river, it is not only important to know the current state of the river, but what will be the eventual stable morphological form. With consistent documented, objective observations, stream types can relate much more than what may be initially assumed from a casual inspection.

Equilibrium Interpretations

Some researchers have developed a visual "image" of the equilibrium channel, where the morphologic variables have mutually adjusted to a "stable" condition. For example, Bull, 1978 in describing "stream power" as the power expenditure per unit bed area ($\omega = \rho g Q s / w$), found such values to be high in a narrow channel, so the channel may be unstable and subject to widening until a smaller value of ω is reached and the channel stabilized. (Where: ω = stream power; ρ = water density; g = gravitational acceleration; Q = stream discharge in cms; s = slope; and w = stream width). This statement about a channel naturally widening to balance available stream power is appropriate for some stream types such as the A3-A6, B1-B6, and G3-G6 stream types. However, a stable "E" stream type has a comparatively low width/depth ratio (narrow and deep) and accommodates a high unit stream power by naturally developing an increased form resistance, as observed in the high values of meander width ratio (belt width/bankfull width) of 20-40, and a high sinuosity. In the case of the E streamtype, a widening of the channel to decrease stream power in order to "stabilize" the stream would in actuality initiate de-stabilization. The classification of rivers helps to stratify the morphological types so that interpretations of adjustment processes from generalized statements can be directed appropriately to stream types where these conditions apply. The channel widening in an "E" stream type would lead to: (1) an increase in width/depth ratio; (2) an increase in channel slope; (3) a shift to a higher boundary shear stress in the near-bank region; (4) an increase in sediment supply due to increasing bank erosion; (5) an increase in bar deposition which adds to the boundary shear stress imposed on the banks; (6) an increase in belt width, and (7) and an increase in width/depth ratio that would lead to a decrease in meander width ratio. These imposed changes acting in combination tend to accelerate lateral channel extension with a continuing increase in width/depth ratio and a corresponding shift in stream type. The systematic channel adjustments would, in effect, increase the sediment supply but decrease sediment transport capacity which would then lead to instability.

LEVEL III FIELD PARAMETERS: THE STREAM CHANNEL INFLUENCE VARIABLES

Introduction

The Level III field inventory uses ten additional parameters to more fully describe stream condition beyond the fundamental Level II morphological template. The additional parameters are: (1) riparian vegetation, (2) streamflow regime, (3) stream size and stream order, (4) organic debris and/or channel blockage, (5) depositional patterns, (6) meander patterns, (7) streambank erosion potential, (8) aggradation/degradation potential, (9) channel stability rating, and (10) altered channel materials and dimensions. Parameter Items 7-10 are additions to an earlier classification system (Rosgen, 1985).

Each of the listed parameters exerts a strong influence on existing stream condition and future operational potential. The parameters are not incorporated into the stream typing process until Level III to limit the sheer number of potential combinations of variables that would otherwise lead to an unworkable number of stream types. For example, incorporating flow regime directly as a classification parameter would increase the number of stream types from 41 to 492. The attributes of the flow regime are clearly important for interpreting both physical and biological stream characteristics, but can be more directly evaluated through a series of simple descriptors that provide additional information. The other Level III parameters are similarly handled and each provide a basis for additional interpretive capability.

In addition to exerting an important influence on stream potential and condition, Level III parameters are especially suited for integration with a number of other related stream system and watershed analyses as previously illustrated in *Figure 6-1*. The related analyses may include assessments of riparian vegetation, aquatic habitat and fisheries condition, non-point pollution sources, sediment budgets, cumulative watershed effects, and establishing restoration priorities.

Description of Level III Channel Influence Variables

Riparian Vegetation

Riparian vegetation has a marked and significant influence on the stability of certain stream types. The controlling influence of riparian vegetation that can change morphology and stability was rated by stream type, and found to vary from low for A stream types to very high for C stream types (Rosgen 1994). Changes in the composition, vigor, and density of riparian vegetation produce corresponding changes in rooting depth, rooting density, shading, water temperature, physical protection from bank erosion processes, terrestrial insect habitat, and contribution of detritus to the channel. Water quality and esthetic values are also affected by changes in riparian vegetation. Examples of the key role of vegetation in maintaining channel stability and function for certain stream types was depicted previously in *Figure 6-3*.

Stream morphology influences the potential for establishment of riparian vegetation communities. The determination of stream types is a process which integrates channel bed particle size, the presence or absence of floodplains, and other channel morphology parameters, which directly influence the presence and amount of riparian vegetation. Thus, stream typing can significantly assist in determining the potential succession of riparian vegetation communities. For example, the width/depth ratios of stream types C, D, E, and F generally increase when

TABLE 6-1. Riparian vegetation inventory/condition survey.

RIPARIAN VEGETATION

Existing Vegetation:

Composition: _____

Vigor, Density: _____

Potential: _____

Summary Categories (Identify individually and/or in combination)

1.	Bare		RV 1
2.	Forbs only -	Low density	2a
		Moderate density	2b
3.	Annual grass with forbs -	Low density	3a
		Mod. density	3b
		High density	3c
4.	Perennial grass -	Low density	4a
		Mod. density	4b
		High density	4c
5.	Rhizomatous grasses (bluegrass, grasslike plants, sedges, rushes)	Low density	5a
		Mod. density	5b
		High density	5c
6.	Low brush	Low density	6a
		Mod. density	6b
		High density	6c
7.	High brush -	Low density	7a
		Mod. density	7b
		High density	7c
8.	Combination grass/brush -	Low density	8a
		Mod. density	8b
		High density	8c
9.	Deciduous overstory -	Low density	9a
		Mod. density	9b
		High density	9c
10.	Deciduous with brush/ grass understory	Low density	10a
		Mod. density	10b
		High density	10c
11.	Perennial overstory -	Low density	11a
		Mod. density	11b
		High density	11c
12.	Wetland vegetation community		12a
		Bog	12b
		Fen	12c
		Marsh	RV 12d

the riparian woody species are converted to grass/forb communities. The resultant channel instability is often followed by a degradation of fish habitat. The increase in width/depth ratios generally sets in motion a series of channel adjustments seen as an increase in sediment deposition, bank erosion, sediment supply, and channel slope; along with a decrease in meander width ratio and sinuosity. Additionally there is a concurrent redistribution of channel bed particle sizes. In step/pool or rapids

LEVEL III: ASSESSMENT OF STREAM CONDITION AND DEPARTURE

dominated stream types (A,B,G), the practice of clearcutting within riparian areas and the subsequent changes from woody to herbaceous vegetation communities frequently reduce the availability and the natural recruitment of large woody debris needed for fisheries habitat and stream-bed stability.

Vegetation mapping can be used to identify current patterns and historical trends in riparian communities and overlain on Level II stream classifications to identify stream reaches most vulnerable to disturbance. Mapping procedures and riparian vegetation scorecards to document riparian condition are available from federal and state resource agencies (USFS 1992, USDI BLM 1993). Alternatively, a general descriptive riparian evaluation such as shown in *Table 6-1* can be used. The alpha-numeric descriptor for riparian vegetation should be added following the stream type designation.

An example is shown in *Figure 6-9*, of the same stream type (B3) found in three separate hydro-physiographic regimes in North Carolina, California, and Arizona, involving different riparian vegetation communities ranging from eastern hardwoods to desert shrub to montane coniferous forest with a willow/alder understory.

Flow Regime

There are many distinct streamflow regimes that can be determined from streamflow records (Poff and Ward 1989), and it is well recognized that streamflow exerts a strong influence on channel morphology, aquatic habitat, and riparian vegetation. *Table 6-2* lists the categories of streamflow recommended for documenting a Level III condition. Streamflow categories include: (E) ephemeral, (I) intermittent, (P) perennial, and (S) subterranean; with specific additional notations for streamflow patterns dominated by (1) snowmelt, (2) stormflow, (3) glacial melt, (4) spring-fed, (5) ice flows, (6) tidal influence, (7) regulated streamflow, and (8) streamflow patterns altered by development. Thus, the lower reach of the Simms Bayou in Texas would be designated as a C5c-, (P:2,6) to indicate that its hydrologic regime is described as perennial flow, stormflow dominated, and tidal influence (*Table 6-2*).

Level III uses the streamflow categories to further compare the probable response of flow regimes for streams having the same morphological type. For example, similar C4 stream types that differ in the dominant source of runoff (such as snowmelt vs. spring-fed) in their flow regime will also differ in regard to macro-invertebrate diversity, water temperature, timing or pattern of flows, ratio of low to high flows, and the efficiency of sediment transport. Similarly, analyses of hydraulic geometry from gaged sites indicates that ephemeral streams normally have greater channel widths for the same discharges than perennial streams of the same type. The increased channel width is due primarily to greater rates of bank erosion resulting from significant differences in flow duration and magnitude, combined with poor vegetative cover, shallow rooting depth, and low root density—characteristic of arid areas. Hydraulic geometry relationship values that lie beyond the ranges predicted from regional drainage area curves are frequently the result of stream regulation or watershed alteration (subcategories 7 and 8, respectively). In particular, reservoirs and diversions are known to affect the total or annual hydrograph, with obvious, attendant effects on river condition.

Glacial fed streams not only have a unique hydrology, but the water quality parameters of temperature and suspended sediment are considered unique to this type of flow regime.

The photographs shown in *Figure 6-9* depict three B3 stream types in North Carolina, California and Arizona for three different flow regimes. Interpretations for physical and biological purposes would vary by local conditions and would be accounted for with documentation of specific channel influence variables at the sub-category level of analysis.

Size and Stream Order

Both stream size and stream order are used to further describe the state of a given stream type. Bankfull width is primarily used to describe stream

LEVEL III: ASSESSMENT OF STREAM CONDITION AND DEPARTURE

FIGURE 6-9a. North Carolina

FIGURE 6-9b. California

FIGURE 6-9c. Arizona

FIGURE 6-9. Three B3 stream types showing variation in vegetative communities and flow regime for the same stream type.

LEVEL III: ASSESSMENT OF STREAM CONDITION AND DEPARTURE

FLOW REGIME

General Category

E. Ephemeral stream channels - flows only in response to precipitation. Often used in conjunction with intermittent (USDA SCS, 1982).

S. Subterranean stream channel - flows parallel to and near the surface for various seasons - a subsurface flow which follows the stream bed.

I. Intermittent stream channel - one which flows only seasonally, or sporadically. Surface sources involve springs, snow melt, artificial controls, etc. Often this term is associated with flows that re-appear along various locations of a reach, then run subterranean.

P. Perennial stream channels. Surface water persists year long.

Specific Category

1. Seasonal variation in streamflow dominated primarily by snowmelt runoff.

2. Seasonal variation in streamflow dominated primarily by stormflow runoff.

3. Uniform stage and associated streamflow due to spring fed condition, backwater, etc.

4. Streamflow regulated by glacial melt.

5. Ice flows, ice torrents from ice dam breaches.

6. Alternating flow/backwater due to tidal influence.

7. Regulated streamflow due to diversions, dam release, dewatering, etc.

8. Altered due to development, such as urban streams, cut-over watersheds, vegetation conversions (forested to grassland) that changes flow response to precipitation events.

TABLE 6-2. Categories of flow regime for specification in level III inventories.

size because bankfull width is the most directly observable stream dimension (Leopold 1994) and because of the many hydrologic and geomorphic interpretations that can be derived from width measurements. Hydraulic geometry relationships should be stratified not only by stream type but also by stream size. Similarly, fisheries biologists stratify streams by size to improve their ability to interpret estimates of parameters such as population dynamics, standing crop, usable area, and fishing pressure.

Table 6-3 lists thirteen channel bankfull width categories that vary in dimension from less than 1 foot to greater than 1000 feet, which provide a required perspective for interpreting hydraulic processes, sediment transport, and biological processes. For example, the relative roughness ratio is much less and the average velocity greater for bankfull stage on a relatively large C4, S-12 stream type (i.e., bankfull width of 500-1000 feet) than that of a C4, S-3 stream type (i.e., bankfull width of 5-15 feet).

Stream order has long been used by hydrologists to develop quantitative relationships such as the bifurcation ratio (the ratio of the number of

6-17

LEVEL III: ASSESSMENT OF STREAM CONDITION AND DEPARTURE

TABLE 6-3. Categories of stream size as indicated by bankfull surface width and stream order.

STREAM SIZE

S-1	Bankfull width less than .305 m (1 foot)
S-2	Bankfull width .3-1.5 m (1-5 feet)
S-3	Bankfull width 1.5-4.6 m (5-15 feet)
S-4	Bankfull width 4.6-9 m (15-30 feet)
S-5	Bankfull width 9-15 m (30-50 feet)
S-6	Bankfull width 15-22.8 m (50-75 feet)
S-7	Bankfull width 22.8-30.5 m (75-100 feet).
S-8	Bankfull width 30.5-46 m (100-150 feet)
S-9	Bankfull width 46-76 m (150-250 feet)
S-10	Bankfull width 76-107 m (250-350 feet)
S-11	Bankfull width 107-150 m (350-500 feet)
S-12	Bankfull width 150-305 m (500-1000 feet)
S-13	Bankfull width greater than 305 m (1000 feet)

STREAM ORDER

Add categories in parenthesis for specific stream order of reach. For example a third order stream with a bankfull width of 6.1 meters (20 feet) would be indexed as: S-4(3).

streams of a given order to the number in the next lower order), and relationships between stream length, stream density, and drainage area. Stream order can be simply described as a numbering sequence which starts when two first order channels join—they form a second order stream and so on. Stream order is often used to describe stream size; however, such inferences are often misleading due to differences among hydro-physiographic provinces, geologic characteristics, and the scale of the maps being used for stream order determinations. Thus, records of stream size and stream order are kept separate to improve the interpretations associated with of Level III analyses. The categories shown in *Table 6-3* are used in combination with Strahler's (1952) stream order for specific reaches.

Stream order is also important as a biological stratification at Level III since it provides another index of the diversity in the food web (a first order channel being less diverse than fourth order channel of the same stream type).

Depositional Patterns (sediment)

Depositional patterns are easily observed features that are helpful for interpreting stream condition. Categories of different in-channel bar features are shown in *Table 6-4* and are illustrated in *Figure 6-10*. Initial descriptions of depositional forms were presented by Mollard (1973) and Galay et al. (1973) and modified by Rosgen (1985). Various depositional patterns can be used to illustrate the effects of past land management on sediment supply and sediment storage and subsequent effects on channel form and stability. For example, delta bars indicate a subdrainage water source contribution of coarse bedload to the receiving streams. Channel adjustments caused by floods, direct disturbances, changes in riparian vegetation, or flow regime are

TABLE 6-4. Categories of depositional features (bars) in a channel reach.

DEPOSITIONAL FEATURES (BARS)

B-1	Point Bars
B-2	Point Bars with Few Mid Channel Bars
B-3	Many Mid Channel Bars
B-4	Side Bars
B-5	Diagonal Bars
B-6	Main Branching with Many Mid Bars and Islands
B-7	Mixed Side Bar and Mid Channel Bars Exceeding 2-3X Width
B-8	Delta Bars

LEVEL III: ASSESSMENT OF STREAM CONDITION AND DEPARTURE

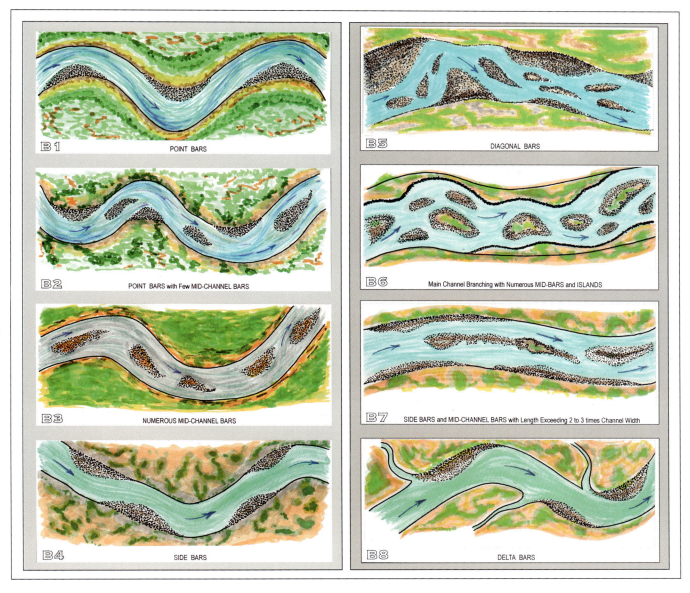

FIGURE 6-10. Illustrations of various depositional features as modified from Galay et al. (1973).

frequently reflected in directly observable depositional features. Excessive sediment deposition is associated with corresponding increases in width/depth ratios, stream slope, lateral migration, and decreases in sinuosity and meander width ratio. Depositional features often indicate channel aggradation. Photographs representing various categories of depositional patterns are shown in **Figure 6-11**.

Aerial photographs can be used to assess channel stability related to depositional features. Cross-sections at gaging stations and other benchmark sites can be re-surveyed to verify interpretations of vertical and lateral channel stability.

Meander Patterns (channels)

Channel meander patterns provide a plan-view of lateral adjustments, meander width ratios, and lateral containment characteristics for various stream types. The manner in which a stream adjusts its slope to that of its valley can be estimated from an analysis of meander patterns.

Categories of meander patterns are listed in **Table 6-5** and are illustrated in **Figure 6-12**. The initial descriptions of meander patterns as presented were described by Mollard (1973) and Galay et al. (1973) and modified by Rosgen (1985). Photographs representing various categories of meander patterns

LEVEL III: ASSESSMENT OF STREAM CONDITION AND DEPARTURE

FIGURE 6-11a. Point Bars (B1).

FIGURE 6-11b. Channel branching with numerous mid-channel bars and islands (B6).

FIGURE 6-11c. Delta Bar (B8).

FIGURE 6-11d. Numerous mid-channel bars (B3).

FIGURE 6-11e. Diagonal and transverse bars (B5).

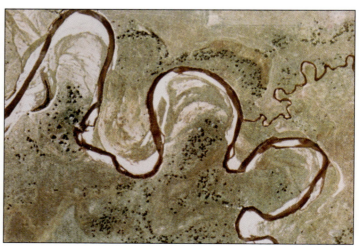

FIGURE 6-11f. Side and mid-channel bars (B7).

LEVEL III: ASSESSMENT OF STREAM CONDITION AND DEPARTURE

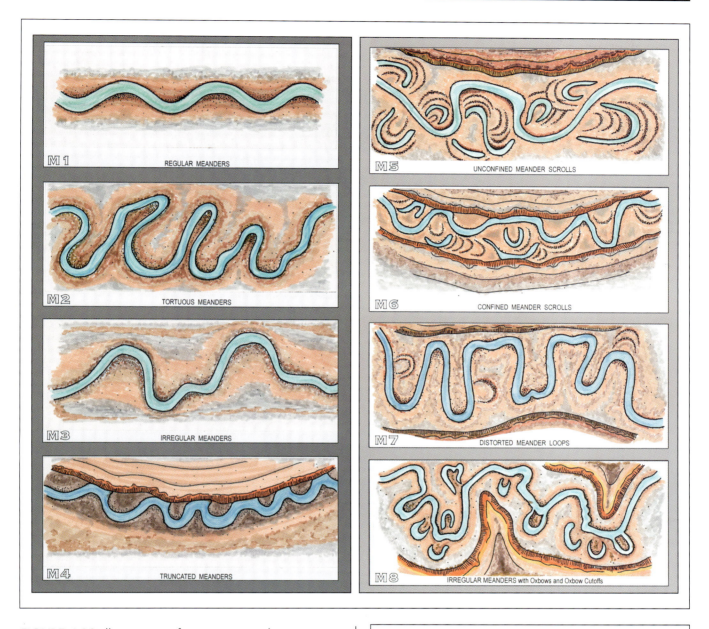

FIGURE 6-12. Illustrations of various meander pattern descriptions as modified from Galay et al. (1973)

MEANDER PATTERNS

- M-1 Regular Meander
- M-2 Tortuous Meander
- M-3 Irregular Meander
- M-4 Truncated Meanders
- M-5 Unconfined Meander Scrolls
- M-6 Confined Meander Scrolls
- M-7 Distorted Meander Loops
- M-8 Irregular with Oxbows, Oxbow Cutoffs

TABLE 6-5. Categories representing various meander patterns of alluvial rivers.

are shown in *Figures 6-13a* through *6-13f*. The primary meander categories are further described in terms of confined (i.e., laterally contained) or unconfined channel reaches. Examples of confined and unconfined meandering stream types are shown in *Figures 6-14* and *Figure 6-15*, respectively. Valley characteristics which limit lateral channel extension include structurally controlled valleys, terraces, colluvial slopes, and alluvial fans impinging on the channel. For example, resistant side slope or valley wall materials can produce truncated channel

LEVEL III: ASSESSMENT OF STREAM CONDITION AND DEPARTURE

FIGURE 6-13a. Truncated Meander (M4).

FIGURE 6-13b. Unconfined Meander Scrolls (M5).

FIGURE 6-13c. Tortuous Meanders (M2).

FIGURE 6-13d. Irregular Meanders (M3).

FIGURE 6-13e. Regular Meanders (M1).

FIGURE 6-13f. Irregular Meanders with Oxbows and Oxbow Cutoffs (M8).

LEVEL III: ASSESSMENT OF STREAM CONDITION AND DEPARTURE

FIGURE 6-14. Examples of various confined meandering stream types.

LEVEL III: ASSESSMENT OF STREAM CONDITION AND DEPARTURE

FIGURE 6-15. Examples of unconfined rivers (C and E) stream types.

LEVEL III: ASSESSMENT OF STREAM CONDITION AND DEPARTURE

STREAM CHANNEL DEBRIS/BLOCKAGES		
DESCRIPTION/EXTENT		Materials, which upon placement into the active channel or floodprone area may cause an adjustment in channel dimensions or conditions, due to influences on the existing flow regime.
D1	NONE	Minor amounts of small, floatable material.
D2	INFREQUENT	Debris consists of small, easily moved, floatable material; i.e. leaves, needles, small limbs, twigs, etc.
D3	MODERATE	Increasing frequency of small to medium sized material, such s large limbs, branches and small logs that when accumulated effect 10% or less of the active channel cross-sectional area.
D4	NUMEROUS	Significant build-up of medium to large sized materials, i.e. large limbs, branches, small logs or portions of trees that may occupy 10 to 30% of the active channel cross-section area.
D5	EXTENSIVE	Debris "dams" of predominantly larger materials, i.e. branches, logs, trees, etc., occupying 30 to 50% of the active channel cross-section; often extending across the width of the active channel.
D6	DOMINATING	Large, somewhat continuous debris "dams," extensive in nature and occupying over 50% of the active channel cross-section. Such accumulations may divert water into the floodprone areas and form fish migration barriers, even when flows are at less than bankfull.
D7	BEAVER DAMS - FEW	An infrequent number of dams spaced such that normal streamflow and expected channel conditions exist in the reaches between dams.
D8	BEAVER DAMS - FREQUENT	Frequency of dams is such that backwater conditions exist for channel reaches between structures; where streamflow velocities are reduced and channel dimensions or conditions are influenced.
D9	BEAVER DAMS - ABANDONED	Numerous abandoned dams, many of which have filled with sediment and/or breached, initiating a series of channel adjustments such as bank erosion, lateral migration, evulsion, aggradation and degradation.
D10	HUMAN INFLUENCES	Structures, facilities, or materials related to land uses or development located within the floodprone area, such as diversions or low-head dams, controlled by-pass channels, velocity control structures, and various transportation encroachments that have an influence on the existing flow regime, such that significant channel adjustments occur.

TABLE 6-6. Debris and channel blockages categorized by size and extent.

meander patterns. The degree of valley confinement affects meander geometry such as sinuosity, meander length, meander radius of curvature, and meander width ratio. Meander geometry relationships are useful for designing channel restoration and interpreting channel adjustment processes, such as avulsion and lateral accretion rates. Analyses of meander patterns can be used to indicate the potential onset of disequilibrium and evolutionary adjustments in an appropriate stream type. Meander geometry interpretations can be used to assess the effects of changes in width/depth ratios, bank erosion estimates, sediment supply, and changes in pattern, dimension, and slope on channel stability.

Debris and Channel Blockages

Woody or organic debris can profoundly affect stream channel stability, sediment storage, width/depth ratio, bank erosion, aggradation/degradation processes, and fish habitat. As shown in *Table 6-6,* debris is categorized by relative size and extent along a reach. Beaver dams, check dams, fish habitat, and irrigation structures are included since they also obstruct streamflow.

A, B, and G stream types generally respond to reduced input of woody debris by an increase in channel scour and increases the spacing of step/pool features, as a function of bankfull width and slope gradient. An increase in step-pool spacing reduces the characteristic mode of natural energy dissipation for A, B, and G stream types, and frequently corresponds to a reduction in fish habitat. Many of the A, B, and G step/pool stream types have evolved in the presence of large woody organic debris, which exerts control on the alternating step/pool bed features. The stability and biological function of A, B, and G stream types is directly linked to the type, amount, and extent of large, woody organic debris. The spacing of the channel step features is proportional to stream width, and inversely related to channel slope. Spacing between pools is highly variable; however, a central tendency exists on riffle/pool streams for an average pool-to-pool spacing of 5-7 bankfull channel widths (or 1/2 of the linear distance associated with a meander wavelength). For streams other than those with riffle/pool bed features, the spacing between pools averages 4-5 bankfull channel widths on entrenched streams with less than 2% slope; 3-4 channel widths on streams with slopes of 2-4 percent; 2-3 channel widths on streams with slopes of 4-6 percent; and 1.5 to 2 channel widths on streams with slopes greater than 6 percent. Exceptions to the average bed feature sequence spacing as a function of gradient and width are found where inputs of large woody organic debris to the system influence the natural sequence of the rock-material bed features. Field observations of debris input tendencies need to be obtained and debris management considerations should incorporate the principles of natural bed-feature spacing. Total removal or excessive introduction of large woody organic debris without regard to the stream type and the related sediment transport characteristics will upset the established balance of processes, with potential adverse consequences.

The presence of large woody debris changes slope, affects potential and kinetic energy, shifts boundary shear stress, creates extremes of velocity, and directly influences sediment storage. Stream types such as the A1-A2 and B1-B6 can support a considerable amount of organic debris and flow blockages without developing adverse impacts. In contrast, the riffle/pool stream types such as C3-C6, E3-E6, and D3-D6 can be adversely affected. The photographs in *Figure 6-16* and *Figure 6-17* illustrate large woody organic debris/channel blockage ratings for various stream types.

Stream Channel Stability

As previously discussed, the stability of streams is key to the full expression of natural stream function. Actual channel stability is best determined by installing permanent channel cross-sections along with bank erosion pins and profiles as discussed in Chapter 7. The channel and bank profile measurements which express lateral and vertical stability are used to verify the rate and direction of change which then indicate the nature and degree of stability inherent in a particular stream system or type.

Pfankuch (1975) developed a system to rate channel stability which has been widely used by the Forest Service and the Bureau of Land Management in the Great Basin, central plains, and the northern, central, and southern Rocky Mountains (*Table 6-7*). Hydrologists use the Pfankuch system to quantitatively describe the potential for sediment material detachment and changes in sediment supply due to changes in streamflow and/or changes in watershed condition. Fisheries biologists have often used the channel stability rating system to generally assess fisheries habitat conditions, and to indirectly assess streambank damage resulting from cattle grazing.

Pfankuch's channel stability rating system was developed prior to the stream classification. Consequently, the original stability rating system uses only an average stability index, invariant of stream type. That is, a rating of 80 is associated with fair channel stability conditions, while a rating of 100 is associated with poor channel stability, regardless of stream type.

The variety of stream types have inherently different channel stabilities, which if expressed in terms of a rating value are masked by use of a single, average value. The good, fair, and poor rating values obtained with the Pfankuch method have been adjusted by stream type, as shown addi-

LEVEL III: ASSESSMENT OF STREAM CONDITION AND DEPARTURE

tionally in *Table 6-7*. The objective of a stream type conversion with the rating values from the original channel stability rating system, is to reflect the naturally inherent and differing value ranges for each stream type. It is important to remember that the values shown are simply an index to channel stability. To determine actual stability, the data collection methods outlined in the Level IV analysis process (Chapter 7) would be implemented. A stability index value is useful to provide a "red flag" or warning system. Once an indication of channel stability is determined from field measurements, the range of index values may be adjusted or "fine tuned". The Wyoming and Colorado state offices of the Bureau of Land Management have evaluated channel stability ratings by stream type in a similar manner, (Mitchell, personal commun. 1994). Similar relations are shown for major groupings of stream types. As always, such ratings are best determined and refined on the basis of field observation and measurement within local and regional hydro-physiographic regions. Stratified channel stability ratings are valuable and have been used for selecting representative sediment rating curves for analyzing cumulative watershed impacts (USEPA, 1980).

Channel stability index value modifications substantially improve the sensitivity of interpretations of the Pfankuch channel stability ratings. Values

FIGURE 6-16a. Example of an A3 "debris torrent" stream where the noticeable lack of large organic debris contributed to bed scour to bedrock. Rated (D1).

FIGURE 6-16b. Example of an appropriate accumulation of debris in a C4 stream type. The rating is (D3).

FIGURE 6-16c. Example of an appropriate amount of debris for this stable A3 stream type. Rated (D4).

FIGURE 6-16d. Example of excessive debris contributing to sediment storage and avulsion. Rated (D6).

LEVEL III: ASSESSMENT OF STREAM CONDITION AND DEPARTURE

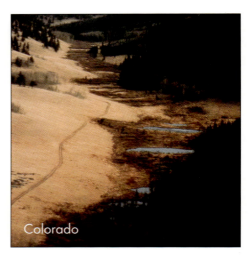

FIGURE 6-17a. Example of stable beaver dams in low bedload, broad valleys with dense willow community in an E3b stream type. Rated (D7).

FIGURE 6-17b. Example of extensive beaver dams. Rated (D8).

FIGURE 6-17c. Example of log weirs installed for fish habitat on B3 stream type. The rating is (D10) which has created backwater between each structure.

FIGURE 6-17d. Example of abandoned beaver dam and adverse adjustment of a C4 stream type which is rated (D9).

greater than the mean values indicate a potential departure from typical stability conditions for a given stream type, and suggest the onset or existence of channel instability. Lower than average ratings indicate a system sensitivity to potential change associated with channel disturbance. Similarly, higher stability numbers indicate an increased system sensitivity and potential for increased erosion/sediment supply if there are commensurate increases of streamflow magnitude and duration.

Data categories have been added to the channel stability field data form to provide for recording interpretations and field observations of sediment supply, aggradation, degradation, and width/depth ratios. Note, however, that both the original and the modified channel stability assessment system require field observations – this is not an office procedure. Additions to the data form used to record revised channel stability ratings are shown in *Table 6-7*, and are used to summarize the interpretations of the field observer. The additional stability categories are:

1. Sediment supply.

   This category is designed to assess the availability of sediment as determined from observed

LEVEL III: ASSESSMENT OF STREAM CONDITION AND DEPARTURE

CHANNEL STABILITY (PFANKUCH) EVALUATION AND STREAM CLASSIFICATION SUMMARY (LEVEL III)

Reach Location_____ Date_____ Observers_____
Stream Type _____

		Category	EXCELLENT	
UPPER BANKS	1	Landform Slope	Bank Slope Gradient <30%	2
	2	Mass Wasting	No evidence of past or future mass wasting.	3
	3	Debris Jam Potential	Essentially absent from immediate channel area.	2
	4	Vegetative Bank Protection	90%+ plant density. Vigor and variety suggest a deep dense soil binding root mass.	3
LOWER BANKS	5	Channel Capacity	Ample for present plus some increases. Peak flows contained. W/D ratio <7.	1
	6	Bank Rock Content	65%+ with large angular boulders. 12"+ common.	2
	7	Obstructions to Flow	Rocks and logs firmly imbedded. Flow pattern without cutting or deposition. Stable bed.	2
	8	Cutting	Little or none. Infreq. raw banks less than 6".	4
	9	Deposition	Little or no enlargement of channel or pt. bars.	4
BOTTOM	10	Rock Angularity	Sharp edges and corners. Plane surfaces rough.	1
	11	Brightness	Surfaces dull, dark or stained. Gen. not bright.	1
	12	Consolidation of Particles	Assorted sizes tightly packed or overlapping.	2
	13	Bottom Size Distribution	No size change evident. Stable mater. 80-100%	4
	14	Scouring and Deposition	<5% of bottom affected by scour or deposition.	6
	15	Aquatic Vegetation	Abundant Growth moss-like, dark green perennial. In swift water too.	1
			TOTAL	

		Category	GOOD	
UPPER BANKS	1	Landform Slope	Bank Slope Gradient 30-40%	4
	2	Mass Wasting	Infrequent. Mostly healed over. Low future potential.	6
	3	Debris Jam Potential	Present, but mostly small twigs and limbs.	4
	4	Vegetative Bank Protection	70-90% density. Fewer species or less vigor suggest less dense or deep root mass.	6
LOWER BANKS	5	Channel Capacity	Adequate. Bank overflows rare. W/D ratio 8-15	2
	6	Bank Rock Content	40-65%. Mostly small boulders to cobbles 6-12"	4
	7	Obstructions to Flow	Some present causing erosive cross currents and minor pool. filling. Obstructions newer and less firm.	4
	8	Cutting	Some, intermittently at outcurves and constrictions. Raw banks may be up to 12"	6
	9	Deposition	Some new bar increase, mostly from coarse gravel.	8
BOTTOM	10	Rock Angularity	Rounded corners and edges, surfaces smooth, flat.	2
	11	Brightness	Mostly dull, but may have <35% bright surfaces.	2
	12	Consolidation of Particles	Moderately packed with some overlapping.	4
	13	Bottom Size Distribution	Distribution shift light. Stable material 50-80%.	8
	14	Scouring and Deposition	5-30% affected. Scour at constrictions and where grades steepen. Some deposition in pools.	12
	15	Aquatic Vegetation	Common. Algae forms in low velocity and pool areas. Moss here too.	2
			TOTAL	

		Category	FAIR	
UPPER BANKS	1	Landform Slope	Bank slope gradient 40-60%	6
	2	Mass Wasting	Frequent or large, causing sediment nearly year long.	9
	3	Debris Jam Potential	Moderate to heavy amounts, mostly larger sizes.	6
	4	Vegetative Bank Protection	<50-70% density. Lower vigor and fewer species from a shallow, discontinuous root mass.	9
LOWER BANKS	5	Channel Capacity	Barely contains present peaks. Occasional overbank floods. W/D ratio 15 to 25.	3
	6	Bank Rock Content	20-40% with most in the 3-6" diameter class.	6
	7	Obstructions to Flow	Moder. frequent, unstable obstructions move with high flows causing bank cutting and pool filling.	6
	8	Cutting	Significant. Cuts 12-24" high. Root mat overhangs and sloughing evident	12
	9	Deposition	Moder. deposition of new gravel and course sand on old and some new bars.	12
BOTTOM	10	Rock Angularity	Corners and edges well rounded in two dimensions.	3
	11	Brightness	Mixture dull and bright, ie 35-65% mixture range.	3
	12	Consolidation of Particles	Mostly loose assortment with no apparent overlap.	6
	13	Bottom Size Distribution	Moder. change in sizes. Stable materials 20-50%	12
	14	Scouring and Deposition	30-50% affected. Deposits & scour at obstructions, constrictions, and bends. Some filling of pools.	18
	15	Aquatic Vegetation	Present but spotty, mostly in backwater. Seasonal algae growth makes rocks slick.	3
			TOTAL	

TABLE 6-7. Channel stability evaluation *(Pfankuch, 1975)* with a conversion of the channel stability rating to a reach condition by stream type.

LEVEL III: ASSESSMENT OF STREAM CONDITION AND DEPARTURE

CHANNEL STABILITY (PFANKUCH) EVALUATION AND STREAM CLASSIFICATION SUMMARY (LEVEL III)

		Category	POOR	
UPPER BANKS	1	Landform Slope	Bank Slope Gradient 60%+	8
	2	Mass Wasting	Frequent or large causing sediment nearly year long or imminent danger of same.	12
	3	Debris Jam Potential	Moder. to heavy amounts, predom. larger sizes.	8
	4	Vegetative Bank Protection	<50% density, fewer species and less vigor indicate poor, discontinuous and shallow root mass.	12
LOWER BANKS	5	Channel Capacity	Inadequate. Overbank flows common. W/D ratio >25	4
	6	Bank Rock Content	<20% rock fragments of gravel sizes, 1-3" or less.	8
	7	Obstructions to Flow	Sediment traps full, channel migration occurring.	
	8	Cutting	Almost continuous cuts, some over 24" high. Failure of overhangs frequent.	16
	9	Deposition	Extensive deposits of predom. fine particles. Accelerated bar development.	16
BOTTOM	10	Rock Angularity	Well rounded in all dimensions, surfaces smooth.	4
	11	Brightness	Predom. bright, 65%+ exposed or scoured surfaces.	4
	12	Consolidation of Particles	No packing evident. Loose assortment easily moved.	8
	13	Bottom Size Distribution	Marked distribution change. Stable materials 0-20%.	16
	14	Scouring and Deposition	More than 50% of the bottom in a state of flux or change nearly year long.	24
	15	Aquatic Vegetation	Perennial types scarce or absent. Yellow-green, short term bloom may be present.	4
			TOTAL	

Stream Width _____ x avg. depth _____ x mean velocity _____ = Q _____ cfs

Gauge Ht _____ Reach Gradient _____ Stream Order _____ Sinuosity Ratio _____

Width $_{bkf}$ _____ Depth $_{bkf}$ _____ W/D Ratio _____ Discharge (Q_{bkf}) _____

Drainage Area _____ Valley Gradient _____ Stream Length _____ Valley Length _____

Sinuosity _____ Entrenchment Ratio _____ Length Meander (Lm) _____ Belt Width _____

Sediment Supply
Extreme _____
Very High _____
High _____
Moderate _____
Low _____
Remarks _____

Stream Bed Stability
Aggrading _____
Degrading _____
Stable _____

TOTAL SCORE for Reach   E____ = G____ + F____ + P____ =

Width/Depth Ratio Condition
Normal _____
High _____
Very High _____

Stream Type [ ]
Pfankuch Rating [ ]
from table [ ] Reach Condition

CONVERSION OF STABILITY RATING TO REACH CONDITION BY STREAM TYPE*

Stream Type	A1	A2	A3	A4	A5	A6	B1	B2	B3	B4	B5	B6
GOOD	38-43	38-43	54-90	60-95	60-95	50-80	38-45	38-45	40-60	40-64	48-68	40-60
FAIR	44-47	44-47	91-129	96-132	96-142	81-110	46-58	46-58	61-78	65-84	69-88	61-78
POOR	48+	48+	130+	133+	143+	111+	59+	59+	79+	85+	89+	79+

Stream Type	C1	C2	C3	C4	C5	C6	D3	D4	D5	D6		
GOOD	38-50	38-50	60-85	70-90	70-90	60-85	85-107	85-107	85-107	67-98		
FAIR	51-61	51-61	86-105	91-110	91-110	86-105	108-132	108-132	108-132	99-125		
POOR	62+	62+	106+	111+	111+	106+	133+	133+	133+	126+		

Stream Type	DA3	DA4	DA5	DA6	E3	E4	E5	E6				
GOOD	40-63	40-63	40-63	40-63	40-63	50-75	50-75	40-63				
FAIR	64-86	64-86	64-86	64-86	64-86	76-96	76-96	64-86				
POOR	87+	87+	87+	87+	87+	97+	97+	87+				

Stream Type	F1	F2	F3	F4	F5	F6	G1	G2	G3	G4	G5	G6
GOOD	60-85	60-85	85-110	85-110	90-115	80-95	40-60	40-60	85-107	85-107	90-112	85-107
FAIR	86-105	86-105	111-125	111-125	116-130	96-110	61-78	61-78	108-120	108-120	113-125	108-120
POOR	106+	106+	126+	126+	131+	111+	79+	79+	121+	121+	126+	121+

Generalized relations ... need additional Level IV data to expand data base for validation.

TABLE 6-7. Channel stability Evaluation (Pfankuch, 1975)

LEVEL III: ASSESSMENT OF STREAM CONDITION AND DEPARTURE

deposition, transport, and storage features within a river reach. Examples of very low to extremely high sediment supply conditions are shown in *Figure 6-18.*

2. Bed Stability

This category is designed to document locations where scour has led to degradation or lowering of river bed, and where aggradation or the effects of deposition over time have raised bed elevations.

Examples of degradation are shown in *Figure 6-19*. Degradation is usually associated with concurrent channel adjustments, such as: (a) tributary rejuvenation (i.e., over steepening of main stem and tributaries); (b) acceleration of bank erosion ; (c) increase in sediment supply and transport; (d) dewatering of floodplains and changes in vegetative communities; (e) abandonment of floodplains and creation of new terraces; and (f) steepening of water surface slope.

Aggradation processes are illustrated in *Figure 6-20* and are associated with the following channel adjustments: (a) increase in width/depth ratio; (b) increase in sediment storage; (c) increase in bank erosion rates; (d) decrease in pool quality, and other fishery habitat features; (e) increase in the loss of channel-adjacent land and associated sediment availability from bank erosion and lateral migration; and (f) an increase in over-bank flooding with less than flood-flow magnitudes.

Stable stream bed types are shown in *Figure 6-21*. The streams depicted are stable, yet transport a wide range of streamflows and sediment sizes and volume without adverse adjustments. Note the stable characteristic of lower width/depth ratios and high quality habitat components associated normally with stable stream systems.

3. Width/depth ratio shifts

This category is designed to document existing width/depth ratios and place them into groupings that would indicate normal or abnormal channel width conditions.

When the bankfull width of a stream is increased without a corresponding increase in bankfull discharge, the stream's mean depth decreases. Channel geometry changes such as an increase in the width/depth ratio, are an important indicator of channel instability.

Changes in bankfull width and depth can be determined by comparing the bankfull width versus width hydraulic geometry for stable and altered reaches of the same stream type. Alternatively, channel geometry changes can be determined by plotting frequency distributions of width/depth ratios for various stream types and checking for departures from average observed values. Width/depth ratios can also be compared at points above and below or before and after a disturbance.

Stable and unstable channel conditions and their varying width/depth ratios for the same stream types are contrasted in *Figure 6-22* and *Figure 6-23*. High width/depth ratios, beyond the stable range, indicate a sharp reduction in the channel's capacity to transport sediment, due to reduced shear stress associated with the reduced mean depths. Moreover, the distribution of shear stress in streams with a high width/depth ratio is greatest in the near-bank region, thus, accelerating bank erosion and adding to the sediment supply. Increased bank erosion causes aggradation and lateral extension of the channel within its valley, and an increased sediment supply which creates downstream impacts. Loss of fish habitat, is characteristically, a direct result. Sufficiently large increases in width/depth ratio can initiate development of new stream types, usually with an attendant loss in stability, habitat, and aesthetic values. Examples of typical stream type conversions or evolutionary sequences developing from changes in width/depth ratios include conversions from an E to C, from a C to F, or C to D stream type, as previously discussed. (*Figures 6-4, 6-5 and 6-6*)

Ironically, stream channel instability due to altered width/depth ratios are frequently the direct result of well-intentioned attempts to restore streams. One common restoration design places logs or check dams in over widened streams in an attempt to increase depth by creating a plunge pool below the dam. Unfortunately, the check-dam design often adds to channel instability by decreasing sediment transport capacity, increasing the width/depth ratio, accelerating headward aggradation, initiating lateral migration, and accelerating streambank erosion. An example is shown in *Figure 6-24*, where a state highway construction project altered the width/depth ratio of a stable B3 stream type by placing a gabion mattress and rip-rap in a misguided attempt to stabilize the banks, and installing a check dam in an attempt to re-establish pools. The high width/depth ratios created by application of the check-dam design will initiate a condition of disequilibrium due to the inability of the stream to efficiently transport sediment. The biological attributes of the altered stream have also been compromised, as compared to its potential, shown by comparison with the reach immediately downstream. In this case, significant resource values have been compromised, and at a high cost in engineering design and construction.

LEVEL III: ASSESSMENT OF STREAM CONDITION AND DEPARTURE

FIGURE 6-18a. Example of very low sediment supply on a B3 stream type.

FIGURE 6-18b. Example of a high sediment supply with extreme embeddedness and bi-modal distribution of bed materials. (B4 stream type)

FIGURE 6-18c. Example of an extreme sediment supply condition A4 stream type (note natural sorting of materials into a step/pool sequence.)

FIGURE 6-18d. Example of very high sediment supply of an F4 stream type incised in a delta due to the stage controlled by a reservoir.

FIGURE 6-18e. Example of tributary (G4) contribution of sediment and delta bar formation.

LEVEL III: ASSESSMENT OF STREAM CONDITION AND DEPARTURE

FIGURE 6-19a. Example of degradation of a meadow (C4) stream, converting to a G4 stream type.

FIGURE 6-19b. Example of a degradational stream (G4) downcutting through a debris fan.

FIGURE 6-19c. Examples of degradational process (G4 stream type).

FIGURE 6-19d. Example of degradation following a major flood (F4 stream type).

FIGURE 6-19e. Degradation, leading to entrenchment (F5 stream type).

FIGURE 6-19f. Degradation of meadow from an E5 to G5c to F5. Rock is failed attempt at bed stabilization with loose rock check dam.

LEVEL III: ASSESSMENT OF STREAM CONDITION AND DEPARTURE

a. C4/D4 Stream type (Idaho).

b. C4 Stream type (Colorado).

c. C5 Stream type (Colorado).

d. C5 Stream type (California).

FIGURE 6-20. Examples of aggradation process showing increases in width/depth ratio, decrease in meander width ratio, deposition, and habitat loss.

Streambank Erosion Potential

Recent studies indicate that the contribution of streambank erosion to total sediment yields has been greatly underestimated (Rosgen 1973, Rosgen 1976, Rosgen 1993c). Bank erosion is a natural river adjustment process. Lateral migration rates of the stream channel can be accelerated, however, when the variables controlling bank erosion processes are altered, especially those variables affecting detachment of bank material and flow stresses in the near-bank region.

Bank erosion occurs as a result of a number of processes, such as dry ravel, mass wasting, surface erosion, liquification, freeze-thaw, fluvial entrainment, and ice scour. The ability of streambanks to resist erosion is primarily determined by:

(1) the ratio of streambank height to bankfull stage;

(2) the ratio of riparian vegetation rooting depth to streambank height;

(3) the degree of rooting density;

(4) the composition of streambank materials;

(5) streambank angle (i.e., slope);

(6) bank material stratigraphy and presence of soil lenses; and

(7) bank surface protection afforded by debris and vegetation.

Hazard rating procedures that characterize various streambank conditions into numerical

LEVEL III: ASSESSMENT OF STREAM CONDITION AND DEPARTURE

FIGURE 6-21. Examples of stable bed stream types that are neither aggrading or degrading.

indices of bank erosion potential are shown in **Figure 6-25** and **Table 6-8** (Rosgen, 1993a).

Streambank erosion is also related to the distribution of streamflows in the near-bank region, defined as the one-third portion of channel cross-section nearest the streambank being evaluated. Near-bank stress is determined primarily by an analysis of flow stage, local energy slope, width/depth ratio, and position in the channel planform (i.e., meander bend or straight reach). In addition, depositional features determine the distribution of local flow velocity gradients, stream power, and shear stress. Extensive bar development creates high velocity gradients, with high boundary (or near-bank) shear stress, due to the influences of bed topography that increases depth and local slope.

Velocity isovels developed from vertical velocity profiles for a C3 stream type for a given stage and cross-section are shown in **Figure 6-26**. These relations show a contrast of the strong velocity gradients toward the left bank at cross-section 2+79, with the depositional process on the right bank. Although just 10 feet apart, these cross-sections show a major shift in velocity distribution for the same discharge.

Table 6-9 converts numerical stress indices using velocity gradients and ratio of stress in the near-bank region with qualitative ratings of near-bank stress. Examples of integrated streambank erodibility and near-bank stress indices used to estimate bank erosion rates are shown in **Table 6-11**.

LEVEL III: ASSESSMENT OF STREAM CONDITION AND DEPARTURE

FIGURE 6-22. Comparisons of stable width/depth ratios to unstable (high values) of width/depth ratios for various stream types.

LEVEL III: ASSESSMENT OF STREAM CONDITION AND DEPARTURE

FIGURE 6-23. Comparisons of stable stream types with appropriate width/depth ratios to unstable stream types showing high width/depth ratios.

LEVEL III: ASSESSMENT OF STREAM CONDITION AND DEPARTURE

a. Stable B3 stream type located immediately downstream of re-constructed reach "b" (Maryland).

b. High width/depth ratio converted F4 stream type "stabilized" (Maryland).

c. Grade control structure to "stabilize" bed and create scour pool for fish habitat and grade control (Maryland).

FIGURE 6-24. Example of channel changes on highway construction which violates the "rules of the river" — loss of natural stability, biological function and visual values — at a great cost.

The relations shown in the erodibility and "stress in the near-bank region" indices (*Table 6-8, Table 6-9,* and *Figure 6-27*) were developed from research the author conducted on the Front Range of Colorado for the USDA Forest Service (Rosgen, 1993a) and for the USDI Park Service in Yellowstone National Park (Rosgen, 1993c). The indices were developed as part of field work conducted by USDA Forest Service and NPS personnel for these projects. Estimates of potential stream bank erosion using the established indices were developed in conjunction with the measurement of stream bank erosion pins that were installed in the early spring of 1989, prior to snowmelt runoff. Following snowmelt runoff, the erosion pins were again measured to determine actual erosion rates. The snowmelt runoff for 1989 was well below normal. The Front Range of Colorado experienced bankfull discharges which were 60 percent to 70 percent of normal, thus the bank erosion rates measured were below the rates experienced during normal runoff seasons. The Colorado and Wyoming data sets were kept separate for analysis purposes. Although there was a large wildfire in Yellowstone Park in 1989, the post runoff bank erosion pin data was obtained early in the season of 1989, prior to the fire.

An analysis of variance was conducted using the collected data with streambank erodibility and stress in the near-bank region defined as independent variables and measured erosion rates as the dependent variable. The coefficient of determination (R^2) for the Colorado data was 0.93 and for the Wyoming data 0.87 both were highly significant at the 95 percent level (*Figure 6-27*). These data represent a wide range of stream types in various erodibility stress, and hydro-physiographic provinces.

A similar research study was conducted by the author in conjunction with Newmont Gold Company in north central Nevada on Maggie Creek in 1992 and 1993. During the course of this study, a rain-on-snow event produced lateral bank erosion rates of 12 to 25 feet during one event, in G4 and F4

LEVEL III: ASSESSMENT OF STREAM CONDITION AND DEPARTURE

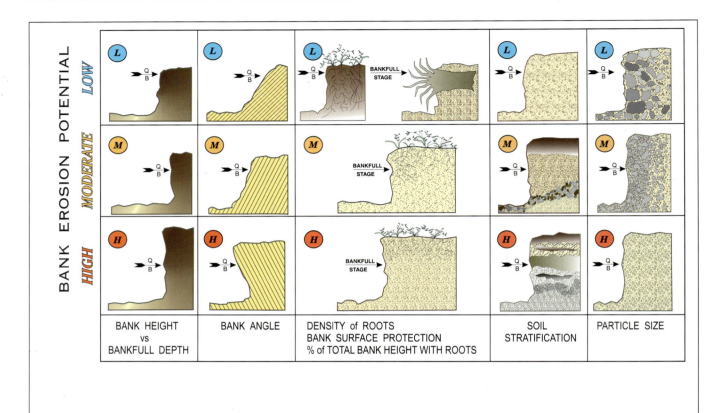

FIGURE 6-25. Stream bank erodibility factors. (Rosgen, 1993a)

stream types, respectively. The G4 and F4 stream types studied in Nevada had extreme erodibility and very high near-bank stress ratings which are obviously much larger than the data shown in *Figure 6-27*. Measurement methods used in all of the described research studies are discussed in Chapter 7 depicting Level IV determinations.

The mythical "flushing flow" is often proposed as a solution to sediment deposition problems usually caused by upstream changes within a watershed. The commonly held belief is that a "flushing flow" will re-distribute the excess sediment in the bars and restore the stream. In fact, for many stream types, such flows will increase channel instability by accelerating bank erosion. However, flushing flows can be effective at removing fines on stream types A1, A2, B1, B2, B3, F1, F2, E2, and E3 due to the stable nature of their bed and banks since the higher flows generally increase the distribution of stress away from the deposition features and toward the banks, as illustrated by the velocity isovels shown in *Figure 6-26*. The high velocity gradients and distribution of the boundary stress in the near-bank region on unstable banks generally accelerates bank erosion due to "flushing flows." The depositional features often intended to be removed by flushing flows frequently become permanent islands or part of the floodplain, as point bars are abandoned through lateral migration. Unfortunately, restoring the natural distribution of bed materials and improving channel stability in streams overloaded with sediment requires much more of a prescribed flow regime than a "flushing flow."

Changes in riparian vegetation, rooting depth, root density and the percent of the bank surface area that is protected by vegetation are reflected in the erodibility ratings. Examples of erodibility ratings for low, moderate, high and extreme for a variety of stream types are shown in *Figures 6-28* through *6-30*.

LEVEL III: ASSESSMENT OF STREAM CONDITION AND DEPARTURE

TABLE 6-8. Streambank characteristics used to develop Bank Erosion Hazard Index (BEHI)

Adjective Hazard or risk rating categories		Bank Height/ Bankfull Ht	Root Depth/ Bank Height	Root Bank Height	Bank Angle (Degrees)	Surface Protection %	Totals
Very Low	Value	1.0-1.1	1.0-0.9	100-80	0-20	100-80	
	Index	1.0-1.9	1.0-1.9	1.0-1.9	1.0-1.9	1.0-1.9	5-9.5
Low	Value	1.11-1.19	0.89-0.5	79-55	21-60	79-55	
	Index	2.0-3.9	2.0-3.9	2.0-3.9	2.0-3.9	2.0-3.9	10-19.5
Moderate	Value	1.2-1.5	0.49-0.3	54-30	61-80	54-30	
	Index	4.0-5.9	4.0-5.9	4.0-5.9	4.0-5.9	4.0-5.9	20-29.5
High	Value	1.6-2.0	0.29-0.15	29-15	81-90	29-15	
	Index	6.0-7.9	6.0-7.9	6.0-7.9	6.0-7.9	6.0-7.9	30-39.5
Very High	Value	2.1-2.8	0.14-0.05	14-5.0	91-119	14-10	
	Index	8.0-9.0	8.0-9.0	8.0-9.0	8.0-9.0	8.0-9.0	40-45
Extreme	Value	>2.8	<0.05	<5	<119	<10	
	Index	10	10	10	10	10	46-50

For adjustments in points for specific nature of bank materials and stratification, the following is used:
Bank Materials: Bedrock (very low), Boulders (low), cobble (subtract 10 points unless gravel/sand > 50%, then no adjustment), gravel (add 5-10 points depending on % sand), sand (add 10 points), silt/clay (no adjustment).
Stratification: Add 5-10 points depending on the number and position of layers.

TABLE 6-9. Velocity gradient and near-bank stress indices

Bank Erosian Risk Rating	Velocity Gradient*	Near-Bank Stress/ Shear Stress**
Very Low	Less than 0.5	Less than 0.8
Low	0.5-1.0	0.8-1.05
Moderate	1.1-1.6	1.06-1.14
High	1.61-2.0	1.15-1.19
Very High	2.1-2.4	1.20-1.60
Extreme	greater than 2.4	greater than 1.60

* Velocity gradient in ft/sec/ft is the difference in velocity from the core of the velocity isovel along the orthogonal length to the near-bank region in feet.

** Near-bank shear stress/mean shear stress
  where shear stress = (mean depth) (slope) (specific weight of water)

LEVEL III: ASSESSMENT OF STREAM CONDITION AND DEPARTURE

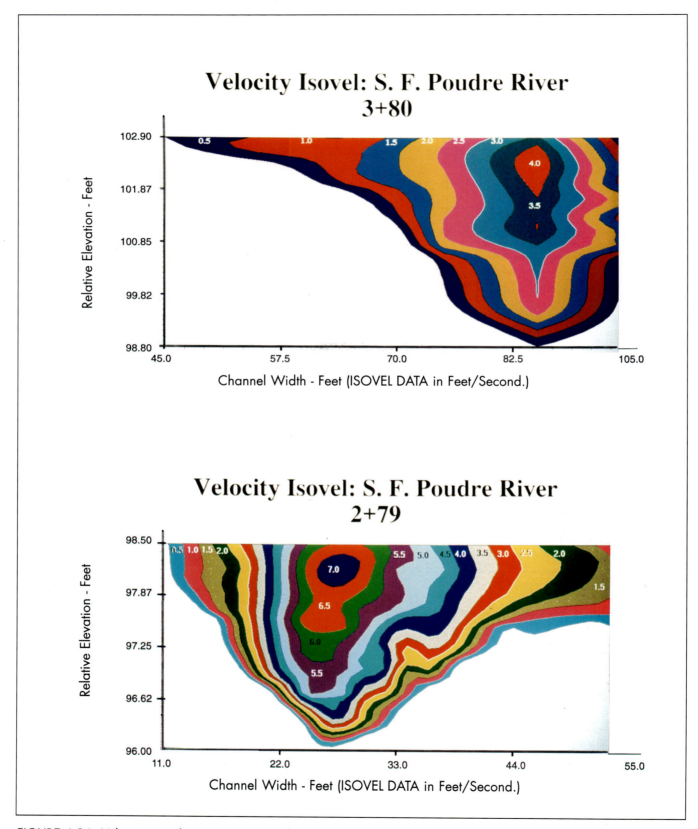

FIGURE 6-26. Velocity isovels at two separate locations on a "C3" streamtype reach; showing variation in stress velocity distribution. *(Nankervis, USDA Forest Service, 1989)*

LEVEL III: ASSESSMENT OF STREAM CONDITION AND DEPARTURE

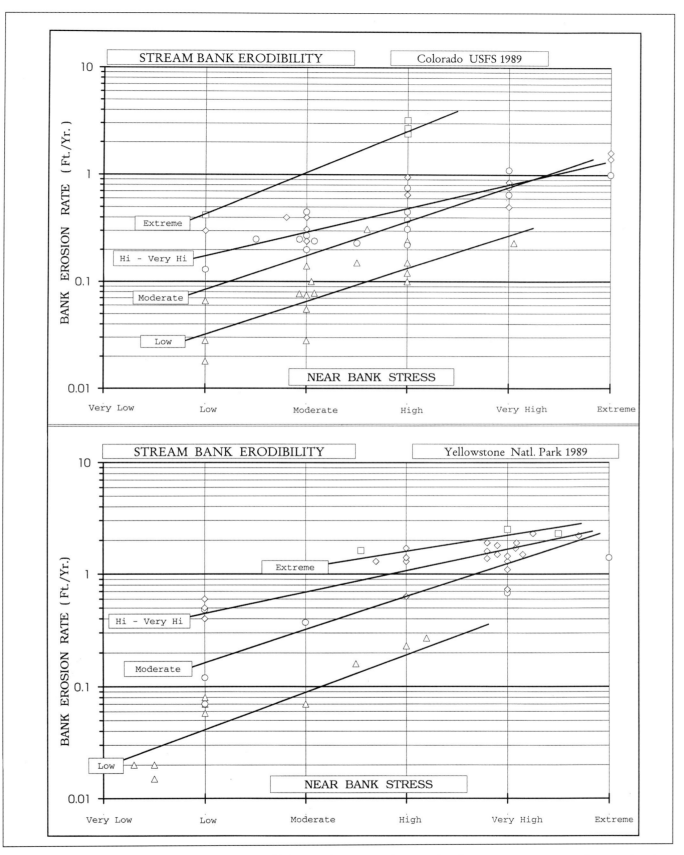

FIGURE 6-27. Relationship of streambank erodibility and stress in the near-bank region vs. measured streambank erosion rates, (USFS, Colorado and NPS, Yellowstone, 1988-1989) (Rosgen 1993c)

LEVEL III: ASSESSMENT OF STREAM CONDITION AND DEPARTURE

FIGURE 6-28. Examples of low erodibility and low stress stream types.

LEVEL III: ASSESSMENT OF STREAM CONDITION AND DEPARTURE

FIGURE 6-29. Examples of moderate erodibility streambank conditions.

LEVEL III: ASSESSMENT OF STREAM CONDITION AND DEPARTURE

FIGURE 6-30. Examples of high/very high bank erodibility potential (a-c) and extreme erodibility potential (d-f).

LEVEL III: ASSESSMENT OF STREAM CONDITION AND DEPARTURE

Altered stream channels

Stream channels have been straightened, deepened, widened, lined, reshaped, relocated, routed through pipes, tunnels, turbines, and trans-basin diversions with profound effects on the stability and integrity of natural systems. Examples of a few altered streams are shown in *Figure 6-31*. Hydrologists and Engineers have frequently been blamed for stream degradation, however, their traditional designs were primarily implemented in response to public requests for protection from floods, as more cities, farms, and homes were located on the floodplains. Floodplain encroachment, in turn, caused river adjustments, increasing the perceived need for additional "hard" engineering controls that have changed the natural function and values of many rivers. Societal values have changed, and hydrologists and engineers now have a new challenge: to restore the natural stability and natural function of rivers.

A format for documenting alterations to rivers is provided in *Table 6-10*. Data for undisturbed bankfull channel dimensions, patterns, and profiles can be obtained from: (1) historical aerial photography, (2) streams within similar valley types in nearby unimpaired watersheds, or (3) regional relationships developed by drainage area and stream types. Information and data in the form of relationships is used to determine the departure of existing conditions from previous conditions and to determine the channel dimensions that need to be restored.

SUMMARY OF LEVEL III ASSESSMENTS OF STREAM CONDITION

A summary form for documenting Level III assessments of stream condition is provided in *Table 6-11*. The form summarizes the important conclusions regarding stream condition, including: stream type, stream size and order, flow regime, riparian vegetation, meander pattern, modified channel stability rating, depositional pattern, debris or other blockages, sediment supply, vertical stability, streambank erosion potential, near-bank stress, width/depth ratios, meander lengths, radius of curvature, belt width, sinuosity, water surface slope, valley slope, and bedform features.

As the above parameter list indicates, the stepwise progression from Level I to Level III analyses yields a great deal of data tailored to quantitatively assess existing and potential stream condition, and to evaluate the significance of instability thresholds. Because the analysis is hierarchical, the more labor-intensive data required for Level III analyses need only be collected at a relatively small fraction of representative stream reaches. The fundamental stratification by stream types that forms the basis of the hierarchy enables the valuable extrapolation of Level III results and interpretations to similar Level II reaches.

LEVEL III: ASSESSMENT OF STREAM CONDITION AND DEPARTURE

FIGURE 6-31. Altered states of streams where the dimension, pattern, slope, and materials have been modified.

LEVEL III: ASSESSMENT OF STREAM CONDITION AND DEPARTURE

TABLE 6-10. Altered stream state or "condition" categories due to direct modification of dimension, pattern, profile, and channel materials.

VARIABLES ALTERED

Channel dimensions
- Width   Previous _____   Existing _____
- Depth   Previous _____   Existing _____
- Width/depth ratio   Previous _____   Existing _____

Channel Patterns (* show as a function of Bankfull Width):
- Meander Length *   Previous _____   Existing _____
- Radius of curvature *   Previous _____   Existing _____
- Belt width *   Previous _____   Existing _____
- Sinuosity   Previous _____   Existing _____

Longitudinal Profile
- Water surface slope   Previous _____   Existing _____
- Valley slope   Previous _____   Existing _____
- Bed features: Previous: _____   Existing _____
    - Riffle/pool _____   Riffle/pool _____
    - Step/pool _____   Step/pool _____
    - Conver/Diverg _____   Conv./div. _____
    - Plane bed _____   Plane bed _____
    - Other: _____   Other _____

Spacing of bed features (as function of bankfull width):
_____   _____

Describe channel alteration:
_____
_____
_____
_____
_____

Existing stream type _____   **Potential stream type** _____

PHOTOGRAPHS
(Looking upstream and downstream)

LEVEL III: ASSESSMENT OF STREAM CONDITION AND DEPARTURE

TABLE 6-11. Field summary form for Level III inventory for stream classification.

SUMMARY OF "CONDITION" CATEGORIES FOR LEVEL III INVENTORY

Stream Name _____
Location _____
Riparian Vegetation _____
Stream Size, Stream order _____
Meander pattern _____
Channel stability rating (Pfankuch) ____
    Sediment supply (check appropriate category):
        Extreme_____
        Very high_____
        High_____
        Moderate_____
        Low_____
    Streambed (vertical) stability
        Aggrading_____
        Degrading_____
        Stable_____
    Width/depth ratio condition:
        Normal (stable)_____
        High_____
        Very high_____
    Streambank erosion Potential:
        Bank erodibility: Near-bank stress:
        Extreme_____   Extreme_____
        High_____   High_____
        Moderate_____   Moderate_____
        Low_____   Low_____

Observers _____
Stream Type _____ Date _____
Flow regime _____
Depositional pattern _____
Debris/channel blockages _____
Altered Channel State: _____
Dimension/shape:
    Width_____
    Depth_____
    Width/depth ratio_____
Patterns: (*show as funct. of W_{bkf}):
    Meander length*_____
    Radius of curve*_____
    Belt width*_____
    Sinuosity_____
Profile:
    Water surface slope_____
    Valley slope_____
    Bed features:
        Riffle/pool_____
        Step/pool_____
        Conver./divrg._____
        Plane bed_____
        Other_____
        Spacing*_____
    Describe alterations:_____

General Remarks
_____
_____
_____
_____
_____
_____
_____
_____

Attach photographs taken mid-stream looking up and downstream. Make site map.

Attach vicinity map of reach and/or aerial photo for specific location.

Note any permanent cross-sections for level IV verification of cross-section stability, actual erosion rates, change in pebble counts, deposition studies, sediment sampling, etc.

Attach copy of: stream classification field form, channel stability rating form, bank erosion rating form, profiles, cross-sections, pebble counts, etc.

CHAPTER 7

LEVEL IV: FIELD DATA VERIFICATION

"Field data is the best cure for a precarious prediction."

INTRODUCTION

Stream inventory Level IV analyses are conducted to verify process-based assessments of stream condition, potential, and stability as predicted from preceding analyses. The verification is achieved through reach-specific observation and analyses of sediment condition, streamflow, and stability measurements (*Figure 7-1*). After reach conditions have been verified, these data are also used to establish empirical relationships for testing, validating, and improving the prediction of velocity, hydraulic geometry, sediment transport characteristics, bank erosion rates, and channel stability.

Great variability occurs amongst models designed to predict similar processes, even for the same data set. The common question is, "What model is the most appropriate for a given stream type?", and "How can I adjust the model to make it work?" There are also wide and divergent opinions of what constitutes a "stable channel," which leads to the age-old question of "is this channel stable?" These questions are not new and without site-specific field observations, they will continue to arise and pose their complex challenges to river managers and scientists. Detailed field observations that measure the correct variables can provide the data for a better understanding of these complex systems and a basis to answer the questions posed.

This also is the level where detailed measurements are made from companion inventories such

LEVEL IV: FIELD DATA VERIFICATION

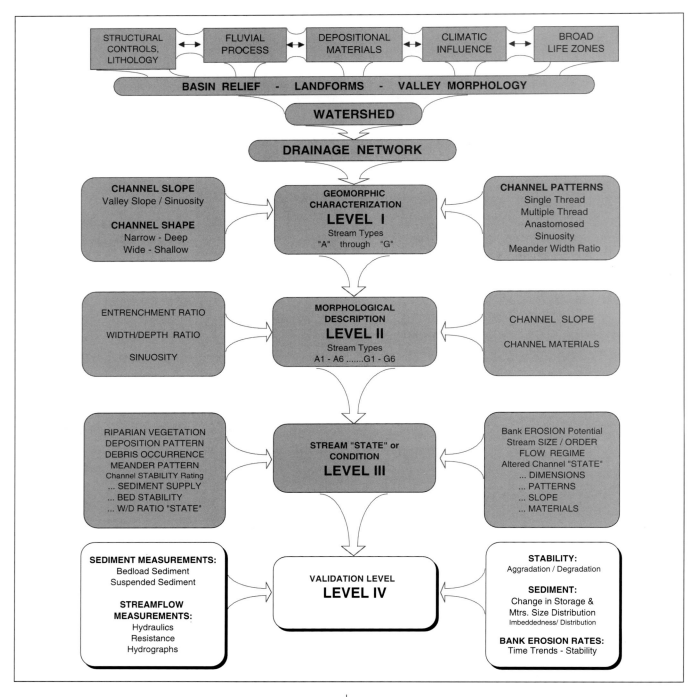

FIGURE 7-1. Validation level (Level IV) for stream classification in relation to the hierarchical river inventory.

as the assessment of the biological system. Specific data on fish populations and food chains can often be related to the morphology and condition of the physical system.

Empirical relationships can be developed for specific stream types, for a given state (condition), and as such, can often be extrapolated to other reaches of the same stream type and state for which Level IV data are not available. As noted in Chapter 2, it is not possible to extrapolate Level IV data and relationships without the geomorphic context provided by the previous levels of analysis. Thus, full use of Level I through Level IV analyses enables the use of existing data that could not otherwise be applied. In contrast, attempts to shortcut verification analyses can result in misinterpretations and for

LEVEL IV: FIELD DATA VERIFICATION

restoration applications could easily result in unintended de-stabilization of stream systems.

Chapter 7 is presented in 3 parts that discuss: (1) streamflow measurements; (2) sediment analyses; and (3) verification of stream stability.

Streamflow Measurements

Streamflow measurements are obtained to establish relations of basin hydrology/channel dimension, hydraulic geometry, and flow resistance characteristics. The streamflow data also validates various models that estimate velocity, streamflow changes in the basin, time of concentration of streamflows, peak flows, minimum flows and altered flow regimes. Actual streamflow data is useful to develop local and regional curves of bankfull dimension/drainage area relations as shown by the analysis of streamflow measurement data for a New Mexico stream as plotted by Jackson, (1994) and shown in *Figure 7-2*. These measured values were plotted onto regional curves as previously referenced in chapter 5, as described by Dunne and Leopold (1978). The regional curves are useful for the initial estimate of bankfull dimensions in the absence of streamflow data, providing they represent the hydro-physiographic province. Bankfull dimensions and drainage area relations further demonstrate how to utilize existing streamflow measurement data and present it in a form useful for extrapolation. The estimate of bankfull dimensions by drainage area can be improved by stratifying the bankfull dimensions and drainage area values by stream type and stream size (bankfull width). Additional basin parameters and morphological indices of streamflow as developed from measured values are described in Carlson, (1963), Stall and Fok, (1967), and Marston, (1978).

Hydraulic geometry relations as described by Leopold and Maddock (1953) allow for validation of velocity and other hydraulic variables by plotting measured values as a simple power function of discharge. An example of at-a-station hydraulic geometry relations is presented in *Figure 7-3*, where the author plotted values from the stream discharge notes for the East Fork San Juan River at the U.S. Geological Survey stream gage location. The hydraulic geometry relations were extrapolated to determine bankfull dimensions for natural channel design on a large river restoration project due the similarity of stream types. Instructions for applying the same procedure is described in the forms provided in *Table 5-1* and *Table 5-2*, Chapter 5. Similar analysis should be additionally stratified by stream type.

Sediment Analyses

Sediment analyses can be divided into measurements of bedload and suspended sediment, changes in sediment storage, size distributions, and source areas. Sediment plays a major role in the influence on channel stability and morphology. As such, many prediction methodologies are applied to determine sediment supply, sizes, and load. The U.S. Environmental Protection Agency has recently determined that sediment is the number one non-point water quality pollutant problem. Thus, monitoring should emphasize field measurements of both suspended and bedload sediment. In practice, however, sediment is predicted much more often than it is ever measured, often resulting in peculiar values. Suspended sediment often is used to determine sediment sources and is often associated with a supply limitation in relation to its transport. Bedload, while influenced by supply, is more often associated with an energy limitation rather than a supply limitation. Bedload is also more critical for stability assessment since the coarse sediment is more sensitive to an energy requirement for transport. Because of the obvious importance to channel stability, the measurement of bedload, not only provides an insight into channel processes, but it also is valuable to validate the great variety of sediment prediction models currently used.

Due to the great diversity of stream types, there is a similar corresponding diversity of sediment models and/or calibration coefficients for each type. Calibration of existing models or improving empirical models with actual sediment data by stream type obviously will improve the prediction. Validation will also help select the most appropriate model by stream type.

LEVEL IV: FIELD DATA VERIFICATION

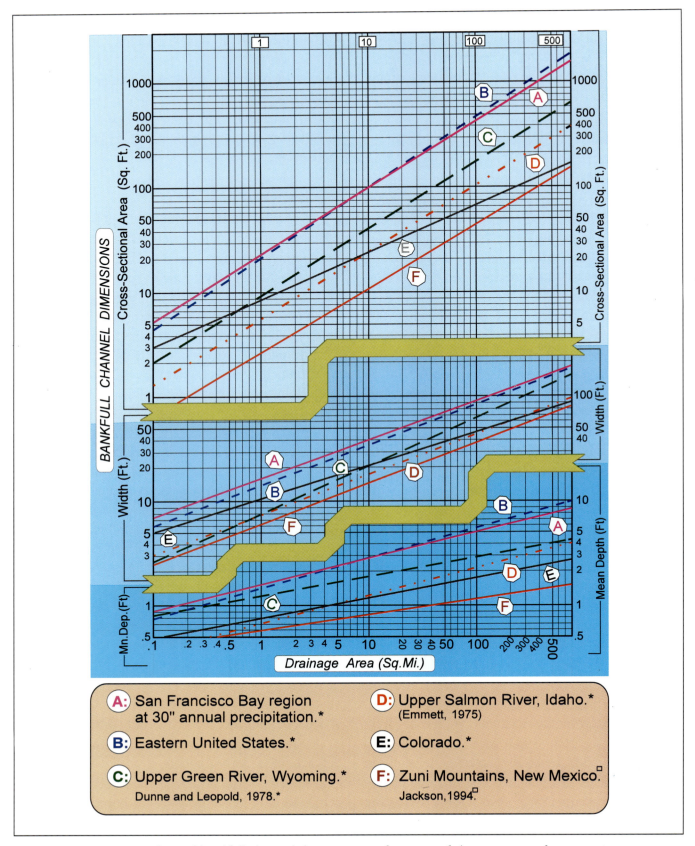

FIGURE 7-2. Average values of bankfull channel dimensions as functions of drainage area for six regions. *(Dunne and Leopold, 1978, Emmett, 1975 and Jackson, 1984)*

LEVEL IV: FIELD DATA VERIFICATION

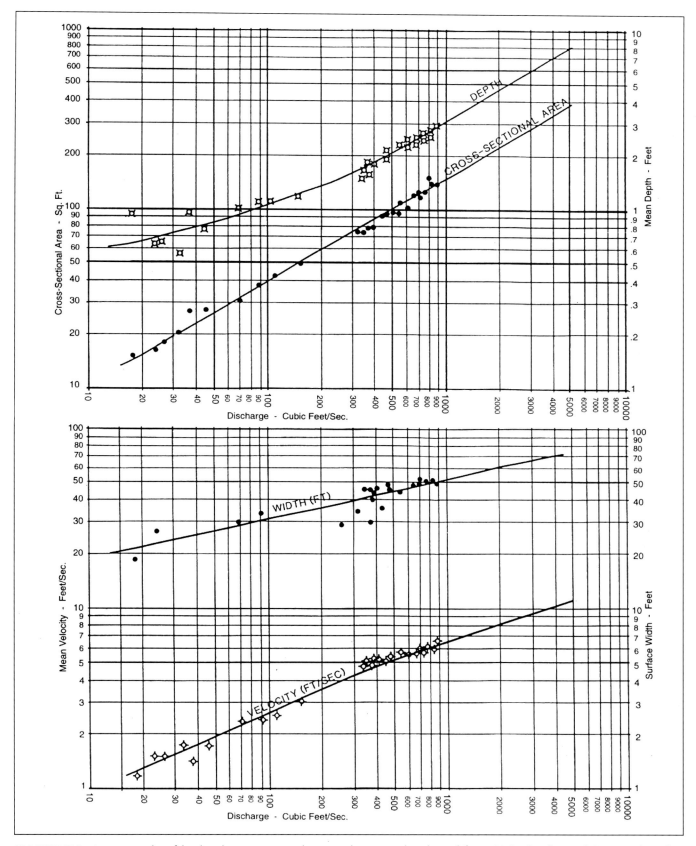

FIGURE 7-3. An example of hydraulic geometry data analyses as developed from U.S. Geological Survey data for the East Fork of the San Juan River, Colorado.

LEVEL IV: FIELD DATA VERIFICATION

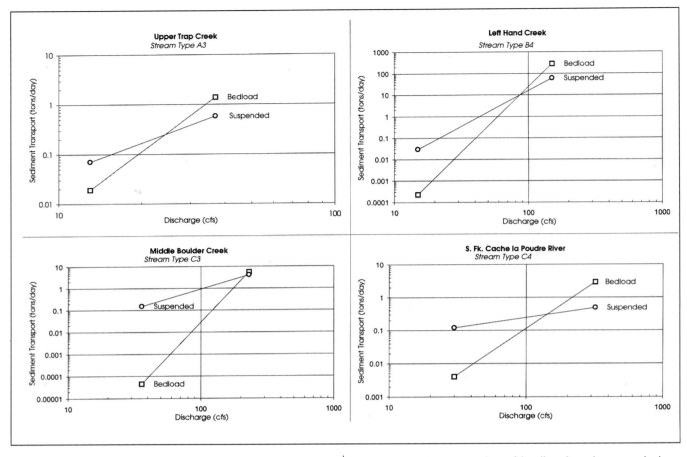

FIGURE 7-4a. Examples of bedload and suspended sediment rating curves for various stream types. (Data from USDA Forest Service as reported by Rosgen, 1993a)

Ratio of Bedload to Total Load

Since both suspended and bedload sediment are important in sediment analysis and/or interpretations by stream type, the measurement of the ratio of bedload to total sediment load can assist in data analysis and the validation of prediction methodologies. An example of concurrently measured values of bedload and suspended sediment as stratified by stream type is shown in **Figure 7-4a** and **Figure 7-4b** (Rosgen, 1993a). Data show that there is not a constant relation to ratios of bedload to suspended load either by flow or for all stream types. As shown in this data set, bedload ratio to suspended sediment varies considerably both from low flow to high flow and by stream type. Large rivers with characteristic high wash loads (large silt and/or clay component to the suspended load) and large discharges generally have a lower proportion of bedload to total load. The Tanana River in Alaska, with suspended sediment concentrations of 3,250 mg/l for flows of 52,970 cfs, contributes 1-2 percent of total load as bedload (Emmett et al., 1978).

It is much easier to measure suspended sediment than bedload in many rivers, and as a result, bedload is often estimated rather than measured. In the literature, it is common to see estimates of contributions of bedload to total load of less than 5-10%. In the absence of bedload data, often estimates based on these ratios are used in sediment budget predictions. Analysis of measured data from Williams and Rosgen (1989) has shown C3 stream types that are cobble bed, gentle gradient streams with relatively high width/depth ratios to have less than 5% bedload to total sediment load. However, in the same data set, values greater than 75% bedload to total load at the bankfull stage for an F4 stream type (Horse Creek near Westcreek, Colo.) have also been measured. With the exceptions of the high wash load at the large rivers described above, the stream types which have a significant contribution

LEVEL IV: FIELD DATA VERIFICATION

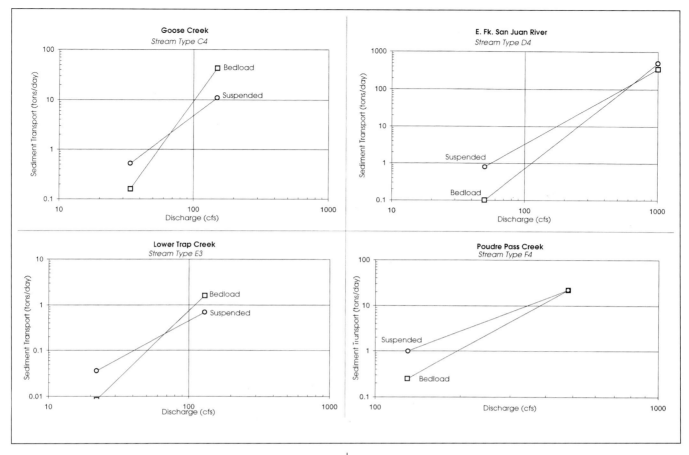

FIGURE 7-4b. Examples of bedload and suspended sediment rating curves for various stream types.

of bedload to total load are the A3, A4, A5, D3, D4, D5, F4, F5, G3, G4, and G5 stream types.

Bedload Size Distribution at or Near the Bankfull Discharge

Since bedload transport and sizes of bedload are often estimated rather than measured, it is necessary to suggest some field techniques which may assist in these estimates. Often prediction of bed material entrainment sizes using Shields criterion (Shields, 1935) underestimate actual sizes in gravel and cobble bed streams. Since the Shield's criterion is based on channels of homogeneous bed material and a rigid boundary, conditions in natural channels often do not match the assumptions of this prediction model. Entrainment computations will be discussed in more detail in Chapter 8; however, these procedures need to be improved by field verification of measured bedload at various flow stages. When bedload cannot be measured at the bankfull stage, a field procedure was developed by the author to determine the particle size distribution of bedload being transported at the bankfull stage. Core samples were obtained on depositional features such as point and central bars during low flow periods and were compared to bedload measurements at or near the bankfull stage for each river sampled. The core sample is obtained from an open bottom bucket not to exceed a depth of twice the largest particle size diameter observed at this specific location. The material obtained from the core sample is then sieved and weighed in order to establish the particle size distribution from the point bars and/or central bars, half-way between the bankfull stage and the channel thalweg on the downstream side of the bar feature. The photograph in ***Figure 7-5*** shows a typical location of the core site sample on a bar as indicated by the 6 inch ruler located at the sample site. Particles are generally coarser on the very top of the bar than are the sizes measured both from bedload at bankfull and on the downstream

LEVEL IV: FIELD DATA VERIFICATION

FIGURE 7-5. Example of a bar site on a C4 stream type where a core sample is obtained to relate bar size distribution to size distribution of bedload at bankfull.

face of this bar. It took several years of trial and error to locate a representative site on depositional features whose particle size distribution was similar to those transported as bedload at the bankfull discharge. The relations between the size distribution of bedload at or near the bankfull stage, bar samples, and bed material for a variety of stream types are shown in *Figure 7-6a* and *Figure 7-6b*. The bed material is obtained from the pebble count method described in Chapter 5. The ratios that display differences between bedload at bankfull and bed material sizes vary by stream type. For example the D-84 bedload size for the A2 stream type is the same size as the D-10 of the bed material; whereas, the D-84 bedload sizes at bankfull for an F4 stream type is the same size as the D-50 of the bed material (*Figure 7-6*). For a sand bed channel (C5), obviously the bar material, bed material, and bedload at bankfull is of the same size distribution (*Figure 7-6a* and *Figure 7-6b*). The relations shown in these figures also show that the bar sample, as described, is a fairly good indicator of the bedload sizes that are available and are being transported at bankfull or the normal high flows. It is often a good field procedure to check the largest size transported as measured from the bar sample with computed values using the Shields criterion for entrainment conditions of the bankfull discharge. Sub-pavement samples, as described by Andrews (1983), also indicate the sizes of bedload available for transport, once mobilization of the bed material occurs. Accurate entrainment computations have applications in channel design, where the competence of the channel to move sediment needs to be accurately quantified.

Fish habitat structures often fail or become non-functional due to induced sediment deposition in the structure. "Gravel traps" are structures used to improve spawning reds; however, many times these structures fill with sand rather than gravel since this is the size of bedload being transported and/or the sizes deposited with a change in stream power and shear stress. In the absence of measured bedload data at high flows, this field method would indicate the sizes of bedload being transported and/ or deposited for the bankfull stage shear stress and stream power.

Sediment Rating Curve Relations

When the slope and intercept of a sediment rating curve is changed, there is generally a corresponding change in the "state" of a stream and, if significant enough, can potentially lead to a change in stream type. As sediment supply increases, it is often associated with a shift in the intercept and/or slope of the regression line of the sediment rating

LEVEL IV: FIELD DATA VERIFICATION

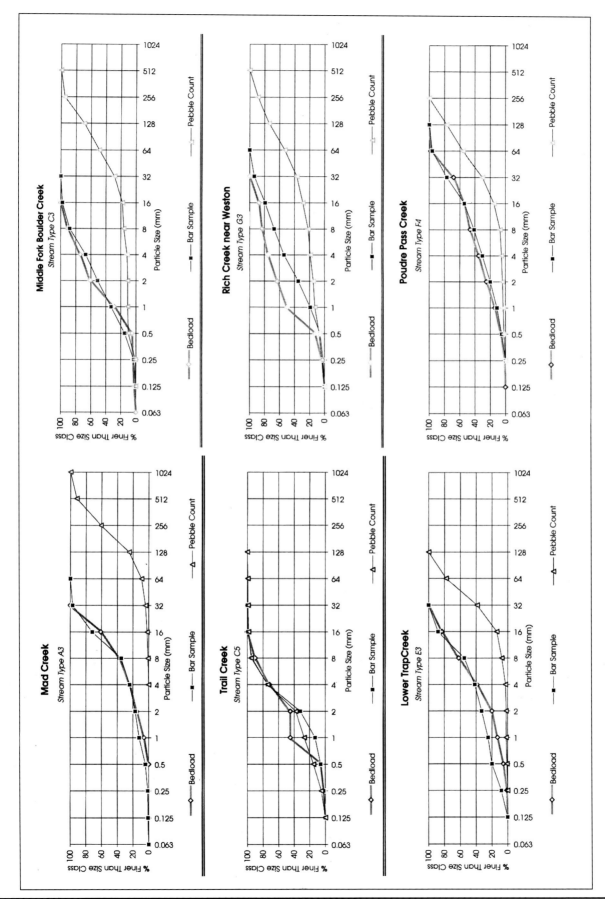

FIGURE 7-6. Relationship of particle size distributions for bedload @ bankfull discharge, bar material, and bed material for various stream types. (Rosgen, 1993a)

LEVEL IV: FIELD DATA VERIFICATION

curve. An example is presented involving a change in stability, morphology, and sediment supply of Joe Wright Creek (Colorado) as determined by field measurements by the author. A shift occurred in the suspended sediment rating curve by an order of magnitude due to construction activities and was responsible for a change in stream type from a B4c to an F4 (Rosgen, 1979) in *Figure 7-7*. The morphological change from a B3c to an F4 stream type was associated with a loss of 100% of the adult fish and corresponding fish habitat, as verified by a concurrent fishery study in the same reach of the physical channel assessment. (Cline 1979).

The sediment rating curve approach has been used to predict flow related changes in sediment supply/transport from hydrograph changes such as from timber harvest (USEPA, 1980). Changes in increased flow and introduced sediment from roads and other surface disturbance activities also need to be predicted. The change in streamflow calculations also relate to similar processes in urban watershed development. Direct measurements are important not only to validate the prediction methodology, but to verify if water quality goals have been met (or violated) and the effectiveness of "best management practices" for mitigation.

Nationally, there still is not an acceptable water quality "standard" for suspended and/or bedload sediment. Existing standards for suspended sediment do not take into account (1) natural geologic rates, (2) acceptable ranges of departure for given uses or values, (3) sensitivity of sediment changes for various stream types and, (4) consequence of sediment change. Often a "one value fits all" criteria is set such as 25 mg/l increase, which is obviously inappropriate. To correct these deficiencies, measured values by stream type under various imposed "states" or conditions may assist in setting appropriate standards. Another consideration is to set goals for maintenance of channel stability. Sediment amounts often exceed "geomorphic thresholds" due to size and/or load of introduced sediment. When sediment monitoring is done in conjunction with channel stability assessment, sediment loading limits may be observed using the sediment rating curve approach by stream type and condition based on channel response and stability to imposed

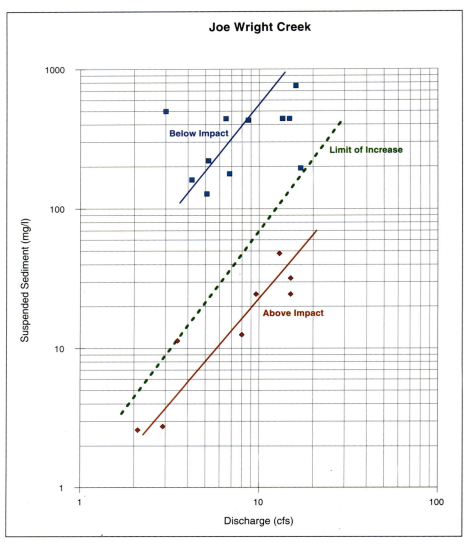

FIGURE 7-7. Above v.s. below-impact sediment yield rating curves for Joe Wright Creek (Colorado) showing sediment yield shift due to construction activities. (Rosgen, 1979)

LEVEL IV: FIELD DATA VERIFICATION

change. These or similar approaches may be used as "sediment standards" to answer the question of "how much is too much?" Biological response to imposed changes in both suspended and bedload sediment also needs to be considered in determining allowable or appropriate exceedance standards. The influence of sediment on habitat conditions, spawning gravels, fry emergence, intra-gravel oxygen levels, food chains, habitat features and other possible limiting factors needs to be better understood. An "integrative" research approach is needed for this type of ecosystem response determination.

Stream Stability Validation

Stream stability is a commonly used term with engineers, hydrologists, fisheries biologists, and others working with river systems. Unfortunately there are more words spoken about "stream channel stability" than data collected to validate observations.

Channel stability, as defined previously, is the ability of the stream, over time, to transport the flows and sediment of its watershed in such a manner that the dimension, pattern and profile of the river is maintained without either aggrading nor degrading. This definition allows for *measurement* of the variables influencing the definition of a "stable channel."

Stream channel monitoring can indicate the current stability of a stream by evaluating whether the stream is: (a) aggrading (building up of the bed elevation or a local base level change by deposition over time), (b) degrading (down cutting or lowering of the bed elevation over time or lowering of local base level due to scour), (c) shifting of particle sizes of stream bed materials, (d) changing the rate of lateral extension through accelerated bank erosion, and (e) changing morphological types through evolutionary sequences. A quantitative assessment of the stability information as stated in 1-5 above is essential to understand the processes of channel adjustment and response to watershed change and/or direct disturbance. A thorough understanding of natural stability tendency of a given river is key for proper design and implementation of proposed restoration, mitigation, and fish habitat enhancement projects. For example, prior to implementing the installation of grade control structures to stabilize a reach, the importance of knowing the vertical stability of a river reach is obvious. As simple as this statement appears, there are, unfortunately, hundreds of grade control structures installed annually in North America without knowing if there is a grade control problem. A validation or quantitative knowledge of the vertical stability of the river is critical so that proposed structures appropriately address channel processes.

To predict the potential impact of an activity on a river, it must be known first what is the current state or stability of the river. Another objective of monitoring is to determine departures from existing conditions or changes in stability due to management activities. Mitigation measures through "best management practices" are often proposed to offset potential adverse consequences of channel instability. Monitoring can determine the effectiveness of such mitigation.

Induced change in channel stability, can be determined by implementing a quantitative, comparative analysis approach that identifies the differences due to climate, geology, and morphology from management impacts on streams. Monitoring programs to assess these impacts can be implemented by using one of the following three approaches:

1. Reference or control reach:

    A stream reach is measured for comparison purposes that is not influenced by management activities, but that is of the same stream type and state or condition as the reach being monitored for management impacts. This reference reach acts as the control or represents the least disturbed condition.

    Reference and study reaches need to be in neighboring watersheds. The assumption for this design is that the same lithology, sediment regime, weather patterns, and morphometric parameters of each basin that influence precipitation/runoff and sediment transport relations are reflected similarly in both the control and study reach (*Figure 7-8*).

LEVEL IV: FIELD DATA VERIFICATION

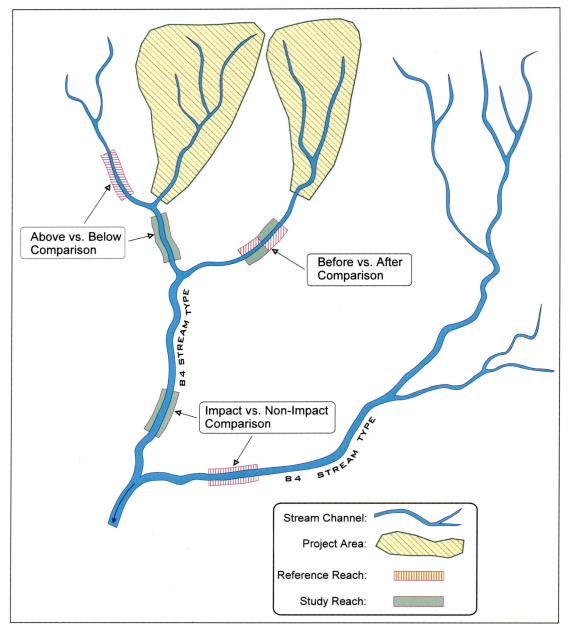

FIGURE 7-8. A monitoring design layout to measure stream impacts, by comparing for the same stream types:
a) "above" v.s. "below" impacts,
b) "before" v.s. "after" impacts, and
c) "impacted" v.s. "non-impacted" systems.

2. Comparison of measurements above versus below impact area of the same stream type:

    Reference reaches on the same stream above a management activity are compared to study reaches below. Reference and study reaches need to be in close proximity for proper comparison of all parameters of the same type and state (*Figure 7-8*).

3. Comparison of measurements *before* versus *after* management activity.

    This approach establishes a "baseline" or calibration of individual reach conditions prior to implementation of a potentially impacting activity (*Figure 7-8*). Ideally, the period of evaluation will span similar weather patterns (dry versus wet years, and unique storms).

Field Methods for Monitoring Channel Stability

Specific monitoring methods are designed to evaluate channel stability and sediment supply that affects bed material distribution and stream bank erosion:

I. Channel Stability

    A. Vertical or bed stability (aggradation/ degradation)

LEVEL IV: FIELD DATA VERIFICATION

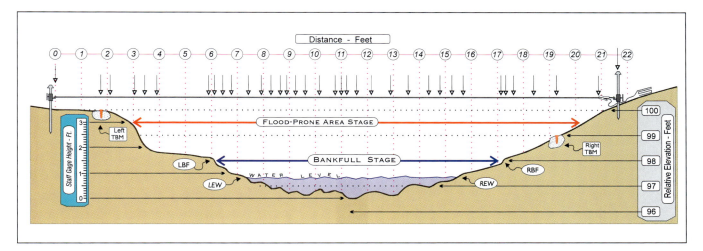

FIGURE 7-9. An example of a permanent channel cross-section with benchmark locations and points of measurement.

Objective: To determine if the stream is either down cutting (degrading), filling (aggrading), or is stable. The rate, magnitude, and direction of vertical change is determined.

B. Lateral stability

Objective: To determine the rate and magnitude of lateral migration through bank erosion processes.

C. Bed material size distribution

Objective: To observe a shift in bed material size distribution in various locations and for various bed features.

II. *Methods*

A. Vertical or bed stability

<u>Monumented cross-sections</u>. This method employs the use of permanently monumented cross-sections located on a riffle and pool segment of a reach (or step versus pool segment if in a step/pool stream). This provides an elevation reference to depict changes. Bench marks need to be located at the cross-section on a stable site above and away from the bankfull channel. An elevation cross-section is often needed if the left or right side of the cross-section is located on an unstable slope. An elevation bench mark is established and often does not represent a true elevation, but rather a relative elevation set at 100 feet. A permanent bench mark is often made by digging a hole which will hold a half of a bag of "Sacrete" at a site for a replicate survey. A 10 inch stove bolt is placed in the concrete before it sets up, flush with the concrete at ground level. This does not interfere with livestock or fisherman and avoids vandalism damage.

The stream cross-section is measured from the intercept of the rod with the "leveled" tape line (***Figure 7-9***). A re-survey of the cross-section should be done annually and/or following stormflow/snowmelt runoff events. Any "high water" levels need to be marked and indicated on the cross-section. The cross-section needs to be plotted for each measurement and compared to previous cross-sections. Measurements must include the floodplain, terraces and stream adjacent slopes. The measurements are obtained by one of several methods including; a level/rod survey using either auto or laser levels, total station surveys, or use of a horizontal tape and a series of vertical rod measurements. The measuring tape and elevation rod method should utilize the following procedure (***Figure 7-9***):

1. Locate the permanent bench mark on both sides of stream (or, if on one side, a bearing for the transect is needed)

LEVEL IV: FIELD DATA VERIFICATION

2. Stretch the tape very tight with spring clamp and level tape
3. Locate tape at same elevation as reference bolt on bench mark
4. Read distance and elevation reading of rod intercept with tape
5. Measure major features such as:
   - Left bench mark (LBM)
   - Left terrace/floodplain (LT, LFP)
   - Left bankfull (LBF)
   - Left bank (LB)
   - Left edge of water (LEW)
   - Various bed features, bars, etc.
   - Thalweg (TW)
   - Inner berm features (IB)
   - Right edge of water (REW)
   - Right bank (RB)
   - Right bankfull (RBF)
   - Right terrace/floodplain (RT, RFP)
   - Right benchmark (RBM)

A vicinity map and detailed site map needs to be prepared so that the monumented cross-section can be located in the future by others not familiar with the site. An upstream and downstream photograph is also recommended for site documentation. Channel dimensions for stream classification need to be collected in order to document morphological comparisons for extrapolation.

Scour Chains. The use of chains, installed vertically in the stream bed, will provide scour depths for various storm events. Often the stream bed will scour, then if the channel is stable, it will return to the pre-flood elevation. Using a combination of scour chains and cross-sections can provide key data not only for vertical stability but also for sediment transport relations and biological interpretations. This method not only validates the entrainment size, but also concurrently determines scour depth in various features of the channel bed.

B. Lateral stability

To determine the rate and magnitude of bank erosion, the installation of bank "pins" and/or bank profiles are installed in sites representative of the stream banks of the river (*Figure 7-10*). Monumented cross-sections are also established concurrently to determine changes in deposition (point bars) and stream width for sediment budget computation. Re-survey of the pin and/or profile sections following runoff events will provide measured stream bank erosion rates. These study sites should be stratified and installed on the outside of bends and on straight reaches of different stream bank erodibility conditions along reaches of the same stream type. Measured stream bank erosion rates can be expressed in feet/year, cubic yards year, or total tons/stream reach for a given flow or for a runoff season. Monumented cross-sections, maps, and photos of this site need to be implemented as described in II.A above. Stream classification measurements need to be taken to provide for extrapolation and comparison purposes. The stream bank erodibility/stress in the near-bank region method described for Level III stream inventory prediction need to be conducted at this reach. The actual measured bank erosion data allows for validation of this prediction method.

1. Bank pin installation

   The erosion pin method involves the installation of 2 or 3 smooth rods (4-5 feet in length with a diameter of 0.3 to 0.5 inch), horizontally into the bank per location. Periodically or as a minimum before or after storm periods, the length of exposed pin should be measured along with the permanent cross-section. This procedure not only provides a rate of lateral migration, but in combination with a permanent cross-section, indicates if dimensions such as bankfull

LEVEL IV: FIELD DATA VERIFICATION

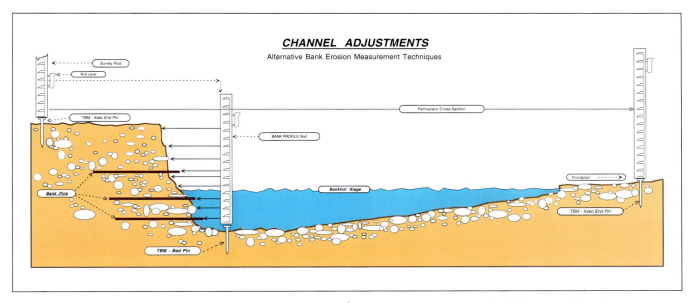

FIGURE 7-10. Example of the use of bank pins, a bank profile, and a monumented cross-section to determine bank erosion rates.

width, bankfull depth, and/or elevation are changing.

2. Bank profile procedures involve:
   a. Installing a permanent cross-section at the bank profile site.
   b. Installing a permanent toe pin (rod) offset and directly adjacent to the study bank (*Figure 7-10*).
   c. Place rod level on survey rod that is set on toe pin.
   d. Stabilize rod with either a tripod or a frame attached to bank to hold rod "plumb."
   e. Measure horizontally with tape rule from vertical rod to bank. Measure at frequent intervals to describe bank dimensions and features.
   f. Plot data to display profile for each survey.
   g. Compare with previous surveys annually following storm events. Compute mean erosion/deposition rate (*Figure 7-10*).

C. Bed material size distribution procedure.
   1. Use a permanent transect to obtain existing and departures of a frequency of particle size distribution by measuring 100 particles at specified locations. These locations are often selected at riffles, pools, glides, runs, spawning redds, and special features of interest. The Wolman pebble count method that has been modified for this purpose is recommended as described in Chapter 5. Permanent transects are preferred for monitoring to decrease spatial and temporal variability and to make replication possible. Permanent transects should be established on riffles separate from pools or other features selected. One hundred samples should be obtained from each transect. Re-survey should collect pebble count data for the same locations for each particle sample.

   2. Embeddedness or the percent fines in various bed features. The determination of the amount of fine material surrounding coarser particles in selected bed features is determined using a variety of field methods. Sample cores taken from the bed, then weighed and sieved is the most accurate. However, frequency distribution of particle sizes for various bed feature provide an index to the fine material distribution. These studies have

significance not only for biological interpretations, but infers imbrication, which can affect particle entrainment conditions for bedload transport (see Embeddedness Field Methods as summarized in McDonald et al., 1991). A method developed by Lisle and Hilton (1992) estimates volume of fine sediment in pools as an index to sediment supply in gravel bed rivers. A stratification by stream type and an index to the streamflow associated with the transport may strengthen this method for widespread application. Similar methods are also included in Lisle (1987) and Eads (1991).

SUMMARY

These measurements are designed to help identify and evaluate physical characteristics of existing conditions and to be used to predict the response of the river to imposed change. The relative stability of the bed, banks, and materials of the river provide valuable interpretations and assessments. Discrete measurements validate the "condition" inventories as discussed in the Level III procedures. Upon implementation of a good monitoring plan, excellent data can be obtained in relatively few days each year. A well designed plan will identify the rate, magnitude, consequence and nature of change of the river character due to human intervention. A comparison of pre-versus-post disturbance, as well as conditions above-versus-below locations of impacts, provides a quantitative measure of change. A well designed data collection effort can provide insight into:

1. Causes, rates, magnitude and direction of river adjustment.
2. Effectiveness of mitigation measures.
3. Accuracy of prediction methodologies.
4. Development of effective mitigation/restoration.
5. Validation of prediction models.
6. Development of empirical relations.
7. Consequence of change.
8. An approach to set limits for channel change and corresponding sediment loads.

Biological monitoring and water quality monitoring—physical, biological and chemical—all need to be studied and integrated by stream type to obtain an ecological assessment and to identify the biological significance and consequence of change.

The Vigil Network, a program proposed by Emmett and Hadley (1968), to measure and permanently document the physical channel characteristics is similar to that previously described in the level IV analysis. A wealth of knowledge of river behavior, over time, would have been obtained if the Vigil Network program had been expanded and continued since its conception by the U.S. Geological Survey. The knowledge of stream system response to imposed change is a key to prevention, prediction, and restoration. This information will help us be more effective at properly managing our lands and rivers for the future. It would now be timely to participate in the Vigil Network and initiate this validation type data collection and analysis to carefully document our observations of the river. Only through a permanent measurement network can the consequence of our actions in the watershed and on the river system be understood.

The key to setting proper watershed management guidance lies in the understanding of cause/effect relations. Since the river is a barometer of the health of its watershed – it is imperative to effectively "read the river."

APPLICATIONS

CHAPTER 8
APPLICATIONS

*"The acid test of our understanding is not whether we can take ecosystems
to bits and pieces on paper, no matter how scientific,
but whether we can put them together into practice and make them work."*
BRADSHAW (1983)

INTRODUCTION

The information in this chapter is going to focus on ways to:

- Utilize existing data and place into usable form
- Develop empirical relations by stream type
- Extrapolate data from research findings
- Analyze sediment and hydraulic data by stream type
- Provide management interpretations of stream types
- Evaluate the suitability of proposed fish habitat structures for various stream types
- Evaluate livestock grazing criteria related to various stream types
- Apply principles for river restoration
- Implement stream inventories utilizing stream classification
- Incorporate river morphology applications in cumulative watershed impacts analysis

HYDRAULIC GEOMETRY RELATIONS

The original work of Leopold and Maddock (1953) was a landmark contribution to the development of hydraulic geometry relationships. They developed simple power functions to describe the variables of depth, velocity, and cross-sectional area as a function of discharge.

APPLICATIONS

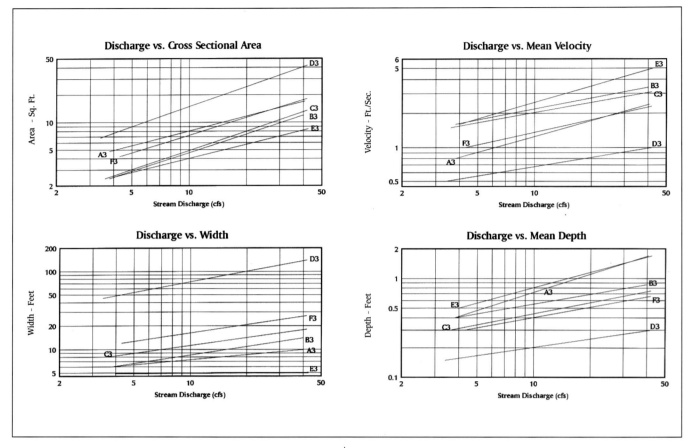

FIGURE 8-1. Hydraulic geometry relations for selected stream types of uniform size.

To investigate whether stream classification might be used to refine hydraulic geometry relationships and exponent values for these power functions, Rosgen (1994) assembled stream dimensions, slopes, and hydraulic data for six different stream types (i.e., A3, B3, C3, D3, E3, and F3) having predominantly cobble-sized bed material. The objective was to determine how width/depth ratio, gradient, sinuosity, and meander geometry affect hydraulic geometry relationships. Streamflow values for a variety of stream types from the baseflow discharge of 4 cfs to the bankfull discharge of 40 cfs were compared for each cross-section and their corresponding widths, depths, velocities, and cross-sectional areas (*Figure 8-1*). These results indicate the slopes of the hydraulic relationships changed little, except for the E3 stream type. This agrees with earlier findings by Leopold and others (1964). The intercept values however, varied significantly suggesting that stratification by stream type can be used to improve estimation of the hydraulic geometry regression variables.

A set of hydraulic geometry equations were developed by Osterkamp, et al. (1983) which were stratified by width/depth ratio as an expression of shear stress distribution and sediment characteristics. These relations allowed a wider range of exponents of the power function to be used as a result of integrating channel morphology relations. Since width/depth ratio is also a key delineative criteria for stream classification, similar results were observed comparing streams of very small to very large width/depth ratios.

The intent of the curves in *Figure 8-1* is not to be extrapolated directly, but rather to introduce a method of a morphological stratification in the development of empirically derived hydraulic relations obtained from gaged records. It would be advantageous, for more widespread applications, to develop dimensionless ratios of the hydraulic geom-

APPLICATIONS

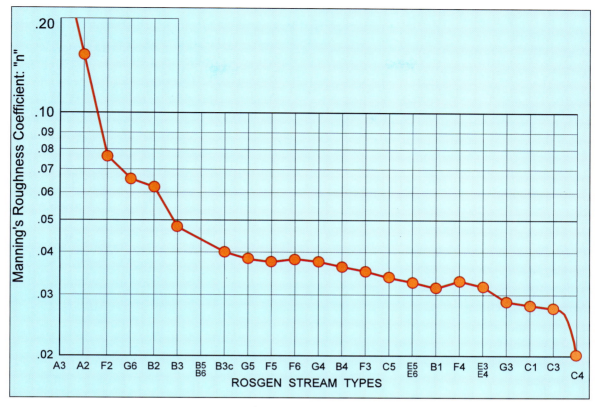

FIGURE 8-2. Manning's "n" value computed at bankfull stage for selected stream types.

etry relations by stream type to help account for differences in morphology and river size.

FLOW RESISTANCE

Determining the resistance to flow for a given channel is required to: (a) estimate stream discharge at stations for which there is no velocity measurements available; (b) determine channel capacity, for computing particle size entrainment for a given flow; and (c) determine sediment transport functions.

Empirically derived hydraulic relationships describing resistance of channels to streamflow are often extrapolated to streams that are not morphologically similar to the streams upon which the equations were established. Similarly, the assumptions upon which the relationships are based may not apply to the stream in question.

Manning "N" Roughness Coefficient

Hydrologists and engineers routinely use Manning's equation to predict mean velocity which was first presented in 1889 (in: Chow, 1964).

$$\overline{U} = \frac{1.49(R)^{2/3}(S)^{1/2}}{n}$$

where: $\overline{U}$ = mean velocity
R = Hydraulic radius (cross-sectional area/wetted perimeter)
S = slope of stream
n = Manning's roughness coefficient

The application of Manning's equation requires the use of a roughness coefficient "n" in the computation. The lack of consistent criteria for selecting "n" values, however, creates great variability in the subsequent estimates of flow velocity. Barnes (1967) and Hicks and Mason (1991) sought to improve consistency in velocity estimates by creating two compendia of photographs and stream data to provide a visual reference to assist in selecting "n" values. Together their compendia covered 128 streams that included a great diversity of stream types.

On the basis of the photography and stream data provided, each of these streams were classified and the bankfull "n" values were plotted by stream type (Rosgen, 1994). The results, shown for bankfull stage only, are illustrated in *Figure 8-2*. The remarkable similarity of "n" values by stream type, derived from two data bases from two separate con-

APPLICATIONS

tinents, suggests that stream classification may help provide more consistent estimates of roughness for use in Manning's equation for estimating stream discharge.

Rather than considering each morphological variable independently, stream classification integrates these and other variables that affect roughness characteristics such as gradient, shape, (width/depth ratio) form resistance, sediment relations, entrenchment, particle size, and relative roughness (depth/D84).

Jarrett (1984, 1990) developed specific relationships between rough- ness and velocity for mountain streams. The relationships he developed were only for steep slopes and cobble/boulder channel materials, and incorporated hydraulic radius and slope as functional parameters. His results are significant and produced values much different than most published equations. Extending his work to include additional stratification by stream type and stream size would further the state-of-the-art for application of these equations.

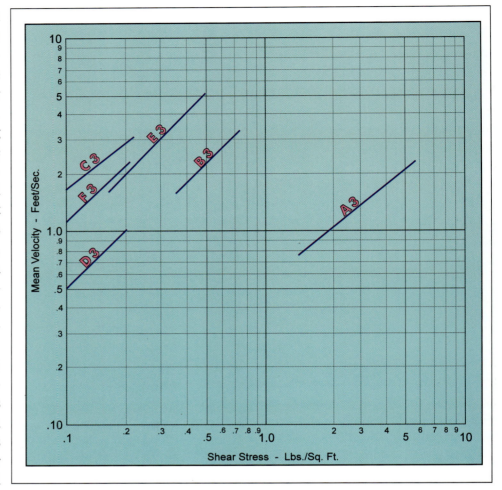

FIGURE 8-3. Relationship of mean velocity v.s. shear stress for six stream types, from base flow (3-4 cfs.) to bankfull discharge (40-41 cfs.). (Rosgen, 1994)

Shear Stress and Velocity Relationships

Shields (1936) described shear stress as:

$$\tau = \gamma RS$$

where:
- τ = shear stress
- γ = specific weight of water,
- R = hydraulic radius, and
- S = channel slope.

To test the relationship between shear stress and mean stream velocity at multiple flow levels, Rosgen (1994) used an aggregated data set for the same six stream types as shown in *Figure 8-1*. As expected, a significant relationship was not found for the aggregated data set. *Figure 8-3* shows relations of shear stress and velocity as stratified by stream type. As shown in this figure, there is a promising trend, suggesting that velocity may be determined empirically from this relation by stream type. While more data are needed to establish mathematical and statistical relationships, the potential applications of such relationships may deserve investigation.

Critical Shear Stress Estimates

Previous investigations of the magnitude of shear stress required to entrain various particle diameters from the streambed material have produced a wide range of values. A number of investigators have assumed the critical dimensionless shear stress values of 0.06 for computations of bedload transport using Shields' (1936) criteria (Baker 1974, Baker and Ritter 1975, Church 1978, Bradley and Mears 1980, Simons and Senturk 1977, Simons et al. 1982). In addition, critical dimensionless shear stress values computed from data compiled by Fahnestock (1963), Ritter (1967), and Church (1978) for the entrainment of gravels and cobbles from a natural riverbed, and as reported by Andrews (1983) showed a range of approximately 0.02 to 0.25. The mean of the computed values was 0.06, which is the value suggested by Shields (1936).

Andrews (1983) described a relationship where the ratio of surface pavement of bed particles to sub-surface (sub-pavement) particles yielded an estimate of critical dimensionless shear stress values (τ_{ci}^*) from 0.02 to 0.28. Additional work using the same equation was applied to several Colorado gravel-bed streams with similar results (Andrews 1984). It is sometimes difficult for many engineers to obtain pavement and sub-pavement data along with the required channel hydraulics information to refine critical dimensionless shear stress estimates using the Andrews (1983, 1984) equation. The use of stream types to help bridge this gap of estimating the critical dimensionless shear stress value (τ_{ci}^*) has potential where these study streams have been analyzed and classified. The classified streams studied by Andrews (1984) were classified by the author and data compiled and the values of τ_{ci}^* (critical dimensionless shear stress for the D-50) were plotted in *Figure 8-4* (Rosgen 1994). Stream types A2 and D4 were obtained from field measurements of bedload sediment and bed-material size distribution for those types (Williams and Rosgen 1989). Stream types and their morphologic/ hydraulic characteristics do not substitute for

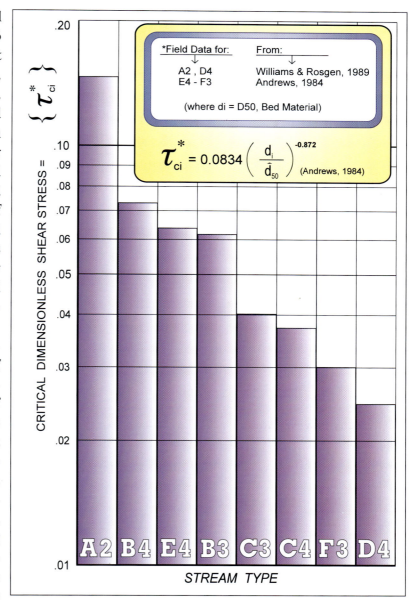

FIGURE 8-4. Relationship of field verified critical dimensionless shear stress values for various stream types. (Rosgen, 1994)

detailed on-site investigations as described by Andrews (1983, 1984); however, calculations of τ_i^* are often made without the benefit of site-specific investigation. Based on the great variability in the estimate of τ_{ci}^*, and the sensitivity of the parameter, sediment transport prediction errors can occur from one to several orders of magnitude. A closer approximation of τ_{ci}^* for stream reaches that cannot be investigated in detail may be possible using this approach.

APPLICATIONS

Sediment Relations

The author has used the sediment rating curve procedure for both suspended and bedload rating curves for management applications. These relationships were initially plotted as a function of channel stability ratings as developed by Pfankuch (1975), and described in Chapter 6. Applications for cumulative effects analysis for non-point sediment sources utilized this approach (USEPA 1980). Subsequent comparisons of data with stream type delineations indicated similar relations. Stream classification can be used to: a) characterize sediment supply in relation to stream discharge; b) to develop rating curves for both suspended and for bedload sediment transport; c) to develop improved estimates of bedload transport from more readily available records of suspended sediment; d) to predict changes in stream state based on shifts in slope and intercept values; and e) to develop families of dimensionless ratio sediment rating curves for extrapolation purposes. Stream types have been used to characterize sediment rating curves that reflect sediment supply in relation to stream discharge. For example, a sediment rating curve regression relation for a low order A2 stream type would have a characteristic flat slope and low intercept. The sediment rating curve for a C4 stream type, however, has a steeper slope and higher intercept - often due to a change in sediment availability from channel derived sources.

Variability in actual sediment values occur in sediment rating curves (sediment vs flow) and in sedi-graphs (sediment vs time) from change in sediment supply due to:

- magnitude, duration, season, source, and nature of runoff
- rising limb vs recession limb of the snowmelt and/or stormflow hydrograph
- first storm of the season vs the last storm

Suspended sediment concentrations which are sensitive to sediment supply are more sensitive to the hysteresis effect than bedload. Bedload is sensitive to energy controls that affect entrainment; however, direct sources of coarse sediment input can be observed from tributaries and from erodible stream banks. These natural variables effect all stream types and there are consistent differences in plotted sediment rating curve by stream types with the same lithology. An example is shown where the integration of stream types have been used to characterize sediment rating curves that reflect sediment supply in relation to stream discharge. A plot of bedload transport rate versus stream discharge for 55 various rivers in Colorado with sediment and flow data from Williams and Rosgen (1989) is shown in *Figure 8-5*. The slope of the line for the regression equation for all streams was 1.2. A large scatter involving 3-6 orders of magnitude was observed for various streamflow ranges for flows less than 1 cfs to over 10,000 cfs. By stratifying individual stream types from this data set, sediment rating curves for a diversity of individual stream types are shown in *Figure 8-6*. When this data is plotted over the "scatter diagram" as shown in *Figure 8-5*, the slope and intercept values of the regression lines reflect relations which typify the interpretations of the various stream types (*Figure 8-6*). For example, F4 stream types are associated with a high sediment supply as shown in *Table 8-1*. As discharge increases, one would expect not only higher sediment transport rates at the lower flows, but also an exponential increase in sediment transport per unit discharge. The steep slope of the sediment rating curve of 2.4 (*Figure 8-6*) is indicative of the very high sediment supply interpretations associated with the F4 stream type (*Table 8-1*). Conversely, the B3 stream types have a very low sediment supply from channel derived sources and are associated with a "flatter" sediment rating curve with a corresponding slope of the sediment rating curve of 0.5. The intercept values also are many orders of magnitude less for the same discharge as the F4 stream types as shown in *Figure 8-6*. The A2 stream type, even though having a much steeper channel gradient, has a low sediment supply (*Table 8-1*), and is similar to the sediment rating curve of the B3 stream type. The C3 stream type is more in the "middle" of the curve and is associated with a moderate to high sediment supply.

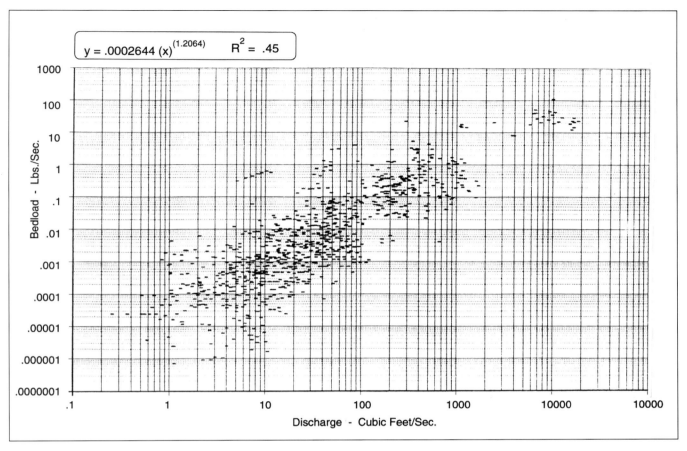

FIGURE 8-5. Measured bedload sediment for 55 various Colorado Rivers. (*From Williams and Rosgen, 1989.*)

The slope of the sediment rating curve for this C3 stream type is 1.9. Thus, for prediction purposes, the application of using sediment rating curves by stream type can provide reasonable estimates of changes in sediment supply when stream types are shifted from one type to another. The sediment supply is often inferred from a comparison of "slope" and "intercept" values of regression coefficients of their respective sediment rating curves. For example, a B4c stream type would have a higher intercept and steeper slope of the sediment rating curve than a B3 due to the entrainment conditions of gravel versus cobble. Sediment supply for the F4 stream type is much higher than either the B3 or B4 as shown in *Figure 8-6*.

Examples of sediment rating curves stratified by channel stability ratings for stream in the northern and central Rockies and for northern California coastal streams are shown in "WRENSS," or water resource evaluation of non-point pollution sources from silvicultural sources (USEPA, 1980).

Sediment rating curves can be effectively used for prediction purposes for a particular lithology. However, to extrapolate outside of the lithology, the sediment and streamflow data are converted into dimensionless ratios. To avoid differences in sediment data resulting from varied geology but for the same stream type and for sediment transport relations for large versus small stream sizes, a dimensionless ratio plot of these "family" of sediment rating curves allows for extrapolation of these procedures. This procedure involves dividing all sediment values by the bankfull sediment value (Q_s/Q_{sbkf}) and the same individual flow values divided by the bankfull discharge (Q/Q_{bkf}). This is done by stream type for the full range of data available. To properly utilize this relation for extrapolation, one must measure actual bankfull sediment transport at the bankfull stage then multiply the bankfull sediment times the Q_s/Q_{sbkf} ratio.

APPLICATIONS

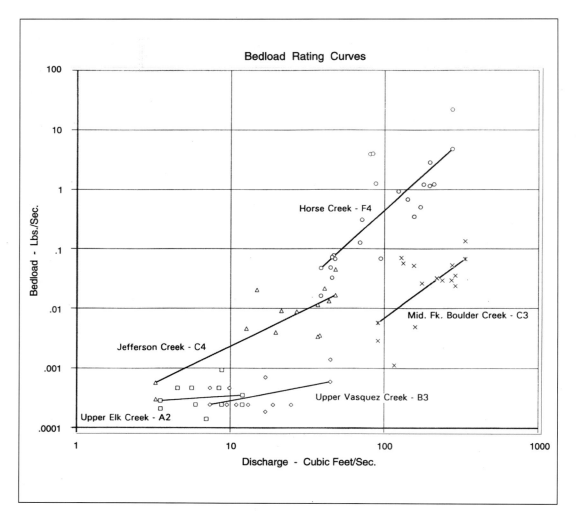

FIGURE 8-6. Bedload sediment rating curves, stratified by stream type. (*From the same data set as used for Figure 8-5*)

The Q/Q_{bkf} ratio is then multiplied by the measured bankfull discharge (Q_{bkf}) for each individual flow measurement to develop local sediment rating curves from bankfull measurements for individual stream types (for Level II and Level III conditions).

GRAZING MANAGEMENT

Grazing management methods designed primarily for upland areas and perennial bunchgrass have altered many riparian areas and their associated stream characteristics throughout the United States and especially in the western states. Improper grazing can change the composition of riparian vegetation communities, and in so doing, also changes rooting depth, rooting character, surface protection, aquatic habitat and aesthetic values. Many of these changes cause adverse stream channel adjustments. These subsequent channel adjustments include : (a) accelerated bank erosion, (b) increased width/depth ratios, (c) altered channel patterns, (d) induced channel instability, (e) increased sediment supply, (f) decreased sediment transport capacity, and (g) damaged fisheries habitat. Poor grazing management practices causes direct mechanical damage that changes the dimensions, pattern, and stability of alluvial channels.

Streams are not similar in terms of the role that riparian vegetation plays in maintaining their stability or their ability to recover from induced damage from over-grazing. When developing or reviewing grazing management plans, it is essential to understand these differences. The response of streams to imposed change is not uniform among stream types. The variability that exists is demonstrated in *Table 8-1*, where interpretations are shown for: sensitivity to disturbance, recovery potential, sediment supply, streambank erosion potential, and vegetation controlling influence. The greatest response in

Stream type	Sensitivity to disturbance[a]	Recovery potential[b]	Sediment supply[c]	Streambank erosion potential	Vegetation controlling influence[d]
A1	very low	excellent	very low	very low	negligible
A2	very low	excellent	very low	very low	negligible
A3	very high	very poor	very high	very high	negligible
A4	extreme	very poor	very high	very high	negligible
A5	extreme	very poor	very high	very high	negligible
A6	high	poor	high	high	negligible
B1	very low	excellent	very low	very low	negligible
B2	very low	excellent	very low	very low	negligible
B3	low	excellent	low	low	moderate
B4	moderate	excellent	moderate	low	moderate
B5	moderate	excellent	moderate	moderate	moderate
B6	moderate	excellent	moderate	low	moderate
C1	low	very good	very low	low	moderate
C2	low	very good	low	low	moderate
C3	moderate	good	moderate	moderate	very high
C4	very high	good	high	very high	very high
C5	very high	fair	very high	very high	very high
C6	very high	good	high	high	very high
D3	very high	poor	very high	very high	moderate
D4	very high	poor	very high	very high	moderate
D5	very high	poor	very high	very high	moderate
D6	high	poor	high	high	moderate
Da4	moderate	good	very low	low	very high
DA5	moderate	good	low	low	very high
DA6	moderate	good	very low	very low	very high
E3	high	good	low	moderate	very high
E4	very high	good	moderate	high	very high
E5	very high	good	moderate	high	very high
E6	very high	good	low	moderate	very high
F1	low	fair	low	moderate	low
F2	low	fair	moderate	moderate	low
F3	moderate	poor	very high	very high	moderate
F4	extreme	poor	very high	very high	moderate
F5	very high	poor	very high	very high	moderate
F6	very high	fair	high	very high	moderate
G1	low	good	low	low	low
G2	moderate	fair	moderate	moderate	low
G3	very high	poor	very high	very high	high
G4	extreme	very poor	very high	very high	high
G5	extreme	very poor	very high	very high	high
G6	very high	poor	high	high	high

a   Includes increases in streamflow magnitude and timing and/or sediment increases.
b   Assumes natural recovery once cause of instability is corrected.
c   Includes suspended and bedload from channel derived sources and/or from stream adjacent slopes.
d   Vegetation that influences width/depth ratio-stability.

TABLE 8-1. Management interpretations of various stream types (*Rosgen, 1994*)

APPLICATIONS

riparian and stream condition would come from placing the highest priority on developing grazing management strategies for those streams that are most sensitive to grazing disturbances and have the highest recovery potential.

For example, the rooting depth of riparian vegetation has diminished effect on entrenched F, A, and G stream types. Many of these banks are collapsed from the lower 1/3 slope position far below existing rooting depths, even of woody species. Conversely, riparian vegetation plays a predominant role in maintaining bank stability for E and C stream types (having characteristically low bank height/rooting depth or bank height/bankfull stage ratios). Myers and Swanson (1992) studied the effects of grazing on the stability of streams in northern Nevada and concluded that "range managers should consider the stream type when setting local standards, writing management objectives, or determining riparian grazing management strategies."

The timing of grazing is also critical in determining the effect that grazing will have on stream condition. This is due to seasonal differences in plant physiology and soil conditions. For example, "E" stream types characteristically support rizomatous grasses and sedges which often have a rooting depth equal to bank height. For "E" stream types, grazing should be limited to mid and late season grazing. This reduces the mechanical damage caused from livestock grazing on highly saturated soils early in the season. Avoiding this damage in turn reduces the collapsing of undercut banks. Another advantage of late season grazing is that many of the Carex and Juncus species associated with E4, E5, and E6 stream types become less palatable as the plants mature, thus providing a natural check on middle to late season grazing utilization.

"C" stream types, however, require a different grazing strategy. "C" stream types are characterized by deep rooted species such as alder, willow, birch and dogwood. Often bank heights are associated with terraces on the outside of bends, requiring riparian species with greater rooting depths. The deeper rooted, woody species, adapted to such riparian sites, are critical to the bank stability of C3, C4, C5, and C6 stream types. For these stream types, grazing should be limited to early season especially for large riparian pastures. This is due to the following reasons:

(1) In early season, the palatability of forage in the upland adjacent to the riparian corridor is very high, thus better utilization is obtained of forage which is often not used. This reduces the concentration of animals and reduces utilization on the streamside vegetation.

(2) Water availability on upland areas is better in the early season with higher plant moisture, both of which provide for easier distribution and less concentrations along the streamside zone.

(3) Air temperatures are often lower early in the season which allows livestock to utilize high energy or exposed slopes adjacent to the streamside zone. This reduces the tendency of concentrations along the riparian corridor.

(4) Nuisance insects are also less problematic earlier in the grazing season which reduce yarding up and wallowing in and adjacent to the stream.

(5) Palatability of critical woody species is higher later in the season; thus, early season grazing in riparian pastures produces less tendency for excessive browsing. This is true under the assumption that stocking rates, livestock distribution, and the carrying capacity is balanced in the riparian pasture.

(6) Rest following early grazing allows for vigor to be replaced throughout the rest of the growing season. The riparian areas respond well to rest during the growing season due to soil moisture availability.

Unfortunately, many grazing strategies in the "C" stream type riparian areas promote late season (mid summer, fall and/or winter) grazing. For the reasons as stated in 1-5 above, this has resulted in degradation of the riparian and stream condition in these "C" stream types. Composition of riparian communities has been altered, generally, by a loss of the woody species in the population. Willows and young cottonwoods are generally the first species to drop out of the vegetative community under late-season grazing due to their sensitivity and loss of vigor under moderate to heavy browsing pressure. This occurs not only under domestic grazing but also under wildlife browsing where concentrations of animals on the winter range exceed carrying capacities.

Studies conducted in the Lamar River basin in the northern portion of Yellowstone have documented a major reduction—and in some cases, total elimination—of willows, aspen, and alder due to heavy late fall and winter browsing by growing populations of elk and bison since the mid-1930's (Kay, 1990; Kay and Chadde, 1991; and Rosgen, 1993c). A conversion from a willow/grass to a grass/forb community can increase bank erosion rates by several orders of magnitude for the same stream type. Many stream types within Yellowstone (e.g., the A3, A3a, A4, A4a, A5, and A5a stream types) contribute great sediment loads from bank erosion and are not greatly influenced by riparian vegetation. The alluvial channel types such as the C, F, D, and G stream types rely on the deeper rooted riparian species to maintain their streambank stability.

Management of grazing, whether by wild animals or domestic livestock, should focus on maintaining the native species that are responsible for maintaining the most stable, natural channel. Grazing standards should focus not only on season of grazing and proper utilization of woody and non-woody plants, but on percentage allowable bank damage seasonally. A good example of the application of this approach is by the USDA Forest Service on the Beaverhead National Forest in Dillon, Montana, where these new riparian grazing standards have been implemented (Bengeyfield, et al., 1994). Monitoring programs (Level IV analysis) should be designed and implemented to verify the vegetative response, stream stability, and "stream health and function." This will document riparian recovery and provide the basis for revised grazing standards and strategies. Some examples of vegetative communities altered by grazing and subsequent effects on streambank stability, fisheries habitat, stream condition, and induced morphological shifts in stream types are shown the photos in *Figure 8-7, Figure 8-8*, and *Figure 8-9*.

Riparian pastures and other grazing strategies should be designed to accommodate the inherent differences in response of various stream types to vegetative changes and potential bank damage. Traditional grazing methods generally have not made any distinction between different stream types, their values and sensitivities. Field evidence has demonstrated that proper grazing methods can be implemented in even the most sensitive of stream types such as the "E"' stream types. Season-long grazing, however, is not one of the grazing strategies compatible with the "E" and "C" stream types. To properly design grazing strategies, it is necessary to:

1. Understand the sensitivity of individual stream types
2. Understand what riparian species, if any, is important for maintaining the stability of the stream type in question
3. Prescribe the grazing season that best favors the key "stability" species
4. Develop acceptable utilization limits for browse and other riparian species
5. Determine allowable, annual bank damage by hoof shear by stream type
6. Design restoration plans concurrent with grazing, i.e., repair the damage to speed up recovery in severely damaged locations
7. Plan water systems to help distribute animals (including construction of ponds in the floodplain for stock water areas to reduce concentrations on the stream)

Although some suggestions for certain grazing strategies have been made, each grazing management plan needs to be specifically designed to meet the grazing objectives while maintaining long term forage production without damage to the riparian and aquatic ecosystem.

APPLICATIONS

FIGURE 8-7a. Conversion of an F4 to C4 stream type due to improved riparian grazing practices (spring grazin riparian pasture). Note sand bar willows on stream banks.

FIGURE 8-7b. Conversion from E4 to C4 stream type due to poor grazing practices to right of fence. South Park, Colorado.

FIGURE 8-7c. Spring grazing allows for good stand of willow along E4 stream type near Chromo, Colorado.

FIGURE 8-7d. Season long grazing downstream of 8-7c which converted willows to grass. Chromo, Colorado.

FIGURE 8-7e. Stable and productive E4 stream type above road crossing, New Fork River, Wyoming.

FIGURE 8-7f. C4 stream type with high width/depth ratio immediately below road crossing from 8-7e, New Fork River, Wyoming.

APPLICATIONS

FIGURE 8-8a. Upstream of Hwy. 185 shows a stable C5 stream type. South Platte River, Colorado.

FIGURE 8-8b. Immediately downstream of 8-8a showing impact of over grazing in riparian. South Platte River, Colorado.

FIGURE 8-8c. Above road crossing showing F5 converted to E5 stream type following major flood. Good grazing practices. Upper Coyote Creek, Colorado

FIGURE 8-8d. Different grazing strategy immediately below 8-8c which has maintained unstable F4 stream type. Upper Coyote Creek, Colorado.

FIGURE 8-8e. Stable C4 stream type below road. Note willow and alders along bank. Mill Creek, Colorado.

FIGURE 8-8f. Unstable C4 stream type immediately upstream of 8-8e. This grazing strategy is not proper to maintain stream stability. Mill Creek, Colorado.

APPLICATIONS

FIGURE 8-9a. Converted E6 stream type from F6. Bed of F6 is now new flood plain of E6. Note F6 downstream of E6. Both reaches are grazed. The F6 below fence line is grazed season long. Tributary near Elko, Nevada.

FIGURE 8-9b. Looking downstream between two other pastures in vicinity of 8-9a. Note E6 stream type downstream of F6. Both pastures are grazed, but not using same strategy. Tributary near Elko, Nevada.

FIGURE 8-9c. Good grazing practices, Maggie Creek Ranch, Nevada - conversion of F4 to E4 stream type - taken same day as downstream reach, Figure 8-9d.

FIGURE 8-9d. Poor grazing practices downstream from 8-9c on other ownership - showing maintenance of high width/depth ratio F4 stream type and ephemeral/subterranean flow.

FIGURE 8-9e. Example of good grazing strategy in riparian pasture - E4 stream type. Upward river condition and trend - O'Neill Creek, Colorado.

FIGURE 8-9f. Example of good grazing practices which are stabilizing E4 stream type in previous F4. Little Navajo, Colorado.

APPLICATIONS

FISH HABITAT

Fish habitat assessment inventories often look very specifically at pool quality, pool spacing, in-stream and over-head cover, substrate composition, and other features of a given stream reach. Identification of limiting factors should precede habitat improvement and enhancement projects. Some recommendations for habitat improvement structures, however, do not consider the ability of the stream to accommodate such structures. There is a desire to increase the number of pools beyond their natural spacing and as a result are often located "out of place". Riffle-pool channels (C stream type) with pool spacing based on 5-7 bankfull widths are often converted to a step-pool channel with the spacing based on 3-4 bankfull widths. This places demands on the stream that is beyond its potential and natural stability. Structures installed must match the natural, stable, characteristics of a given stream. If the inventories and analyses that specify the need for structural improvements do not address the channel morphology and corresponding stable dimension, pattern and profile, then, the effectiveness of the structures is greatly diminished.

Fish habitat improvement projects are enjoying widespread application throughout the west. Trade-offs for proposed development involve mitigation focused on physical habitat enhancement. Flood damage as well as adverse impacts from land management activities are often responsible for deteriorated fish habitat conditions. Restoration plans generally include some type of structural fishery enhancement recommendations.

Often these structures meet with great success on certain streams and are total disasters on others. This occurs for a variety of reasons including, but not limited to: (1) poor understanding of river response as a result of installation of such structures, (2) lack of both field experience and/or documented procedural guidelines (3) economic and time constraints which limit the amount of consultation and pre-project research, (4) the lack of current state-of-the-art knowledge in the applicability of those structures to various field conditions, and (5) Tendency to install the same familiar structure on all stream types (one size fits all). All of these limitations serve to promote the "trial and error" method of application.

Integration of these stream enhancement projects with related disciplines such as hydrology, geomorphology, engineering, and river mechanics has been slow due to the difficulty in the understanding and exchange of technical information. Many biologists have developed an "intuitive feel" after years of experience through trial and error. However, this knowledge has not always been transferred to others less experienced.

A basic understanding of channel relations will assist fisheries biologists in selecting appropriate improvement designs for various streams. Channel patterns are self-developed and self-maintained such that any change in the variables responsible for such patterns sets up mutual adjustments within the channel. Changes in velocity, depth, width, channel materials, discharge, sediment supply, and slope initiates a series of concurrent adjustments between these variables in order to seek a quasi-equilibrium or a "balance." Results of such adjustment often cause aggradation, degradation, lateral channel migration, accelerated bank erosion, increased sedimentation, and substrate material size shifts. These consequences of channel adjustments can often result in actual decreases in habitat quality, even though the initial adjustments were caused by in-channel structures designed to improve the habitat. Since streams follow the basic laws of physics, and habitat development is dependent on physical processes, the marriage of hydrological relationships and habitat enhancement is not only desirable, but essential.

A guideline is presented which is designed to assist the fisheries biologist in evaluating suitability of various proposed fish habitat structures for a wide range of morphological stream "types." This guideline was initially presented by Rosgen and Fittante-Mitchell, 1986 and revised for inclusion in this book. The main objective of the suitability guidelines is to bridge the gap between the "trial and error" methods and detailed hydraulic calculations

APPLICATIONS

for various installations. The guideline is not intended to determine what structure will be the most effective for improving a specific limiting factor, but rather insure the long-term functioning of a particular structure.

The recommended procedure would be to first obtain field inventories which assess the limiting factors for the species and age class of fish favored for management for a given reach or combinations of reaches. The reach should be classified and evaluated at the Level III inventory which determines the state, stability, or "health" of the system. The next step should determine what caused the limiting factors (at the reach and in the watershed) and an analysis of potential remedies that address these causal factors. If through this analysis, a determination is made that structures are the solution, then an initial selection is made to off-set the limiting factors. Analysis of the compatibility of a proposed structure by stream type is then needed before final selection and implementation is made. If some adjustments need to be made in the design to accommodate a particular stream type, then engineering guidelines may be necessary to adjust the design of the structure. Once the design is implemented, the most important part is to evaluate the effectiveness of the structure to its intended purpose. This evaluation also needs to address the adjustment of the channel to the structure. This will not only provide information on meeting the initial objective, but also evaluate the effectiveness of the prediction methods and structures applied.

Habitat Improvement Structures

There is very little standardization for definitions of fish habitat improvement structures. Several names may be used to describe the same structure or a particular structure may have different design criteria in various locales. The descriptions below provide a standard frame of reference for the interpretations provided in the guideline.

Rearing Habitat Enhancement

<u>Low Stage Check Dam.</u>—One of the most common devices installed for fish habitat improvement are check dams (*Figure 8-10*). Low stage dams are check dams that are placed low in the channel profile (generally less than 1/3 of the bankfull stage). These would more appropriately be termed a plunge or ledge rather than a dam because of their low height. These devices are not designed for pool formation above the structure but rather to form a plunge pool below. A variety of structures such as straight log weirs, diagonal log weirs, K dams, wedge dams, and over pour ramps, are evaluated under this description since the channel adjustments are similar. Low stage dams are normally placed in long shallow riffles on straight reaches and meanders.

<u>Medium Stage Check Dam.</u>—Another type of check dam are those placed higher in the channel profile (up to 3/4 bankfull). These are also plunge pool forming structures such as trash catchers, gabion dams, log dams, and rock dams.

<u>Boulder Placement.</u>—A very common method of fish habitat improvement is the placement of boulders intended to provide instream cover (*Figure 8-11*). Velocities are increased such that a scour pool develops around the structure. Boulders are often placed in groups or singly in a "random" fashion. Minimum size rock depends upon maximum velocities at the site but generally two to three feet diameter or larger boulders are utilized. They are generally placed in riffles and glides but are also occasionally placed in pools for added cover.

<u>Bank Placed Materials.</u>—Bank placed materials (boulders, root wads, logs, etc.) are installed for dual purposes, i.e., to provide cover and pools similar to that provided by instream boulder placement and also to protect unstable banks. These structures act as small deflectors diverting high flows away from unstable banks and creating small pockets of back water that provide fish resting areas. Bank placed materials can be placed alone or in series along the bank, generally along the outside bend of meanders. They are keyed into the bank so that high velocity flows cannot scour behind or underneath them.

<u>Single Wing Deflectors.</u>—These commonly used devices are installed to direct streamflows, increase

velocities and form small pools (*Figure 8-12*). They are also used to direct high flows away from unstable banks. The guideline will evaluate deflectors built in a triangular shape. Single wing deflectors are often used in conjunction with other structures such as boulder placement and bank cover structures.

Double Wing deflectors.—The objective of double deflectors is to narrow the channel and increase velocities such that a deep scour pool develops in the center of the channel. They are constructed by installing two single wing deflectors opposite each other reducing channel width by 40 to 80% (*Figure 8-13*).

Channel Constrictor.—This structure is very similar to a double wing deflector in that it is designed to narrow and deepen the channel (*Figure 8-14*). These structures are either paired or placed alone. Channel width is generally reduced up to 80%.

Bank Cover.—Bank cover structures are installed to create an undercut bank effect thus providing hiding cover for adult trout (*Figure 8-15*). They are built along the outside bends for along straight reaches in conjunction with deflectors so that they always have adequate water depth below. They can be built with extensive planking as illustrated by White and Brynildson (1967), the modified version presented by Hunt (1980) or simply with log construction as shown by Seehorn (1985).

Floating Log Cover.—One simple method of cover development is the installation of floating logs (*Figure 8-16*). These structures generally consist of two or more tree boles, fastened together and cabled to the bank, or to a streamside or instream boulder and are free to float and drift with rising and falling stage. They are generally placed over pools, backwater areas, or along meanders to provide overhead protection.

Submerged Shelters.—Submerged shelters such as whole trees, tree tops, shrubs, or brush piles are another simple structure placed in the channel to provide overhead cover (*Figure 8-17*). These structures also provide ideal substrate for aquatic organisms (Seehorn, 1985). The guideline evaluates structures placed on meanders and straight sections separately due to different potential channel adjustments.

Half Log cover.—Half log structures are used to provide overhead cover for adult trout (Hunt, 1977). They are built with 8 to 12 inch diameter logs split lengthwise, placed upon 6 inch spacer blocks and then anchored to the stream bottom (*Figure 8-18*). They are placed parallel or at a slight angle to streamflow and positioned at or adjacent to the thalweg. They are generally placed in a riffle-run (deep glide) or a riffle with adequate depth to keep the structure submerged.

Migration Barrier.—These structures are installed to protect native fish populations in headwater streams from non-native fish populations by blocking upstream fish migration (*Figure 8-19*). They are designed to create an impassable falls generally 4 to 6 feet in height.

Spawning Habitat Enhancement

V-shaped Gravel Traps.—Gravel traps are used where streams have an adequate supply of gravel but have little instream structure such as fallen trees, debris, etc., to trap gravel. Reeves and Roelofs (1982) described these structures used at Coos Bay, Oregon, to retain spawning for anadromous fish (*Figure 8-20*). The "V" shaped structures are placed with the apex downstream in a series of two or more. The upper structure dissipates water velocities and the lower collects and retains gravels.

Log Sill Gravel Traps.—Another type of gravel trap utilized by west coast biologists is the log sill (*Figure 8-21*). These are very similar to low stage dams in design and materials but they are built for gravel accumulation rather than for pool formation. Thus, these structures are generally placed very low (less than 10% of bankfull stage).

Gravel placement.—Another method of spawning habitat enhancement employed by several west coast biologists is the introduction of appropriate size gravel (Reeves and Roelofs, 1982). Clean river gravel is placed in riffles covering at least two square feet. This technique does not utilize any

APPLICATIONS

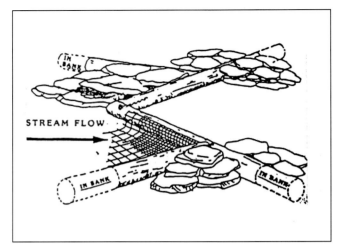

FIGURE 8-10. Low Stage check dam. (*Seehorn, 1985*)

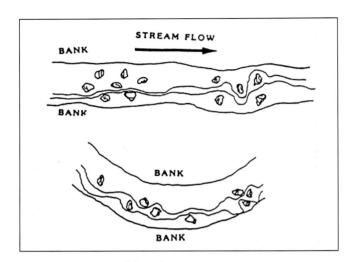

FIGURE 8-11. Boulder Placement. (*USDT, 1979*)

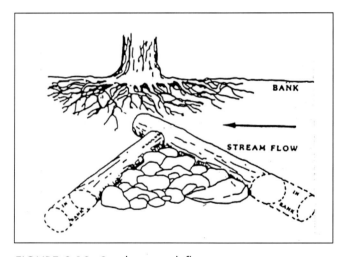

FIGURE 8-12. Single-wing deflector. (*Seehorn, 1985*)

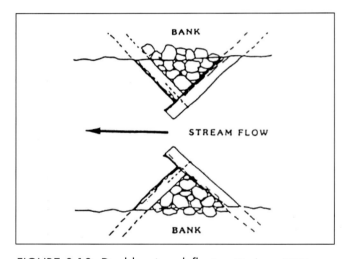

FIGURE 8-13. Double-wing deflector. (*Seehorn, 1985*)

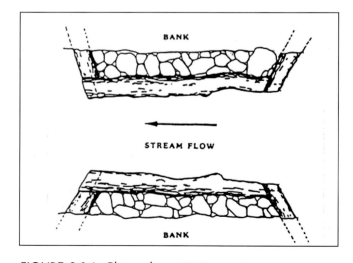

FIGURE 8-14. Channel constrictor. (*Seehorn, 1985*)

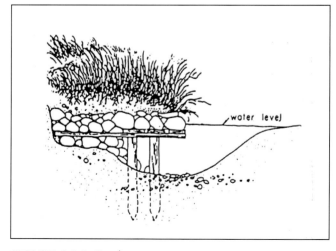

FIGURE 8-15. Bank cover. (*White and Brynildson, 1967*)

APPLICATIONS

FIGURE 8-16. Floating log cover.

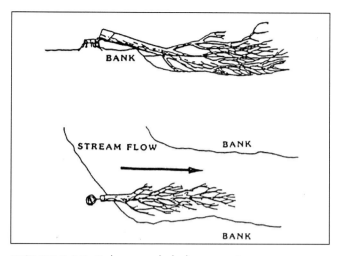

FIGURE 8-17. Submerged shelters. (*Seehorn, 1985*)

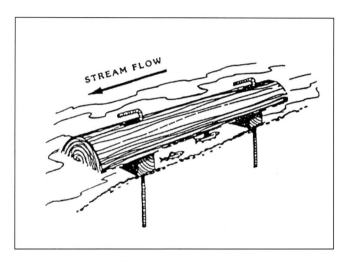

FIGURE 8-18. Half-log cover. (*Hunt, 1977*)

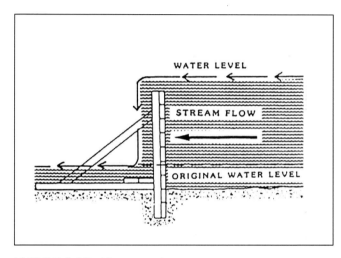

FIGURE 8-19. Migration barrier. (*Culver, 1985*)

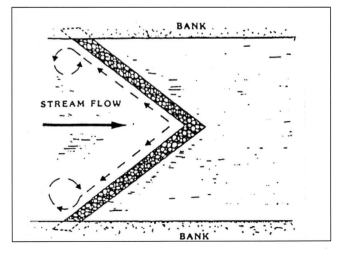

FIGURE 8-20. V-shaped gravel trap. (*Reeves & Roelofs, 1982*)

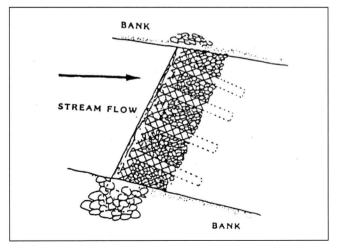

FIGURE 8-21. Log-sill gravel trap. (*Reeves & Roelofs, 1982*)

structural devices but is evaluated in regard to its applicability by channel type.

Channel Stability/Habitat Improvement Structures

Cross-Vane.—This structure was designed to off-set the adverse effects of straight weirs, and check dams, which create backwater and flat slopes. It was also designed to avoid the problems of the downstream pointing weirs which create twin parallel bars and a scour hole which de-stabilizes the structure. The objectives of this structure are to: (1) create instream cover/holding water; (2) take excess shear stress away from the "near bank" region and direct it to the center of the stream to maintain lateral stability; (3) increase stream depth by decreasing width/depth ratio; (4) increase sediment transport capacity; (6) provide a natural sorting of gravel (where naturally available) on the up-welling portion on the downstream side of structure for spawning redds, and; (6) create grade control to prevent down cutting. This design is shown in *Figure 8-22*.

J-Hook Vane.—This structure is designed to re-direct velocity distribution and high velocity gradient in the near-bank region, stabilize streambanks, dissipate energy in deep, wide and long pools created below the structure, create holding cover for fish and spawning habitat in the tail-out of the structure (*Figure 8-23*). Material can vary using native boulders and logs.

Native Material Revetment-Root Wads/Vegetation Transplants.—The objectives of this design are to: (1) protect the streambank from erosion; (2) provide in-stream and overhead cover for fish; (3) provide shade, detritus, terrestrial insect habitat; (4) look natural, and; (5) provide diversity of habitats. Diagrams showing cross-section and plan-views are shown in *Figure 8-24* and *Figure 8-25* (Rosgen, 1993a).

"W" Weir.—This structure is designed for river widths generally greater than 12 m (40 feet). This boulder structure is designed to create in-stream cover and diversity of velocity and depth and more useable area across the channel width. It looks much more natural than straight or curves weirs on wider channels (*Figure 8-26*).

Native Material Transplants and Log Vanes.—Woody riparian vegetation is transplanted along streambanks for stabilization, shade, detritus, terrestial insects and over-head cover for fish habitat. Vanes reduce back-eddy erosion for root wads and transplanted vegetation plus instream cover, up and down-stream of log and/or rock structure (*Figure 8-27*) (Rosgen, 1993a).

Examples of habitat features and various structures installed in a wide range of stream types are shown in the photographs contained in *Figures 8-28* and *Figure 8-29*.

This procedure is designed to evaluate the suitability of a wide range of fishery enhancement structures for various stream types. These guidelines are intended for application in planning and designing enhancement structures over a wide variety of streams to reduce the "error" from the trial and error method. These guidelines are intended as an initial framework for technology transfer that others will improve upon as more data are derived from on-going monitoring and evaluation programs (Rosgen and Fittante Mitchell, 1986).

Application of Guidelines

The generalized rating scheme shown in *Tables 8-2a-c, 8-3a, 8-3b, 8-3c, 8-3d*, and *Table 8-3e* evaluates the potential effectiveness of fish habitat improvement structures based on the morphology of the stream types involved. It is based on actual observations of such structures by stream type, reflecting both good and poor applications of a given structure for a particular stream type. These are only guidelines and are meant to provide general direction or highlight potential problems. They are not intended to be "fixed" or evolve into "hard rules." They in no way substitute for the services of a fisheries biologist and hydrologist in planning enhancement projects. The guidelines may, however, "red flag" some potential problem areas to necessitate more detailed, site-specific analysis prior to design selection.

The interpretations and rationale of this subjective rating scheme of excellent, good, fair and poor are described in *Table 8-3a-e*. These ratings do not reflect on: (1) the biological effectiveness for meeting limiting factors of habitat, (2) costs or difficulty of construction, or (3) cost/benefit relationships.

APPLICATIONS

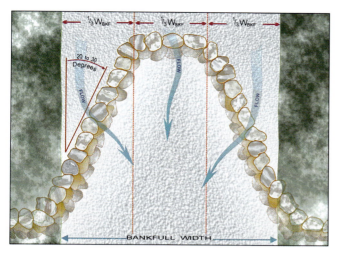

FIGURE 8-22. Cross-Vane Structure.

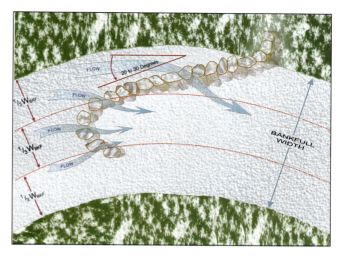

FIGURE 8-23. J-Hook Vane.

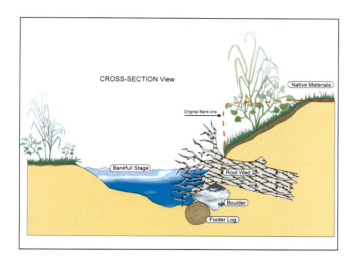

FIGURE 8-24. Native material bank revetment. (*Rosgen, 1993a*)

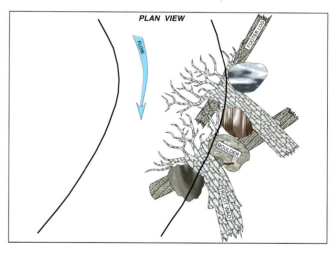

FIGURE 8-25. Native material bank revetment. (*Rosgen, 1993a*)

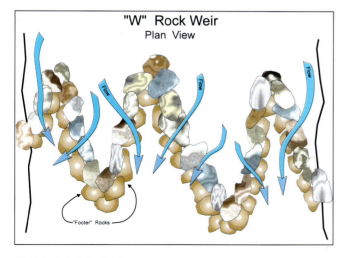

FIGURE 8-26. "W" rock weir. (*Rosgen, 1993a*)

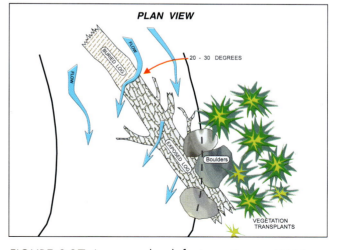

FIGURE 8-27. Log-vane bank feature. (*Rosgen, 1993a*)

8-21

APPLICATIONS

FIGURE 8-28a. A natural "vortex" rock weir on B3 stream type.

FIGURE 8-28b. Stable bed feature in step/pool B2 stream type.

FIGURE 8-28c. Simulation of a step/pool structure in a riffle/pool C4 stream type with a gabion check dam. Note aggradation headward, high width/depth ratio and lateral erosion (right bank).

FIGURE 8-28d. Same location as (c), but looking downstream at gabion check dam. Note channel avulsion on right converting to a "C4" stream type.

FIGURE 8-28e. Log check creating aggradation headward and migration barrier due to scour below dam.

FIGURE 8-28f. Log check dam placed too high creating deposition, high width/depth ratio and accelerating gulley head-cut. An "E4" - "G4" conversion.

APPLICATIONS

FIGURE 8-29a. Random rock placement in "F4" stream type. Need lower width/depth ratio and conversion to a stable morphology such as a "C4."

FIGURE 8-29b. Rock clusters in C4 - high bedload transport - Note bars.

FIGURE 8-29c. Construction of a "central bar" for fish habitat - inappropriate for this "C4" stream type.

FIGURE 8-29d. Central bar construction and rip rap for fish habitat and restoration - inappropriate for this F4 stream type.

FIGURE 8-29e. Log check dam at 80% of the bankfull stage - note high width/depth ratio - C4 stream type.

FIGURE 8-29f. Log check dams on B4 stream type spaced at 1-1/2 bankfull widths.

APPLICATIONS

Previous Types 1985	Channel Type 1994	Low St. Ch. Dams	Med. St. Ch. Dams	Random Boulder Placement	Bank Placed Boulder	Single Wing Deflector	Double Wing Deflector	Channel Constrictor	Bank Cover	Half Log Cover	Floating Log Cover
A1	A1	N/A	N/A	N/A	N/A	N/A	N/A	N/A	N/A	N/A	N/A
A2	A2	N/A	N/A	N/A	N/A	N/A	N/A	N/A	N/A	N/A	N/A
A3	A3	FAIR	POOR	POOR	GOOD	POOR	FAIR	N/A	POOR	POOR	FAIR
A3	A4	FAIR	POOR	POOR	GOOD	POOR	FAIR	N/A	POOR	POOR	FAIR
A4	A5	FAIR	POOR	POOR	GOOD	POOR	FAIR	N/A	POOR	POOR	FAIR
A5	A6	FAIR	FAIR	POOR	GOOD	POOR	FAIR	N/A	POOR	FAIR	FAIR
B1-1	B1	POOR	POOR	POOR	EXC	POOR	POOR	POOR	EXC	GOOD	GOOD
B1	B2	EXC	EXC	N/A	N/A	EXC	EXC	EXC	EXC	N/A	N/A
B2	B3	EXC	GOOD	EXC	EXC	EXC	EXC	EXC	EXC	EXC	EXC
	B4	EXC	GOOD	EXC	EXC	EXC	EXC	EXC	EXC	EXC	EXC
	B5	GOOD	FAIR	FAIR	EXC	GOOD	GOOD	GOOD	EXC	GOOD	EXC
	B6	GOOD	FAIR	FAIR	EXC	GOOD	GOOD	GOOD	EC	EXC	EXC
C1-1	C1	POOR	POOR	POOR	EXC	POOR	POOR	POOR	EXC	EXC	EXC
C2	C2	GOOD	FAIR	N/A	N/A	GOOD	GOOD	GOOD	GOOD	N/A	GOOD
C2	C3	GOOD	FAIR	GOOD	EXC	GOOD	GOOD	GOOD	GOOD	GOOD	GOOD
C3	C4	FAIR	POOR	POOR	GOOD	FAIR	FAIR	FAIR	GOOD	FAIR	GOOD
C4	C5	FAIR	POOR	POOR	GOOD	POOR	POOR	POOR	FAIR	POOR	GOOD
C5	C6	FAIR	POOR	POOR	GOOD	POOR	POOR	FAIR	GOOD	FAIR	GOOD
D1	D3	POOR	POOR	POOR	FAIR	FAIR	FAIR	FAIR	POOR	POOR	POOR
D1	D4	POOR	POOR	POOR	FAIR	FAIR	FAIR	FAIR	POOR	POOR	POOR
D2	D5	POOR	POOR	POOR	FAIR	FAIR	FAIR	FAIR	POOR	POOR	POOR
B6	E3	N/A	POOR	POOR	GOOD	POOR	FAIR	N/A	N/A	N/A	N/A
C6	E4	N/A	POOR	POOR	GOOD	POOR	FAIR	N/A	N/A	N/A	N/A
C6	E5	N/A	POOR	POOR	GOOD	POOR	FAIR	N/A	N/A	N/A	N/A
C6	E6	N/A	POOR	POOR	GOOD	POOR	FAIR	N/A	N/A	N/A	N/A
F1	F1	POOR	POOR	POOR	GOOD	FAIR	POOR	POOR	FAIR	FAIR	FAIR
F2	F2	FAIR	POOR	N/A	N/A	FAIR	FAIR	FAIR	FAIR	FAIR	FAIR
F3	F3	FAIR	POOR	FAIR	GOOD	GOOD	GOOD	FAIR	FAIR	FAIR	FAIR
F3	F4	FAIR	POOR	POOR	POOR	GOOD	FAIR	FAIR	FAIR	FAIR	FAIR
F4	F5	FAIR	POOR	POOR	GOOD	FAIR	FAIR	FAIR	FAIR	FAIR	FAIR
F5	F6	FAIR	POOR	FAIR	GOOD	FAIR	FAIR	FAIR	FAIR	FAIR	FAIR
	G1	N/A	N/A	POOR	N/A	N/A	N/A	N/A	POOR	FAIR	FAIR
	G2	N/A	N/A	N/A	N/A	N/A	N/A	N/A	POOR	FAIR	N/A
	G3	FAIR	POOR	POOR	GOOD	POOR	FAIR	N/A	POOR	POOR	FAIR
B3	G4	FAIR	POOR	POOR	GOOD	POOR	FAIR	N/A	POOR	POOR	FAIR
B4	G5	FAIR	POOR	POOR	GOOD	POOR	FAIR	N/A	POOR	POOR	FAIR
B5	G6	FAIR	POOR	POOR	GOOD	POOR	FAIR	N/A	POOR	POOR	FAIR

TABLE 8-2a. Fish habitat improvement structures - suitability to stream types.

APPLICATIONS

Submerged Shelter			Gravel Traps					Native Material Vegetation		
Meander	Straight	Migration Barrier	"V" Shaped	Log Sill	Spawning Gravel Placement	Cross Vane	"W" Weir	Bank Placed Root Wads	J-Hook, Log & Rock Vanes	Stream Type
N/A	N/A	EXC	N/A	N/A	POOR	N/A	N/A	N/A	N/A	A1
N/A	N/A	EXC	N/A	N/A	POOR	N/A	N/A	N/A	N/A	A2
POOR	POOR	FAIR	POOR	POOR	POOR	FAIR	POOR	GOOD	POOR	A3
POOR	POOR	FAIR	POOR	POOR	POOR	FAIR	POOR	GOOD	POOR	A4
POOR	POOR	POOR	POOR	POOR	POOR	FAIR	POOR	GOOD	POOR	A5
POOR	POOR	FAIR	POOR	POOR	POOR	FAIR	POOR	GOOD	POOR	A6
GOOD	EXC	GOOD	GOOD	GOOD	FAIR	GOOD	GOOD	N/A	N/A	B1
N/A	N/A	GOOD	EXC	EXC	FAIR	N/A	N/A	N/A	N/A	B2
GOOD	EXC	GOOD	GOOD	GOOD	GOOD	EXC	EXC	EXC	EXC	B3
GOOD	EXC	GOOD	N/A	N/A	N/A	EXC	EXC	EXC	EXC	B4
GOOD	EXC	FAIR	POOR	POOR	POOR	GOOD	EXC	EXC	GOOD	B5
GOOD	EXC	FAIR	POOR	POOR	POOR	GOOD	GOOD	EXC	EXC	B6
EXC	EXC	POOR	GOOD	FAIR	FAIR	GOOD	GOOD	EXC	GOOD	C1
N/A	N/A	POOR	GOOD	GOOD	GOOD	N/A	N/A	EXC	GOOD	C2
EXC	EXC	POOR	GOOD	GOOD	GOOD	EXC	EXC	EXC	EXC	C3
FAIR	GOOD	POOR	N/A	N/A	N/A	EXC	GOOD	EXC	EXC	C4
FAIR	GOOD	POOR	POOR	POOR	POOR	GOOD	FAIR	EXC	GOOD	C5
FAIR	GOOD	POOR	POOR	POOR	POOR	GOOD	GOOD	EXC	GOOD	C6
POOR	POOR	POOR	POOR	POOR	POOR	POOR	POOR	FAIR	FAIR	D3
POOR	POOR	POOR	N/A	N/A	N/A	POOR	POOR	FAIR	FAIR	D4
POOR	POOR	POOR	POOR	POOR	POOR	POOR	POOR	FAIR	FAIR	D5
POOR	POOR	POOR	POOR	POOR	POOR	POOR	POOR	FAIR	FAIR	D6
GOOD	GOOD	POOR	FAIR	FAIR	FAIR	GOOD	N/A	GOOD	GOOD	E3
GOOD	GOOD	POOR	N/A	N/A	N/A	GOOD	N/A	GOOD	GOOD	E4
GOOD	GOOD	POOR	POOR	POOR	POOR	GOOD	N/A	GOOD	GOOD	E5
GOOD	GOOD	POOR	POOR	POOR	POOR	GOOD	N/A	GOOD	GOOD	E6
GOOD	GOOD	POOR	POOR	POOR	POOR	N/A	N/A	N/A	N/A	F1
N/A	N/A	POOR	FAIR	FAIR	FAIR	N/A	N/A	N/A	N/A	F2
GOOD	GOOD	POOR	FAIR	FAIR	FAIR	GOOD	FAIR	GOOD	GOOD	F3
GOOD	GOOD	POOR	N/A	N/A	N/A	GOOD	FAIR	GOOD	GOOD	F4
GOOD	GOOD	POOR	POOR	POOR	POOR	GOOD	FAIR	GOOD	GOOD	F5
GOOD	GOOD	POOR	POOR	POOR	POOR	GOOD	FAIR	GOOD	GOOD	F6
FAIR	FAIR	GOOD	N/A	N/A	POOR	N/A	N/A	N/A	N/A	G1
N/A	N/A	GOOD	N/A	N/A	POOR	N/A	N/A	N/A	N/A	G2
FAIR	FAIR	POOR	POOR	POOR	POOR	GOOD	POOR	GOOD	FAIR	G3
FAIR	FAIR	POOR	POOR	POOR	POOR	GOOD	POOR	GOOD	FAIR	G4
FAIR	FAIR	POOR	POOR	POOR	POOR	GOOD	POOR	GOOD	FAIR	G5
FAIR	FAIR	POOR	POOR	POOR	POOR	GOOD	POOR	GOOD	FAIR	G6

STREAM TYPES:
A1 → A6   Typically no fish habitat structures are installed in these stream types
DA4 → Da6

TABLE 8-2b. Fish habitat improvement structures - suitability to stream types. (cont.)

APPLICATIONS

LOW STAGE CHECK DAM			MEDIUM STAGE CHECK DAMS		
Rating	Channel Types	Limitation/Discussion	Rating	Channel Types	Limitation/Discussion
Exc	B2,B3,B4	No limitations	Exc	B2	No limitations
Good	B5,B6 C2,C3	Bank erosion due to lateral migration will occur unless bank stabilization is utilized	Good	B3, B4	Stage increase will result in floodplain encroachment. Limit dam height to less than 75% of bankfull stage and select sites with high stable banks
Fair	A3-A6 C4-C6 F2-F6 G3-G6	Low dams must be constructed in conjunction with bank stabilization in these channel types. Use in conjunction with confinement measures and bank stabilization to reduce lateral migration	Fair	A6, B5, B6, C2, C3	Banks must be adequately protected both up and downstream of structure
Poor	B1,C1	Bedrock streambed limits the development of pools	Poor	A3-A5 C4-C6 D3-D6 F2-F6 G3-G6	Increased stream aggradation accelerated bank erosion, slope rejuvenation and floodplain encroachment can result. Extensive bank stabilization measures must accompany installation. Exceptions are on headwater streams in ephemeral channels to stop gully headcuts, which rate fair.
	D3-D6	High width/depth ratio and high sediment yields makes ineffective. Increases bank erosion.		B1, C1, F1	Bedrock streambed limits pool scour depth, anchoring difficult.
N/A	A1,A2,F1E3-E6 F1,G1,G2	Pools not limiting in these stream types	N/A	A1,A2,G1,G3 E3-E6	Pools not limiting factor in these channel types. These stream types provide excellent fish habitat and generally do not require modification.

RANDOM BOULDER PLACEMENT			BANK PLACED BOULDER		
Rating	Channel Types	Limitation/Discussion	Rating	Channel Types	Limitation/Discussion
Exc	B3,B4	No limitations	Exc	B1,B3 B6 C1,C3	No limitations
Good	C3	Lower gradient provides more opportunity for bar development up and downstream of rock - unless placed on meander points (See Bank placed rock). Use in conjunction with deflectors to increase velocity sufficient to create pools	Good	A3-A6 C4-C6 E3-E6 F1,F3 F6 G3-G6	Boulders must be keyed into the bank.
Fair	B5,B6 F3	Potential bar deposition and lateral migration can be offset by stabilizing the banks and by strategic placement. Due to bed armor and flatter gradients, it is advantageous to create deep pools with a combination of deflectors, boulders and/or rock clusters	Fair	D3-D6	Difficult to locate thalweg channel and where the banks will be inundated from one year to another
			Poor		
Poor	B1,C1,F1, G1 A3-A6 C4-C6 D3-D6 E3-E6 F4,F5,F6 G3-G6	Bedrock limits bed scour, difficult to stabilize in place with high flows. The high sediment supply and highly unstable banks limit the effectiveness of boulders placed in the channel (other than along banks). Bar deposition up and downstream of boulder and excessive bank erosion often occur. Deflectors can reduce sediment deposition, but stress banks.	N/A	B2,C2,F2,G1 G2, A1,A2	Bank rock and streamside boulders naturally occur and banks are naturally stable. Cover and pools not limiting in this channel type
N/A	A1,A2, B2,C2 F2,G2	Large boulder and/or pools are not a limiting factor in these channel types			

TABLE 8-3a. Limitations and discussions of various fish habitat improvement structures by stream types.

APPLICATIONS

	SINGLE WING DEFLECTOR			DOUBLE WING DEFLECTOR	
Rating	Channel Types	Limitation/Discussion	Rating	Channel Types	Limitation/Discussion
Exc.	B2-B4	No limitations.	Exc.	B2-B4	No limitations.
Good	B5,B6 C2,C3,F3	May need bank stabilization.	Good	B5,B6 C2,C3,F3	May need bank stabilization in conjunction with double deflector.
Fair	C4,D3 D6 F1,F2, F4-F6	Must be done with corresponding bank protection. Extensive construction may be needed to gain stabilization of the channel.	Fair	A3-A6 C4,D3 D6 E3-E6, G3-G6,F2, F4-F6	Need bank stabilization. Extensive construction may be needed to gain stabilization.
Poor	A3-A6 C5,C6 E3-E6 G3-G6 C1,B1	Channel instability and high sediment supply reduces effectiveness. Increases W/D ratio. Increase bank erosion. Bedrock bed limits effectiveness.	Poor	C5,C6 F1,C1,B1	Channel instability and high sediment supply reduces effectiveness. Bedrock bed limits effectiveness.
N/A	A1,A2 G1,G2	Pools not a limiting factor.	N/A	A1,A2 G1,G2	Pools not a limiting factor.

	CHANNEL CONSTRICTOR			BANK COVER	
Rating	Channel Types	Limitation/Discussion	Rating	Channel Types	Limitation/Discussion
Exc.	B2-B4	No limitations.	Exc.	B1-B6 C1	No limitations.
Good	B5,B6,C2, C3		Good	C2-C4,C6	Locate on mid-lower 1/3 of outside bank only.
Fair	C4-C6 F2-F6 D3-D6	Need bank protection downstream from constrictor. Need to reshape channel to W/D ratio less than 40 to use - high bedload	Fair	C5,F1-F6	Lateral migration may result in undermining the structure.
Poor	C5 C1,F1,B1	Bank and bed instability and high sediment supply limits effectiveness. Bedrock bed limits effectiveness.	Poor	A3-A6 G1-G6 D3-D6	Channel instability limits effectiveness. Structure loss common. Change in annual thalweg position makes these structures impractical.
N/A	E3-E6 G1-G6 A1-A6	Not limiting due to existing low width/depth ratios.	N/A	E3-E6 A1,A2	Good cover generally available within these channel types - low W/D ratio promotes good bank cover.

TABLE 8-3b. Limitations and discussions of various fish habitat improvement structures by stream types. (Cont.)

APPLICATIONS

	HALF LOG COVER			FLOATING LOG COVER	
Rating	Channel Types	Limitation/Discussion	Rating	Channel Types	Limitation/Discussion
Exc.	B3,B4,B6 C1	No limitations.	Exc.	B3-B6 C1	No limitations.
Good	B1,B5,C3	Will have to use anchoring techniques compatible with substrate.	Good	B1,C3 C6	Overlapping logs reduces bank erosion.
Fair	A6,C4,C6 F1-F6 G1,G2	Increased sedimentation may cause bar formation which results in decreased channel capacity and increased bank erosion. Key is the use low deflectors in conjunction with half log structures.	Fair	F1,F3-F6 A3-A6, G1, G3-G6	Undercutting will cause undermining of the anchor and eventual loss of the structure. Take extra precautions to protect banks.
Poor	A3-A5 C5,D3-D6 G3-G6	Extremely unstable bed conditions - degrading and aggrading reaches which limit the effectiveness of this structure.	Poor	D3-D6	Shifting active channel makes this structure infeasible.
N/A	C2,B2 E3-E6 A1-A2	Cover generally not limiting in these stream types.	N/A	A1 E3-E6,A2, B2,C2,F2,G2	Steep, bedrock, high velocity. These stream types provide excellent fish habitat and generally do not require modification.

	SUBMERGED SHELTERS LOCATED ON MEANDERS			SUBMERGED SHELTERS LOCATED ON STRAIGHT REACHES	
Rating	Channel Types	Limitation/Discussion	Rating	Channel Types	Limitation/Discussion
Exc.	C1,C3	No limitations.	Exc.	B1,B3-B6 C1,C3	No limitations.
Good	B1,B3-B6 E3-E6 F1,F3-F6	Because structures are located on meanders (high velocity areas of the channel), these channel types may be subject to some bank erosion.	Good	C4-C6 E3-E6 F1, F3-F6	Submerged shelters can be placed on straight reaches in these channel types.
Fair	C4-C6	Need bank stability measures on opposite bank to prevent accelerate bank erosion and lateral migration. Done in conjunction with bank stabilization, this structure can deepen and narrow C1, C4, C5, and C6 channels in particular.	Fair	G3-G6	High bedload transport and high stream power/high sediment supply and channel instability limits effectiveness.
Poor	G3-G6	Unstable channel, high sediment supply, and bank erosion.	Poor	D3-D6	Shifting active and thalweg channel makes this structure ineffective.
	D3-D6	Shifting active and thalweg channel makes this structure ineffective.		A3-A6	Steep gradient makes difficult to stabilize, structure loss common.
	A3-A6	Very high sediment supply, Highly unstable channel	N/A	A1,A2,C2,B2, F2,G2,G1,G2	Cover naturally available.
N/A	A1, A2 B2,C2,F2, G1,G2	Cover naturally available.			

TABLE 8-3c. Limitations and discussions of various fish habitat improvement structures by stream types. (Cont.)

APPLICATIONS

	V-SHAPED GRAVEL TRAP*			LOG SILL GRAVEL TRAPS	
Rating	Channel Types	Limitation/Discussion	Rating	Channel Types	Limitation/Discussion
Exc.	B2	No limitations.	Exc.	B2	No limitations.
Good	B1,B3, C1-C3		Good	B1,B3, C2,C3	
Fair	E3, F2,F3	Higher sediment yields make invasion of fines possible. Use with pervious trap so intra-gravel flow rate is maintained.	Fair	C1,F2,F3,E3	Frequent bed scour or bank erosion may inundate gravel with fines.
Poor	G3-G6 B5,B6, C5,C6, D3,D5,D6, E5,E6,F1, F5,F6, A3-A6	Entrench. and low w/d struct. unstable Unstable bank and bed with high sediment supply limits effectiveness and/or no source for suitable spawning gravel. Stream too steep and/or unstable	Poor	B5,B6,C5,C6, D3,D5,D6,F1 F5,F6,E5,E6, A3-A6 G3-G6	High transport of fine sediments. gravel size bedload unavailable. Unstable bed and banks, hi. bedload, Channel too steep and/or unstable for structures.
N/A	B4,C4,E4, F4, A1,A2, G1,G2	Stream types that have gravel sizes for spawning potential. Not associated with spawning habitat	N/A	B4,C4,E4,F4, D4 A1,A2, G1,G2	Gravel bed stream types that have spawning potential. Not associated with spawning habitat

*Note: Downcutting often occurs at the apex which can undermine the structure.

	GRAVEL PLACEMENT			MIGRATION BARRIER	
Rating	Channel Types	Limitation/Discussion	Rating	Channel Types	Limitation/Discussion
Exc.		No limitations.	Exc.	A1,A2	No limitations.
Good	B3,C2,C3	Must select lower velocity areas within the reach - transition zones between pool and riffle.	Good	B1-B4 G1,G2	Proper site selection must be made within the reach where banks are high and stable.
Fair	B1,B2,C1 E3,F2 F3	May not be effective considering the limited area where critical shear velocities would not be exceeded. Can cause capacity reduction and increase bank erosion. Treat smaller percentage of the channel area and/or stabilize banks. Potential for fine sediment invasion with minimal disturbance due to frequent bed shifts.	Fair	A3-A6 B5,B6	Erodible banks and moderate confinement limit barrier placement.
Poor	B5,B6,C5, C6,D3,D5, D6,F5,F6 E5,E6,F1, A1-A6 G1-G6	Will fill in with finer bed load transported material. Channel too steep, deeply incised and/or unstable for spawning channel.	Poor	C1-C6 D3-D6 E3-E6 F1-F6 G3-G6	Bank and bed instability can result in structure failure. Low banks - cannot create adequate height for falls.
N/A	B4,C4,D4, E4,F4	Gravel bed stream type.	N/A		

TABLE 8-3d. Limitations and discussions of various fish habitat improvement structures by stream types. (Cont.)

8-29

APPLICATIONS

CROSS-VANE STRUCTURE			"W" WEIRS		
Rating	Channel Types	Limitation/Discussion	Rating	Channel Types	Limitation/Discussion
Exc.	B3,B4, C3,C4	No limitations	Exc.	B3,B4,B5 C3	No limitations
Good	B1,B5,B6, C1,C5,C6, E3-E6, F4-F6, G3-G6	Minor limitations, need bank vegetation to optimize habitat. Use on E types when starting to incise-bank height ratio >1.0.	Good	B1,B6, C1,C4,C6	Minor limitations
Fair	A3-A6	Steep banks need cut-off trench, difficult to install, debris flows add high risk-need extra footers/larger rock	Fair	F3-F6 C5	High width/depth ratio along with erodible banks limits effectiveness. Deposition of sand bedload occurs in backwater.
Poor	D3-D6	High width/depth ratio makes structure impractical.	Poor	A3-A6 G3-G5 D3-D6 A1,B1,C1, F1,G1,	Low width/depth ratio, unstable, channel makes this structure inappropriate. Braided pattern not conducive for this structure. Bedrock bed could not stabilize boulders, nor scour to make deep pools or holding water.
N/A	A1,A2, B2,C2. F1,F2, G1,G2	Cross-vane structure not required for Bedrock and boulder channels.	N/A	B2,C2, F2,G2 E3-E6	These stream types provide excellent fish habitat and generally do not require modification.

ROOT WADS/VEGETATION TRANSPLANTS			J-HOOK VANES, ROCK AND LOG VANES		
Rating	Channel Types	Limitation/Discussion	Rating	Channel Types	Limitation/Discussion
Exc.	B3-B6 C1-C6	No limitations	Exc.	B3,B4,B6, C3,C4	No limitations
Good	A3-A6 F3-F6 G3-G6 E3-E6	Need to use extensive re-vegetation above bankfull on entrenched stream types. Use log vane to stop back eddy bank erosion.	Good	B5, C1,C2,C5,C6, E3-E6, F3-F6	Need woody vegetation for optimum function. Need extra footers for rock vanes in sand.
Fair	D3-D6	Need to decrease W/D ratio to improve depth of flow in near bank region.	Fair	G3-G6 D3-D6	Steep banks make vanes difficult to install, need extensive transplants and cut-off trench. Braided channels need to reduce high width/depth ratio to be effective.
Poor			Poor	A3-A6	Steep banks make difficult to install, entrenched, high bedload, rejuvenated slopes, debris torrents/flows create high risk.
N/A	A1,A2 B1,B2 F1,F2 G1,G2	Bedrock/boulder banks do not require stabilization.	N/A	A1,A2 B1,B2 F1,F2 G1,G2	Bedrock/boulder channels would not generally require this structure. Pools and cover are normally not limiting habitat components in these stream types.

TABLE 8-3e. Limitations and discussions of various fish habitat improvement structures by stream types.

This procedure is designed to evaluate the suitability of a wide range of fishery enhancement structures for various stream types. These guidelines are intended for application in planning and designing enhancement structures over a wide variety of streams.

APPLICATIONS

RESTORATION

In 1935, the American Forests journal ran an advertisement for DuPont dynamite. Untamed rivers, it seems, were an untapped market for new uses of Dupont's product. The ad intoned:

"Crooked streams are a menace to life and crops in the areas bordering their banks. The twisting and turning of the channel retards the flow and reduces the capacity of the stream to handle large volumes of water. Floods result. Crops are ruined. Lives are lost . . . take the kinks out of crooked streams . . . Dupont dynamite has straightened many thousands of miles of crooked streams . . ."

We are now in an unprecedented era of stream restoration, working to put the kinks back into channelized, over-widened streams. These restoration efforts are much needed but, as with many new programs, restoration efforts run the risk of working counter to natural stability concepts. In an effort to make sure the stream is restored or fixed to its maximum potential, the "structure-mania" as shown in *Figure 8-30* from a recent effort illustrates a growing tendency. The author often refers to this type of layout as getting "nibbled to death by a duck." This work is often not only expensive, but is also often contrary to the river's central tendency for self-stabilization. There is a German term which often applies to such designs called "Schlimmbesserung" which is defined as a so-called improvement that makes things worse.

Stream classification can assist in river restoration by:

1. Enabling more precise estimates of quantitative hydraulic relationships associated with specific stream and valley morphologies.
2. Establishing guidelines for selecting stable stream types for a range of dimensions, patterns, and profiles that are in balance with the river's valley slope, valley confinement, depositional materials, streamflow, and sediment regime of the watershed.
3. Providing a method for extrapolating hydraulic parameters and developing empirical relationships for use in the resistance equations and hydraulic geometry equations needed for restoration design.
4. Developing a series of meander geometry relationships that are uniquely related to stream types and their bankfull dimensions.
5. Identifying the stable characteristics for a given stream type by comparing the stable form to its unstable or disequilibrium condition.

River Restoration Principles

There are basic questions that should be answered before implementing any stream restoration. These are:

What are the observed problems?

It seems obvious, but the first step should be a clear detailed description of the problems observed within the study reach. A practical application is to search for similar stream types that are stable in the same hydro-physiographic province. Comparisons of conditions above and below known problem source areas can also isolate certain conditions. The application of Level III and Level IV stream inventories and described in Chapters 6 and 7 can quantitatively indicate departures from natural stability. These techniques often help in isolating and describing specific problem reaches.

Many restoration projects involve a fish habitat improvement objective. Unfortunately, many projects are implemented without a "limiting factor" analysis. Sometimes we are guilty of "fixing something that is not broke."

What caused the problem?

Current and historical land uses within the watershed should be reviewed. Changes in the watershed that affect the quantity or timing of streamflows (such as changes in vegetation, soil compaction, impoundments, or diversions) also affect the amount and distribution of energy within the stream system. Alluvial stream types obviously respond much differently to changes in flow than do structurally controlled channels such as bedrock or boulder bed stream types.

APPLICATIONS

Before designing restoration methods for an unstable river, it is essential to determine the causes of the disequilibrium conditions. These causes can be complex and interrelated with many watershed variables. Often an attack on the symptom, does not effect the cure! Many restoration attempts have failed, as the processes driving the cause of the problem were not identified, nor understood. As a minimum, it is important to know the bankfull discharge, sediment regime, drainage area, and stream hydrograph characteristics of the watershed.

Changes in the watershed that affect the quantity or timing of streamflows are activities such as vegetation removal, roads, soil compaction, impoundments or diversions, urban development, and drainage alteration. Changes in the nature, size, amounts, and source areas of sediment, influence channel stability. Previous discussions in this book and prediction methods are currently available to quantify these relations.

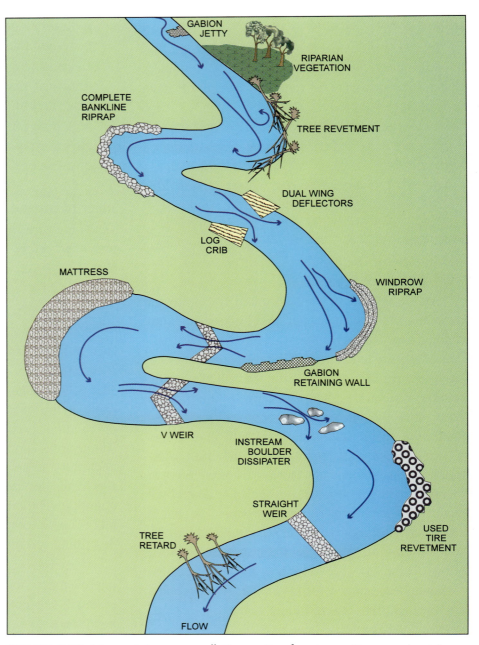

FIGURE 8-30. Non-point source pollution option for stream "improvement."

It is important, however, to understand the interrelations between channel stability and watershed changes. For example, if watershed changes produce an increase in streamflow, alluvial stream channel dimensions will change to accommodate the increases. That is, the channel enlarges and its bankfull dimension grows wider through bank erosion and lateral extension. Since meander geometry is related to bankfull width, other channel dimensions such as meander length, radius of curvature, and meander width ratio will also change. The overall result is channel instability with a corresponding loss of land, increased sediment supply, and the loss of aquatic habitat.

Watershed impacts leading to increased sediment supply can overload the river beyond its carrying capacity. There are many projects which have attempted to restore the river, when the cause was due to excessive sediment introduction from roads and surface disturbance activities. Often, mass wasting is accelerated due to roads and/or sat-

APPLICATIONS

uration due to vegetation alteration and or slope hydrology. To properly restore or repair a stream, one must first, or at least concurrently, repair the watershed or the source of the problem. The importance of stream condition analysis, watershed assessment, and data collection methods as described in chapters 5, 6, and 7 can not be over emphasized.

Stream channel adjustments due to the sediment loading and/or accelerated bank erosion lead to bar deposition and an increased width/depth ratio. Concurrently with this increase in width/depth ratio, the stream slope generally steepens due to a decrease in stream length. If the cause of the problem is understood, the solutions become more evident.

What stream type should this be?

Streams need to be considered often, not in terms of their current degraded state, but in terms of their future potential as conditioned by their watershed and valley features. Applications of the evolutionary stages of stream types presented earlier in this chapter assist in the determination of the appropriate stream type. Similarly, it is essential that restoration goals not be based on conditions—regardless of how desirable—that bear no relationship to the river system.

What is the probable stable form of the stream type under the present hydrology and sediment regime?

The stable dimension, pattern and profile for the identified stream types need to be established. The geometry and hydraulic measurements as stratified by stream type at reference reaches and /or gaged station are often obtained. An example of the existing versus "potential" stream channel relations were previously presented in Figures 6-2a and 6-2b and discussed in Chapter 6 (Level III). Many restoration projects have the right dimensions for the channel, but the pattern (meander geometry) is not matched to this pattern. The profile of the river has to be proportionate to the pattern and dimensions, but often is not designed nor constructed in such a fashion. The design and placement of proper riffle/pool and/or step/pool sequences as a function of the stream width and gradient is critical for natural stability.

Longitudinal profile relations have been presented in the first chapter in this book. The spacing of streambed steps versus pools is a function of stream type that integrates materials, slope, and width. For "A" stream types (having slopes of .04 - .10), spacings between pools are characteristically about 3.5 to 4 bankfull widths. For gradients greater than 0.10 this spacing decreases to about 1.5 to 2 bankfull widths, thus the pool to pool spacing is inversely related to stream gradient. Pool spacing for "B" stream types (rapids dominated bed features) is approximately 4 bankfull widths; increasing to about 4-6 bankfull widths for gradients less than .02 as for Bc stream types. The spacing of riffles and pools in alluvial channels (C, E and F) is described in previous chapters as one half a meander wavelength. Since, on the average, the meander wavelength is 10-14 bankfull widths, then the spacing between pools is 5-7 bankfull widths. This information which has been previously published by Leopold et al. (1964) is very important to integrate into restoration designs for the various alluvial stream types. These are some general examples where the dimension (width), pattern (meander wavelength), and profile (slope and bed features) have to be interrelated.

River works often create an "over-width" channel or a design discharge width which does not meet the dimension, pattern and profile interrelations with the bankfull discharge. These over-width channels, while having larger streamflow capacity, induce sediment deposition, resulting in decreased velocity and loss of channel capacity and competence.

It has been difficult in the past to develop the proper "natural channel" design width. The stream classification system, however can assist in this calculation by establishing the stable (from reference reaches) width/depth ratio for the selected stream type. Along with the cross-sectional area associated with the bankfull discharge obtained from the hydraulic geometry by stream type and regional curves, the following calculation can be made:

$$W_{bkf} = \sqrt{(W/D)(A_{bkf})}$$

Where:

W_{bkf} = bankfull width

W/D = Width/Depth ratio from stable, selected stream type

A_{bkf} = Cross-sectional area of channel at the bankfull discharge

Once the proper bankfull width is selected, the following channel variables can be obtained:

1) the meander length (10-14 bankfull widths);
2) the meander radius of curvature (2-1/2 - 3 bankfull widths);
3) belt width (meander width ratio for selected streamtype multiplied by bankfull width);
4) sinuosity (from layout of meander length, radius of curvature and meander width ratio);
5) slope (from meander geometry layout);
6) spacing of pools (based on stream type, meander geometry and spacing ratios as a function of slope and bankfull width

Flows greater than bankfull are designed to inundate the flood prone area. Flood prone areas should be designed to handled desired specific return period floods rather than the bankfull channel. Thus, the limits of the flood prone area widths can be determined by computing the required flood capacity and using standard engineering methods and the relations shown in *Figures 5-12* through *5-14* (Chapter 5).

River designs need to accommodate the influence of size and sediment supply. For example, from large scale restoration, the author has observed that in high bedload C3 and C4 stream types, the channel pattern as a function of bankfull width varies considerable from average values. The radius of curvature is larger than average. The meander length is shorter, the arc length is greater, and the point bar slopes are steeper. As bank material becomes more cohesive, the radius of curvature for the same width decreases as does the width/depth ratio.

The best blue print for these design concepts are available through close observation of the natural stable form. Stream types help stratify the multiple variables into an integrative combination of dimension, pattern, and profile which can be directly identified and described. These blue prints are unique to each type and help direct restoration design efforts.

Clearly, river restoration designs should be congruent with restoration goals. It is possible, for example, to restore a river to achieve improved fish habitat and still maintain inherent stability. Fortunately, designs based on restoring natural river conditions yield improved biological values and functions as well as improved aesthetics. Too many "river restorations" have been confused with "river stabilization" which focused primarily on flood control works and/or bank stabilization. The use of "hard controls" (concrete and rip rap) associated with these efforts, often creates a need to "restore" the river. An example of this can be demonstrated by a restoration project designed and constructed in 1994 by the author on Wildcat Creek in the Bay Area of California. This restoration project conducted for the East Bay Regional Park District involved the removal of concrete dams and the concrete bed and their replacement with natural stable materials comprised of boulders, root wads, and native riparian plants. Dimensions, patterns, and profiles were altered from existing conditions to construct three distinct stream types along various reaches of this creek. The purpose of this restoration was to secure "natural stability", for both physical and biological function, restore fish migration, and improve aesthetic values. Wildcat Creek experienced a near record 100-year event in January, 1995 and maintained its "natural stability" with minimal post flood maintenance, effecting less than 5 per cent of the restored channel.

Case Studies

Three case studies are used to illustrate the fundamentals of stream restoration; actual designs require more detail and analysis than is presented here.

APPLICATIONS

Weminuche River, southwestern Colorado

In 1987, the study reach of the Weminuche River in southwestern Colorado was a C4 stream type on the way to becoming a G4; i.e., it was broad, shallow, with high sediment supply and excessive bar deposition, and poised to cut a new channel. A review of the local land use and history showed that, as part of a government cost-shared program to increase pasturage in the mountain meadow drained by the Weminuche River, riparian willows had been removed in 1978 and livestock grazing had been increased. Just 12 years later, the riparian vegetation had been converted to a grass/forb community.

The study area for the Weminuche River drains thirty square miles of pristine watershed and had "high geologic sediment supply." The valley slope is 0.01. As a result of the willow removal and livestock grazing in the stream bottoms, the streambanks were de-stabilized. As a result, the following changes ensued:

1. Bank erosion increased and the channel migrated laterally.
2. Width/depth ratio increased from 14 to 35 (increase in bankfull width and decrease in mean bankfull depth).
3. Belt width increased.
4. Meander width ratio decreased from 10 to 2.
5. Down valley meander migration rate increased approximately 8 feet/year.
6. An increase in sediment supply occurred from accelerated bank erosion and land loss.
7. Sediment transport capacity decreased with increased width/depth ratio, with corresponding excess bar deposition leading to stream aggradation.
8. Meander length and radius of curvature increased; sinuosity decreased.
9. Fish habitat and aesthetic values decreased greatly.

As a result of these relatively rapid adjustments, the river was at a geomorphic threshold and it was poised to cut through its banks to create a new main channel that would abandon about 1200 feet of stream length. This would have resulted in a stream slope of 0.055 slope with a valley slope of .01. Obviously, this would have created a shift in stream type by conversion to a gully or G4 stream type. To regain a balance with its valley, the gully would have to advance headward, abandon its floodplain, and laterally erode the stream banks. Downstream of this reach, the river had cutoff over 1700 feet of stream length following a similar process of adjustment. This over-steepened reach was head cutting into the upper reach described. Such a progression would convert the previous C4 reach of the Weminuche River into a G4, then to an F4, and eventually back to a C4, but at a new local base level. The new base level would have triggered rejuvenation and head-cutting of every tributary stream draining into the valley. Moreover, it would require several lifetimes for the Weminuche to develop a new floodplain and balance its slope to be approximately half of the valley slope (sinuosity of 2). A fairly significant adverse channel adjustment occurred over a relatively short time span stemming from this willow removal in the meadow. If this series of predictable consequences had been better understood, the willow removal would undoubtedly not have been implemented.

The restoration design re-created the river's earlier dimension, pattern and profile, rather than piecemeal patching of unstable banks with rip rap. Knowledge of stable features of C4 stream types and a study of pre-disturbance features, developing empirical relations of dimension, pattern, and profile assisted in developing design criteria for this river restoration project. Stream inventories Levels I through Level IV provide the necessary data for design criteria.

The width/depth ratio of the Weminuche River was returned from 35 to 14, and a slope of 0.005 was restored with a corresponding 2.0 sinuosity. A meander wavelength was established at 10 bankfull widths, and a meander radius of curvature at 2.8 bankfull widths. Willows were transplanted along the streambanks to restore their stability, and also to create fish habitat and improved aesthetics. Examples of the restoration is shown in *Figure 8-31a-8-3c.* The landowner, Mr. Robert Lindner of the Weminuche Valley Ranch, financed this restora-

APPLICATIONS

FIGURE 8-31a. Weminuche River showing sediment deposition, channel widening, aggradation, stream slope increasing, sinuosity decreasing and chute cut-offs developing.

FIGURE 8-31b. Weminuche River cross-section showing high width/depth ratio, grass/forb composition of riparian vegetation, fish habitat and loss of land characteristics prior to restoration.

FIGURE 8-31c. Weminuche River, three years following restoration. Note willows and sod transplants.

APPLICATIONS

tion and is presently managing his livestock to maintain the health and function of the river.

East Fork of the San Juan River and Blanco Rivers in Southwestern Colorado

Another example of a restoration application using stream classification involves a conversion of a braided or D4 stream type to a C4 stream type. This work was accomplished in southwestern Colorado in 1986 on the East Fork of the San Juan River (owned by Mr. Dan McCarthy) and in 1987 on the Rio Blanco River (owned by Mr. Robert Lindner). Both of these rivers have had major disturbance dealing with conversion of willows to grass, or channelization, straightening, and sacrificial rip rap levies. Following floods in the late 1960's, flood control works on the Blanco had straightened and over-widened the stream which had accelerated the rate of braiding, increased the rate of bank erosion and land loss, increased sediment supply, and steepened the channel slope. As a result of this, from 2 to 12 feet of bank were lost annually, as well as loss of fish habitat. The valley morphology is associated with relatively steep (.025 slope), glaciated, broad alluvial valleys, with four terraces involving a high glacial terrace and three Holocene alluvial terraces. Drainage areas for both rivers are approximately 52-54 square miles, with elevations from 8,000 ft. to over 13,000 ft. with variation from base flow to peak flows of 2 orders of magnitude due to snowmelt and stormflow runoff events.

A key to this restoration was an understanding of the potential stable channel morphology rather than the existing braided D4 stream type. Since both valleys had a well developed floodplain the question that needed to be answered is what stream type could be stable in this valley and what are the morphological characteristics? In the literature, one would interpret that for stream slopes of .015 to .020 in these coarse alluvial materials, associated with glaciation, that the existing D4 stream type is the natural occurring form. However, upon study of other reaches of the valley that had not undergone willow removal straightening, widening and similar perturbations, a C4 stream type was observed that had been able to transport the sediment of the watershed and maintain a stable pattern following a major flood in 1970 without change in dimension or pattern. This C4 stream type was located downstream on the East Fork San Juan River of the existing D4 stream type. An example of these two different stream types in the same valley morphology for the East Fork San Juan River are shown in *Figure 8-32*. In this Figure, a D4 stream type is shown upstream in the same watershed as the C4 stream type, located approximately 1.5 miles downstream of the braided reach. The C4 reach was located at a US Geological Survey streamgage site, which provided the data as shown in *Table 5-2* and *Figure 7-3*. The verification of bankfull discharge, corresponding bankfull width, and thus, meander geometry could be obtained mathematically and then, verified for the C4 stream type at the location shown in *Figure 8-32b*. Streamflow data was also available for the Blanco River permitting a similar analysis.

A river restoration design was then prepared by the author for the East Fork San Juan River and a 404 permit application was submitted to the Corps of Engineers to allow a conversion from the D4 back to a C4 stream type. This proposal funded by the landowner, Mr. Dan McCarthy of the East Fork Ranch, was not readily accepted by the regulatory agencies due to the designs non-typical "unique" and unusual character. Since the river morphology principles used were initially developed by Dr. Luna B. Leopold, as documented in the previous chapters of this book, then who better to review this design and field investigation? Dr. Leopold was subsequently requested to review the East Fork San Juan River proposal, and field checked the relations described. Dr. Leopold is shown on the banks of the East Fork of the San Juan River mapping and checking the morphological relations at the C4 reach near the streamgage site in 1985 (*Figure 8-33*). As a result of Dr. Leopold's recommendation, the 404 permit was approved. The author was the field superintendent on the project that was constructed in 1986 and also conducted the post-construction

APPLICATIONS

FIGURE 8-32. Comparison of D4 and C4 stream types, East Fork, San Juan River, Colorado 1985.

monitoring for the following three years (Rosgen, 1993a). The restoration project involving conversion of both the East Fork and Blanco River are shown in *Figure 8-34* and *Figure 8-35*. The fish habitat has improved due to increases in over-head and in-stream cover, in-stream mean depth and habitat diversity much beyond what was associated with the braided D4 reach. The author caught and released a 17 inch brown trout female on a dry fly on the restored reach of the East Fork San Juan River two years following construction. Colorado Division of Wildlife personnel documented the lowest biomass ever recorded with their equipment on the same reach two years prior to the construction. The natural stable channel form can often be the best fish habitat as well. The Blanco River Restoration project was discussed in more detail in Restoration of Aquatic Ecosystems: Science Technology and Public Policy (National Research Council, 1991).

Cautionary advice is offered in that the preceeding restoration examples should not infer that it is appropriate to convert all "D" stream types to "C's." There are many "D" stream types that are operating at their potential based on the landtype and valley type associated with the braided channel morphology.

ADDITIONAL RESOURCE MANAGEMENT INTERPRETATIONS

Interpretive information by stream type can also apply to establishment of watershed and streamside management guidelines dealing with silvicultural standards, surface disturbance activities, gravel and surface mining activities, riparian management guidelines, debris management, floodplain management, cumulative effects analysis, and flow regulation from reservoirs/diversions. An example of the implementation of these guidelines by stream type are shown in the Land and Resource Management Plan of the Arapaho/Roosevelt National Forest, Ft. Collins, Colorado (USDA Forest Service 1984).

FIGURE 8-33. Dr. Luna B. Leopold on the East Fork San Juan River.

APPLICATIONS

FIGURE 8-34a. East Fork, San Juan River, D4, prior to construction, 9/86.

FIGURE 8-34b. East Fork, San Juan River, C4, following construction, 8/87.

FIGURE 8-34c. Blanco River D4, prior to construction, 8/87.

FIGURE 8-34d. Same location on Blanco as (c) C4 and E4 stream type and new floodplain showing willow transplants.

FIGURE 8-34e. Aerial view of Blanco River D4, looking upstream prior to construction, 1987.

FIGURE 8-34f. Aerial view of Blanco River looking upstream following conversion to C4 stream type - 10/90.

APPLICATIONS

FIGURE 8-35a. Blanco River (D4) prior to restoration (note location of barn).

FIGURE 8-35b. Blanco River following conversion from (D4) to (C4) stream type (note location of barn for reference).

FIGURE 8-35c. East Fork, San Juan River showing upper restored reach (C4) compared to lower braided (D4) reach.

FIGURES 8-35a-c. Aerial views of conversion from braided (D4) to meandering (C4) stream types - Blanco and San Juan Rivers, Colo.

APPLICATIONS

The management of the riparian areas (USDA Forest Service 1992) have utilized the stream classification system into their recently developed Integrated Riparian Evaluation Guide — Intermountain Region. The classification system is used to help stratify and describe riparian areas associated with individual stream types. In addition, the stream classification system evaluates the potential risks, sensitivities of riparian areas and their potential "desired future condition."

The requirements and procedures of many agencies to assess the proper functioning condition (USDI Bureau of Land Management) and desired future condition (USDA, Forest Service) of their streams can be supplemented with the stream types and hierarchical river inventories presented in Chapters 5 and 6 (Level III) and verified using the procedures and forms outlined in Chapter 7 (Level IV). These procedures address the existing morphology, condition and potential of the river.

SUMMARY OF APPLICATIONS OF THE HIERARCHICAL STREAM INVENTORY LEVELS

Chapters 4 through 7 deal with detailed description of various hierarchical inventory Levels I-IV. The advantages of implementing these levels allows an integration of a large number of variables into an analytical framework for a wide range of planning levels. Management decisions need to be made without the luxury of operating at the most detailed level of information (Level IV). However, decisions made at level I and/or Level II need to follow up with field validation, as well as improve prediction models. *Table 8-4* shows examples of the implementation of the full range of stream inventory levels and how companion inventories may be integrated. Because of space limitation in this table, examples of only sediment and fisheries are demonstrated.

In this chapter, the application procedures presented for hydraulics, sediment, grazing, fish habitat, management interpretations by stream type, and restoration, assists the scientist and land managers in working with the an integration of water resource and river processes.

As there are many conflicting and competing uses of the water resources and as river and aquatic values diminish to the momentum of "development," it is time to take a holistic approach analyzing where have we been?— where are we now?—and where are we going?

The challenge is to integrate these process analysis, data inventories, on a watershed basis towards an ecosystem management approach. Each discipline has been working diligently to understand and to be understood. It is of great benefit to bring all disciplines together for an integrated approach. The "watershed" is one of the key components of an ecosystem assessment and many of the techniques in this book will assist in the watershed analysis.

APPLICATION EXAMPLES

A case history is presented to demonstrate applications of the principals presented in this book. In 1983, the author conducted a watershed analysis on second, third, and fourth order (Strahler's ordering) watersheds that were delineated on the front range of Colorado on the USDA Forest Service Arapaho National Forest. The objective of this watershed analysis was to predict the potential effects of past land management activities and for future resource allocations on these watersheds. Effects of timber harvest grazing, roads, and other activities were quantified. Changes in water yield, timing and introduced sediment from surface disturbance activities were predicted and documented. Streamflow related increases in sediment yield due to channel adjustment processes were quantified. Preceding this analysis was a data collection effort to assess the condition of the creeks, streams, and rivers for 1.2 million acres. The majority of the methods used involved the information contained in this book. Road systems, both existing, and proposed were located in each watershed and the timber harvest areas both from previous harvest and proposed stand changes were mapped by sub-watershed.

APPLICATIONS

Inventory Category	Application Level	Inventory Objectives	Application Examples	Extrapolation Method
Geomorphic Characterization (Level I)	Rapid assessment, overview of large areas for broad level study and associations. Integrates stream types with geology, Landforms, vegetation, broad life zones.	Provide broad level interpretations and general response by valley morphology, relief and land types. Potential habitat types for fisheries, set priorities for study.	Sediment: supply is "moderately high," stream energy low, alluvial deposits. Fisheries: General categories of species adaptability, general life stages and habitat.	Data from detailed inventories used to set averages and interpretations from level III and IV data. General relations set for fish species, life zones and habitat.
Morphological Description (Level II)	Watershed, sub-watershed and individual stream reach.	Management interpretations of morphological potential, sediment supply/energy relations Fisheries habitat type, species, age class and spawning potentials.	Develop regional curves from Level IV data by stream type for sus. and bedload. Fisheries habitat relations by stream type and fishery population relations.	Measured level IV data used to develop dimensionless rating curves. Biomass and benthic studies used to describe general relations for stream types in certain locale.
Stream State or Condition (Level III)	Individual reaches and combinations of channel features such as (Pools, riffles, runs, glides, spawning reds).	Prediction level estimates, stability index, departures from potential, Determine biological values, conditions, migration and habitat problems due to watershed impacts. Evaluation of existing conditions versus potential. Determine fishery impacts from sediment and stability changes.	Apply sediment rating curves for changes in stability and condition by stream type from level IV. Relate prediction index to physical and biological relations. MacroInvert. diversity index, special species, food chains, natural reproduction, habitat quality index applied to stream condition.	Develop relations for extrapolation from level IV that demonstrate departures from potential. Prediction methods that are sensitive to changes in stream condition (debris, flow, sediment, stability) can be extrapolated by condition by stream type to previous levels.
Validation and monitoring (Level IV)	Individual reaches, individual channel features and specific sites. Applied by season, year and sometimes for concurrent locations.	To obtain the most site specific data to quantitatively describe the resource and understand spatial and temporal variation. Data and analysis develop relations, validate prediction models and monitor response of watershed mgt., mitigation and/or restoration.	Establish sediment thresholds for stability, validate prediction models, sed. standards, Study changes in substrate composition, habitat, temperature, reproduction, food chains and populations by changes in stream type, stability, sediment relations. Relate to fishery change the degree and nature of impacts on channels.	Measurement of sediment/channel stability conditions and corresponding fishery conditions all Compared above vs below and/or before vs after impact. Biomass measured, food chains, and limiting factors determined by stream condition by stream type for extrapolation to previous levels.

TABLE 8-4. Example of an integration of fish habitat inventories and sediment/channel evaluation using the hierarchical stream inventory.

The method used for the prediction was adopted from in-service water and sediment yield models previously developed by hydrologists of the Northern Region. These methods were integrated into the Hysed model, adapted for the Central Rocky Mountain hydro-physiographic province and used in this analysis (Rosgen and Silvey, 1978). This analytical system utilized the procedures described previously in this book such as channel stability ratings and sediment rating curves by stream type. A similar, but more detailed procedure, is presented in "WRENSS" (USEPA, 1980).

As a result of the cumulative effects analysis by watershed, it was discovered that eleven fourth order watersheds were potentially at and/or beyond their geomorphic threshold. The proposed timber harvest and road construction would create potential adverse effects. When the watershed analysis results were presented to Gray Reynolds, then Forest Supervisor of the Arapaho and Roosevelt National Forests, he directed the entire Forest Plan be re-analyzed. Management decisions were then made with the benefit of a watershed analysis that informed the management team of the consequence of the proposed actions. This information assisted in the development of standards and guidelines to insure an integrative, ecosystem approach to assist in maintaining resource productivity and values. As a result of this re-analysis, the eleven watersheds were placed in a special management category where watershed restoration and re-forestation were emphasized. Without this integrated watershed analysis, management decisions would have been made without the benefit of a consistent, comparative assessment based on physical processes. The "driving mechanism" for the model and the follow-up monitoring was based on *field data*.

Although complex, the pieces of a watershed can be put into an understandable and useable form to advise those individuals who make policy and eventual resource management decisions. To provide this timely information, field observations are not only recommended—they are essential.

It is the intent of the author that the information and field methods presented in *Applied River Morphology* will assist in the assessment and management of our watersheds, rivers and associated water resources. In so doing —we can all play a key role in the wise management of our water resources and contribute to the promotion of the river ethic.

BIBLIOGRAPHY

ANDREWS, E.D. (1980): Effective and bankfull discharges of streams in the Yampa River Basin, Colorado and Wyoming. Journal of Hydrology 46: 311-330.

―――― (1983): Entrainment of gravel from natural sorted riverbed material. Geol. Soc. Am. Bull. 94: 1225-1231.

―――― (1984): Bed-material entrainment and hydraulic geometry of gravel-bed rivers in Colorado. Geol. Soc. Am. Bull. 95: 371-378.

ANNABLE, W.K. (1994): Morphological relations of rural water courses in southeastern Ontario for use in natural channel design. Masters thesis, Univ. Of Guelph, School of Engineering, Guelph, Ontario, Canada.

BAKER, V.R. (1974): Paleohydraulic interpretation of Quaternary alluvium near Golden, Colorado. Quaternary Research 4: 94-112.

BAKER, V.R. & D.F. RITTER (1975): Competence of rivers to transport coarse bedload material. Geol. Soc. Am. Bull. 86: 975-978.

BARBOUR, M.T., J.B. STRIBLING & J.R. KARR (1991): Biological criteria: streams - 4th draft. EPA Contract No. 68-CO-0093. Washington: U.S. Environmental Office of Science and Technology: 192 pp.

BARNES, H.H. (1967): Roughness characteristics of natural channels. U.S. Geological Survey, Water Supply Paper 1849: 213 pp.

BAUER, S.B. & T.A. BURTON (1993): Monitoring protocols to evaluate water quality effects of grazing management on western rangeland streams. U.S. Environmental Protection Agency, 910/R-93-017, Washington, D.C.

BENGEYFIELD, P.D. SVOBODA, D. BROWNING, & G. GALE (1994): Beaverhead riparian Guidelines. USDA Forest Service Northern Region, unpublished report: 24 pp.

BRADLEY, W.C. & A.I. MEARS (1980): Calculations of flows needed to transport coarse fraction of Boulder Creek alluvium at Boulder, Colorado. Geol. Soc. Am. Bull., Part II. 91: 1057-1090.

BRADLEY, W.C. & D.G. SMITH (1984): Meandering Channel Response to Altered Flow Regime: Milk River, Alberta and Montana. Water Resources Research Vol. 20, No. 12: 1913-1920.

BRICE, J.C. & J.C. BLODGETT (1978): Counter measures for hydraulic problems at bridges. Vol. 1, Analysis and Assessment. Report No. FHWA-RD-78-162, Federal Highway Admin., Washington, D.C.: 169 pp.

BULL, W.B. (1978): The Threshhold of Critical Power. Geol. Soc. Amer. Bull. in Press.

CARLETON, C.W. (1963): Drainage density and streamflow. U.S. Geogical Survey, Professional paper 422-C.

CHOW, V.T. (Ed.) (1964): Handbook of Applied Hydrology, McGraw-Hill, New York.

CHURCH, M. (1978): Paleohydrological reconstructions from a Holocene valley. In A. D. Miall (ed.) Fluvial sedimentology, Canadian Society of Petroleum, Geologists Memoir 5: 743-772.

CHURCH, M. & K. ROOD (1983): Catalogue of alluvial river channel regime data. Natural Sciences and Engineering Research Council of Canada. Dept. Geology Univ. British Columbia, Vancouver, B.C. Edition 1.0.

CLINE, L. (1979): Masters Thesis, Colorado State University, Fisheries Department, Fort Collins, CO.

CULBERTSON, D.M., L.E. YOUNG & J.C. BRICE (1967): Scour and fill in alluvial channels. U.S. Geological Survey, Open File Report: 58 pp.

CULVER, S.R. & K.R. BESTGEN, (1985): Greenback Cutthroat Trout Recovery Project: 1985 Progress Report. Colorado Dept. of Natural Resources, Division of Wildlife: 36 pp.

BIBLIOGRAPHY

DAVIS, W.M. (1899): The geographical cycle. Geographical Journal 14: 481-504.

DUNNE, T. & L.B. LEOPOLD (1978): Water in Environmental Planning. W.H. Freeman and Co., San Francisco, CA: 818 pp.

EMMETT, W.W. (1975): Hydrologic evaluation of the upper Salmon River area, Idaho. U.S. Geological Survey, Professional Paper 870-A, U.S. Govt. Printing Office, Washington, D.C.

EMMETT, W.W. & R. HADLEY (1968): The Vigil Network: preservation and access of data. U.S.Geological Survey Circular 460-C, U.S. Dept. Interior. Govt. Printing Off., Washington, D.C.

EMMETT, W.W., R.L. Burrows & B. Parks (1978): Sediment transport in the Tanana River in the vicinity of Fairbanks, Alaska, 1977 U.S. Geological Survey, Open-File Report 78-290: 28 pp.

FAHNESTOCK, R.K. (1963): Morphology and hydrology of a glacial stream — White River Mount Rainier, Washington. U.S. Geological Survey Prof. Paper 422-A: 70 pp.

FAHNESTOCK, R.K. & W.C. BRADLEY (1973): Knik & Matanuska Rivers, Alaska: A Contrast in Braiding; In Morisawa, M., ed., Fluvial geomorphology: Binghamton, State University of New York, Publications in Geomorphology: 221-250.

FISHER, W.L., L.F. BROWN, JR., A.J. SCOTT & J.H. MCGOWEN (1969): Delta systems in the exploration for oil and gas, a research colloquium. Bureau of Economic Geology, University of Texas at Austin.

GALAY, V.J., R. KELLERHALS & D.I. BRAY (1973): Diversity of river types in Canada. In Fluvial Process and Sedimentation. Proceedings of Hydrology Symposium. National Research Council of Canada: 217-250.

GRANT, G.E., F.J. SWANSON & M.G. WOLMAN (1990): Pattern and origin of stepped-bed morphology in high-gradient streams, Western Cascades, Oregon. Geol. Soc. Am. Bull. 102: 340-352.

HARRELSON, C.C, C.L. RAWLINS & J.P. POTYONDY (1994): Stream channel reference sites: an illustrated guide to field technique. Gen. Tech. Rep. RM-245. U.S. Department of Agriculture, Forest Service, Rocky Mountain Forest and Range Experiment Station: 61 pp.

HICKS, D.M. & P.D. MASON (1991): Roughness characteristics of New Zealand rivers. New Zealand Dept. of Scientific and Industrial Research, Marine and Freshwater, Natural Resources Survey, Wellington, N.Z.: 329 pp.

BIBLIOGRAPHY

HUNT, R.L. (1977): Instream Enhancement of trout habitat. Wisconsin Department of Natural Resources. In Hashagen, K. (ed.). A National Symposium on Wild Trout Management, Proceedings. California Trout Inc., San Francisco, CA.

HUNT, R.L. (1980): Two recent modifications of habitat development techniques in Wisconsin. In Proceedings of the Trout Stream Habitat Improvement Workshop. Asheville, NC. November 3-6: 60-62.

INGLIS, C.C. (1942): Rapid westerly movement of the Kosi River. Central Board of Irrigation, India (Technical): pages 7 & 8.

JACKSON, F. (1994): Plotting bankfull dimensions of a New Mexico stream on Leopold's Regional Curves. In Stream notes. Stream Systems Technical Center, USDA Forest Service, Fort Collins, Co.

JARRETT, R.D. (1984): Hydraulics of high-gradient streams. J. Hydraul. Engr. 110(11): 1519-1539.

JARRETT, R.D. (1990): Hydrologic and hydraulic research in mountain rivers. Water Res. Bull. WARBAQ 26(3): 419-429.

KAY, C.E. (1990): Yellowstone's northern elk herd: a critical evaluation of the "natural regulation" paradigm. Dissertation, Utah State University, Logan, UT.: 490 pp.

KAY, C.E. & S. CHADDE (1991): Reproduction of willow seed production by ungulate browsing in Yellowstone National Park. Symposium on Ecology and Management of Riparian Shrub Communities, Sun Valley, ID.

KELLERHALS, R., C.R. NEILL & D.I. BRAY (1972): Hydraulic and geomorphic characteristics of rivers in Alberta. Research Council of Alberta, River Engineering and Surface Hydrology Report 72-1: 52 pp.

KELLERHALS, R., M. CHURCH & D.I. BRAY (1976): Classification and analysis of river processes. J. Hydraul. Div., ASCE 102(HY7): 813-829.

KHAN, H.R. (1971): Laboratory studies of alluvial river channel patterns. Ph.D. Dissertation, Dept. of Civil Engineering Department, Colorado State University, Fort Collins.

LANE, E.W. (1955): The importance of fluvial morphology in hydraulic engineering, American Society of Civil Engineering, Proceedings, 81, paper 745: 1-17.

LANE, E.W. (1957): A study of the shape of channels formed by natural streams flowing in erodible material. Missouri River Division Sediment Series No. 9, U.S. Army Engineer Division, Missouri River, Corps of Engineers, Omaha, Neb.

BIBLIOGRAPHY

LANGBEIN, W.B. & L.B. LEOPOLD (1966): River meanders – Theory of minimum variance. U.S. Geological Survey Prof. Paper No. 422-H.: 15 pp.

LEOPOLD, L.B. (1994): "A View Of The River". Harvard University Press, Cambridge, Mass.: 298 pp.

LEOPOLD, L.B. & T. MADDOCK (1953): The hydraulic geometry of stream channels and some physiographic implications. U.S. Geological Survey Prof. Paper No. 252. Supt. of Docs,, U.S. Government Printing Office. Washington, D.C.: 57 pp.

LEOPOLD, L.B. & M.G. WOLMAN (1957): River channel patterns: braided, meandering, and straight. U.S. Geological Survey Prof. Paper 282-B.

LEOPOLD, L.B., M.G. WOLMAN & J.P. MILLER (1964): Fluvial processes in geomorphology. Freeman, San Francisco, CA: 522 pp.

LISLE, T.E. (1987): Using "residual depth" to monitor pool depths independently of discharge. Res. Note PSW-394. Berkeley, CA. USDA, Forest Service, Pacific Southwest Forest and Range Experiment Station: 4 pp.

LISLE, T.E. & R. EADS (1991): Methods to measure sedimentation of spawning gravels. Research Note PSW-411.USDA, Forest Service, Pacific Southwest Research Station, Berkeley, CA.

LISLE, T.E. & S.J. HILTON (1992): The volume of fine sediment in pools: an index of sediment supply in gravel-bed streams. Water Res. Bul. 28(2): 371-383.

LOWHAM, H.W. (1976): Techniques for estimating flow characteristics of Wyoming streams. U.S. Geological Survey. Water Resources Investigations: 76-112.

MACDONALD, L.H., A.W. SMART & R.C. WISSMAR (1991): Monitoring guidelines to evaluate effects of forestry activities on streams in the Pacific Northwest and Alaska. EPA/910/9-91-001. Seattle, WA: U.S. Environmental Protection Agency and University of Washington: 166 pp.

MARSTON, R.A. (1978): Morphometric indices of streamflow and sediment yield from mountain watersheds in western Oregon. USDA Forest Service Siuslaw National Forest, Corvallis, Oregon.

MATTHES, G. (1956): River engineering. Chapter 15, American Civil Engineering Practice. Abbott (ed.). John Wiley and Sons, New York.

MELTON, F.A. (1936): An empirical classification of flood-plain streams. Geographical Review 26.

MOLLARD, J.D. (1973): Air photo interpretation of fluvial features. In Fluvial Processes and Sedimentation. Research Council of Canada: 341-380.

BIBLIOGRAPHY

MONTGOMERY, D.R. & J.M. BUFFINGTON (1993): Channel classification, prediction of channel response, and assessment of channel condition, TFW-SH10-93-002, Timber, Fish, and Wildlife Agreement. Department of Natural Resources, Olympia, Washington: 84 pp.

MEYERS, T.J. & S. SWANSON (1992): Variation of stream stability with stream type and livestock bank damage in Northern Nevada. Water Resources Bull. AWRA. 28(4): 743-754.

MITCHELL, B. (1994): Comparisons of channel stability ratings (Pfankuch) with ranges by stream type in Wyoming and Colorado. USDI BLM, Personal Communication. BLM Technical Service Center, Denver, CO.

NANKERVIS, J. (1989): Fluvial studies on South Fork Cache la Poudre River. USDA Forest Service, Fort Collins, CO.

NASH, D.B. (1994): Effective sediment-transporting discharge from magnitude-frequency analysis. Journal of Geology, Vol. 102: 79-95.

NATIONAL RESEARCH COUNCIL (1992): Restoration of aquatic ecosystems. Restoration case studies. National Academy Press, Wash. DC.: 470-477.

NORDIN, C.F. JR. (1977): Graphical aids for estimating general scour in long channel contractions: U.S.Geological Survey Open-File Report 77-837: 12 pp.

OSBORN, J.F. & J.M. STYPULA (1987): New models of hydrological and stream channel relationships. In Erosion and sedimentation in the Pacific Rim. IAHS Publ. No. 165, Oregon State Univ., Corvallis, OR.

OSTERKAMP, W. & L.J. LANE AND G.R. FOSTER (1983): An analytical treatment of channel morphology relations. U.S. Geol. Surv. Prof. Paper 1288: 21 pp.

OSTERKAMP, W. (1994): Personal communication on techniques of bankfull determination for step/pool channel types. U.S. Geological Survey, Denver, CO.

PFANKUCH, D.J. (1975): Stream reach inventory and channel stability evaluation. USDA Forest Service, R1-75-002. Government Printing Office #696-260/200, Washington D.C.: 26 pp.

PLATTS, W.S. (1980): A plea for fishery habitat classification. Fisheries 5(1), 1-6.

POFF, N.L. & J.V. WARD (1989): Implications of streamflow variablility and predictability for lotic community structure: a regional analysis of streamflow patterns. Can. J. Fish. Aquat. Sci., 46: 1805.

BIBLIOGRAPHY

REEVES, G.H. & T.D. ROELOFS, (1982): Rehabilitating and enhancing stream habitat: field applications. In Meehan, W. (ed.). Influence of Forest and Rangeland Management on Andromous Fish Habitat in Western North America. USDA Forest Service. General Technical Report PNW-140.

RICHARDS, K. (1982): Rivers. Methuen & Co. New York: 358 pp.

RITTER, J.R. (1967): Bed-material movement. Middle Fork Eel River, California. U.S. Geological Survey Prof. Paper 575-C: C219-C221.

ROSGEN, D.L. (1979): Sediment measurements and analysis. Joe Wright Creek. USDA Forest Service, Ft. Collins, CO. Data Report, unpublished.

——— (1985): A stream classification system. In Riparian Ecosystems and Their Management. First North American Riparian Conference. Rocky Mountain Forest and Range Experiment Station, RM-120: 91-95.

——— (1993_a): Applied Fluvial Geomorphology, Training Manual. River Short Course, Wildland Hydrology, Pagosa Springs, CO.: 450 pp.

——— (1993_b): River Restoration Utilizing Natural Stability Concepts. In Conference Proceedings, Watershed '93. A national conference on watershed management. USDA, Alexandria, VA.

——— (1993_c): Stream classification, Streambank Erosion and Fluvial Interpretations for the Lamar River and Main Tributaries. Report for USDI, NPS, Yellowstone, N.P., Gardner, Montana.

——— (1994): A classification of Natural Rivers. Catena, Vol 22: 169-199. Elsevier Science, B.V. Amsterdam.

ROSGEN, D.L. & B.L. FITTANTE (Mitchell) (1986): Fish habitat structures - A selection guide using stream classification. 5th Trout Stream Habitat Improvement Workshop. Lock Haven Univ., Lock Haven, PA. Penn. Fish Comm. Publics, Hamsburg, PA.

SCHUMM, S.A. (1963): A tentative classification of alluvial river channels. U.S. Geological Survey Circular 477. Washington, DC.

——— (1968): River adjustment to altered hydrologic regimen - Murrumbidgee River and paleochannels, Australia. U.S. Geological Survey Prof. Paper 598: 65 pp.

——— (1977): The fluvial system. Wiley and Sons, New York: 338 pp.

SEEHORN, M.E. (1985): Fish habitat Improvement Handbook. U.S. Forest Service. Southern Region, Technical Publication R8-TP7: 21 pp.

BIBLIOGRAPHY

SELBY, M.J. (1985): Earth's changing surface: an introduction to geomorphology. Oxford University Press, Oxford.

SHIELDS, A. (1936): Application of similarity principles and turbulence research to bed-load movement. Mitt. Preuss. Verschsanst., Berlin. Wasserbau Schiffbau. In W.P. Ott and J.C. Uchelen (translators), California Institute of Technology, Pasadena, CA. Report No. 167: 43 pp.

SIMONS, D.B. & F. SENTURK (1977): Sediment transport technology. Water Resource Publications. Fort Collins, CO: 807 pp.

SIMONS, D.B., LI, R.M. & SLA (1982): Engineering analysis of fluvial systems. SLA, P. O. Box 1816, Fort Collins, CO.

SMITH, D.G. (1986): Anastomosing river deposits, sedimentation rates and basin subsidence, Northwestern Colombia, South America. Sediment. Geol., 46: 177-196.

SMITH, D.G. & P.E. PUTNAM (1980): Anastomosed river deposits: modern and ancient examples in Alberta, Canada. Can. J. Earth Sci., 17: 1396-1406.

SMITH, D.G. & N.D. SMITH (1980): Sedimentation in anastomosing river systems: examples from alluvial valleys near Banff, Alberta. J. Sed. Pet. 50: 157-164.

STALL, J.B. AND YU-SI FOK (1967): Discharge as related to stream system morphology. International Association of Scientific Hydrology Publication 75: 224-235.

STEWART, B. & L. MAYNARD (1994): Personal Communication, USDA, Forest Service, Helena National Forest, Helena, MT.

STRAHLER, A.N. (1952): Hypsometric (area-altitude) analysis of erosional topography. Bull. Geol. Soc. Am., 63: 1117-42.

THORNBURY, W.D. (1969): Principles of geomorphology. 2nd Edition, John Wiley and Sons, New York.

U.S. DEPARTMENT OF AGRICULTURE, FOREST SERVICE (1984): Land and Resource Management Plan - Arapaho and Roosevelt National Forests and Pawnee Grasslands. Fort Collins, Colorado.

——— (1989): Fisheries habitat surveys handbook. Region 4 - FSH 2609.23. Ogden, Utah.

——— (1989_b): Hydrologic, geomorphic, sediment, and aerial surveys of stream types. Water Division I Data Collection Effort. Rocky Mountain Region. Denver, CO.

——— (1992): Integrated Riparian Evaluation Guide. USDA, Forest Service, Intermountain Region - Ogden, UT.

BIBLIOGRAPHY

U.S. DEPARTMENT OF AGRICULTURE, FOREST SERVICE (1993): Environmental impact statement on management of habitat for late-successional and old-growth related species within the range of the northern spotted owl. Portland, OR: 2 vols. 1 map.

―――― (1995): A Guide To Field Identification of Bankfull Stage in the Western United States (video). Rocky Mountain Forest and Range Experiment Station, Stream Systems Tech. Center, Fort Collins, CO.

U.S. DEPARTMENT OF AGRICULTURE, U.S. DEPARTMENT OF COMMERCE, USDI, ENVIRONMENTAL PROTECTION AGENCY: (1993): Forest ecosystem management: an ecological, economic, and social assessment. Report of the Forest Ecosystem Management Assessment Team. Portland, OR: U.S. Department of Agriculture, Forest Service; U.S. Department of Commerce, National Oceanic and Atmospheric Administration, National Marine Fisheries Service; U.S. Department of the Interior, Bureau of Land Management, Fish and Wildlife Service, National Park Service; Environmental Protection Agency: 1033 pp.

U.S. DEPARTMENT OF THE INTERIOR - BUREAU OF LAND MANAGEMENT (1993): Riparian Area Management Tech. Ref. 1737-9. BLM Service Center, Denver, CO.

U.S. DEPARTMENT OF TRANSPORTATION, FEDERAL HIGHWAY ADMINISTRATION (1979): Restoration of Fish Habitat in Relocated Streams. FHWA-IP-79-3: 63 pp.

U.S. ENVIRONMENTAL PROTECTION AGENCY (1980): An approach to water resources evaluation of non-point silvicultural sources. Chap. VI, Total Potential Sediment, D. Rosgen. EPA-600/8-80-012, Athens, GA: 39-41.

U.S. SOIL CONSERVATION SERVICE (1975): Soil Taxonomy. Agricultural Handbook No. 436. U.S.Govt. Printing Office, Washington, D.C.

WERTZ, W.A. & J.F. ARNOLD (1972): Lands System Inventory. USDA, Forest Service, Intermountain Region, Ogden, UT: 12 pp.

WHITE, R.J. AND O.M. BRYNILDSON (1967): Guidelines for management of trout stream habitat in Wisconsin. Wisconsin Department of Natural Resources. Technical Bulletin 39: 65 pp.

WILLIAMS, G.P. (1978): Bankfull discharge of rivers. Water Resources Research. Vol. 14, No. 6: 1141-1153.

―――― (1978): The Case of the Shrinking Channels - The North Platte and Platte Rivers in Nebraska. U.S. Geological Survey circular 781: 48 pp.

―――― (1986): River meanders and channel size. Journal of Hydrology 88: 147-164.

―――― (1987): Unit hydraulic geometry - An indicator of channel changes. Selected Papers in the Hydrologic Sciences. U.S. Geological Survey Water-Supply Paper 330.

BIBLIOGRAPHY

WILLIAMS, G.P. & D.L. ROSGEN (1989): Measured total sediment loads (suspended loads and bedloads) for 93 United States streams. U.S. Geological Survey Open File Report 89-67. Denver, CO: 128 pp.

WILSON, D. & R. JONES (1994): Personal communication, USDA Forest Service, Clearwater National Forest, Orofino, Idaho.

WOLMAN, M.G. & J.P. MILLER (1960): Magnitude and frequency of forces in geomorphic processes. Journal of Geology 68: 54-74.

WOLMAN, M.G. (1954): A method of sampling coarse river-bed material. Transactions of American Geophysical Union 35: 951-956.

YANG. C.T. & C. SONG (1979): Dynamic Adjustments of Alluvial Channels. In Adjustments of the Fluvial System. Edited by D. Rhodes & G. Williams. A proceedings volume of Tenth Annual Geomorphology Symposium Series. George Allen & Unwin, London, UK.

SUBJECT INDEX

A

"A" stream types: 4-6, 4-12, 4-20, 4-24, 4-29, 5-7, 8-33
—A1: 3-6, 5-36 thru 5-39, 5-60, 5-166, 5-167, 6-26, 6-30, 8-9, 8-24 thru 8-30
—A2: 3-6, 5-30, 5-33, 5-34, 5-40, 5-41, 5-43, 5-171, 6-26, 6-30, 6-44, 7-8, 8-5, 8-6, 8-9, 8-24 thru 8-30
—A3: 3-6, 5-30, 5-34, 5-44, 5-45, 5-47, 5-56, 5-123, 6-13, 6-27, 6-30, 6-46, 7-7, 8-2, 8-9, 8-11, 8-24 thru 8-30
—A4: 3-6, 5-48, 5-49, 5-51, 5-56, 6-30, 6-33, 7-7, 8-9, 8-11, 8-24, 8-25
—A5: 3-6, 5-52 thru 5-56, 5-182, 6-30, 7-7, 8-9, 8-11, 8-24 thru 8-28
—A6: 5-56 thru 5-59, 6-13, 6-30, 8-9, 8-24 thru 8-30
Abandoned floodplain: 2-3, 5-8, 5-19
—*see also terrace*
Accelerated bank erosion: 5-92, 5-96, 6-4, 6-7, 7-11, 8-8, 8-15, 8-26, 8-33, 8-35
Aggradation: 1-3, 2-9, 4-5, 4-6, 4-8, 4-10, 4-11, 6-4, 6-7, 6-11, 6-13, 6-19, 6-25, 6-28, 6-35, 7-12, 8-15, 8-22, 8-26, 8-35, 8-36
Alluvial channels: 2-5, 3-2, 4-5, 5-8, 5-14, 5-84, 8-8, 8-33
Alluvial fan: 4-7, 4-10, 4-11, 4-12, 4-14, 4-20, 4-23, 4-24, 4-29, 5-34, 5-44, 5-48, 5-64, 5-68, 5-72, 5-73, 5-76, 5-77, 5-80, 5-108, 5-109, 5-112, 5-113, 5-116, 5-170, 5-174, 5-175, 5-178, 5-179, 5-183, 5-187, 6-10, 6-21
Alluvium: 4-5, 4-9, 4-12, 4-15, 4-16, 5-52, 5-77, 5-84, 5-93, 5-97, 5-109, 5-113, 5-159, 5-163, 5-175, 5-179, 5-183, 5-187
Alpine glaciation: 4-29
Altered channel pattern: 8-8
Altered stream channels: 6-47
Anastomosed channels: 5-122
Aquatic ecosystem: 8-11, 8-38

B

"B" stream types: 4-6, 4-10, 4-20, 4-24, 4-29, 5-9, 6-10, 8-33
—B1: 5-60 thru 5-63, 6-13, 6-20, 6-26, 6-30, 8-9, 8-24 thru 8-30
—B2: 5-64, 5-65, 5-67, 6-30, 8-9, 8-22, 8-24 thru 8-30
—B3: 5-15, 5-17, 5-68, 5-71, 6-12, 6-15, 6-16, 6-20, 6-28, 6-30, 6-33, 6-37 thru 6-39, 6-44, 6-45, 7-10, 8-2, 8-6, 8-7, 8-9, 8-22, 8-24 thru 8-30
—B4: 5-17, 5-30, 5-33, 5-72, 5-73, 5-75, 6-30, 6-33, 6-36, 6-44, 6-45, 7-10, 8-7, 8-9, 8-23 thru 8-30
—B5: 5-76 thru 5-79, 5-81, 6-20, 6-30, 8-9, 8-24 thru 8-30
—B6: 5-80 thru 5-83, 6-13, 6-20, 6-26, 6-30, 8-9, 8-24 thru 8-30
Bank angle: 6-35, 6-40, 6-41
Bank cover: 8-17, 8-18, 8-27
Bank erodibility factors: 2-5, 6-40
Bank erosion: 2-4, 2-5, 2-9, 3-8, 4-5, 4-6, 4-8, 4-10, 4-11, 5-21, 5-22, 5-44, 5-48, 5-52, 5-56, 5-108, 5-112, 5-116, 5-120, 5-154, 5-158, 5-174, 5-178, 5-182, 5-186, 6-4, 6-7, 6-8, 6-13 thru 6-15, 6-25, 6-26, 6-43, 6-50, 7-1, 7-12 thru 7-15, 8-8, 8-11, 8-26 thru 8-30, 8-32, 8-35, 8-37
Bank erosion hazard index: 6-41
Bank erosion measurements: 6-39, 7-14, 7-15
Bank erosion processes: 5-186, 6-8, 6-14, 7-13
Bank erosion rates: 4-5, 4-6, 4-8, 4-10, 5-21, 5-108, 5-112, 5-116, 5-120, 5-154, 5-158, 6-43, 7-1, 7-14, 7-15, 8-11
Bank erosion sediment source: 6-35
Bank height: 3-1, 5-21, 5-150, 5-154, 5-158, 8-10
Bank height ratio: 8-30
Bank placed boulder: 8-26
—*see also streambank erosion*
Bankfull cross-sectional area: 2-5, 5-11
Bankfull dimensions: 5-18, 7-3, 8-31
Bankfull discharge: 2-2 thru 2-5, 5-1, 5-7 thru 5-14, 5-23, 5-25, 5-124, 7-8, 7-9, 8-2, 8-4, 8-7, 8-8, 8-32 thru 8-34, 8-37
Bankfull discharge calibration: 5-9 thru 5-14
Bankfull discharge drainage area relations: 5-18, 7-4
Bankfull discharge return period: 2-3, 5-11, 5-13, 5-14
Bankfull maximum depth: 5-11
Bankfull mean depth: 5-11, 5-13, 5-16
Bankfull mean velocity: 5-13
Bankfull stage: 2-2 thru 2-5, 2-9, 3-7, 4-6, 4-10, 5-7 thru 5-21, 5-150, 5-154, 5-158, 5-162, 6-17, 7-6 thru 7-8, 8-3, 8-7, 8-10, 8-16, 8-17, 8-23, 8-26
Bankfull stage indicators: 5-8, 5-9, 5-23
Bankfull validation: 5-8 thru 5-14
Bars:
—Delta bars: 6-18, 6-20, 6-33
—Diagonal bars: 6-18
—Mid-channel/central bars: 6-20
—Particle size distribution of bars: 7-7 thru 7-9
—Point bars: 2-5, 4-8, 5-8, 5-14, 5-15, 5-17, 5-96, 5-100, 6-8, 6-18, 6-20, 7-7, 7-14, 8-34
—Transverse bars: 4-5, 5-108, 5-112, 5-116, 5-120, 5-150, 5-154, 5-158, 6-20
—Side bars: 6-18
—*see also depositional features*

Base level: 4-6 thru 4-8, 4-16, 4-20, 5-122, 6-8, 6-9, 7-11, 8-35
Basin relief: 2-1, 2-5, 3-4
Beaver dams: 6-25, 6-28
Bed features: 1-4, 2-2, 2-5, 3-2, 4-5, 4-8, 4-12, 4-20, 4-23 thru 4-25, 5-1, 5-2, 5-25 thru 5-29, 5-40, 5-48, 5-60, 5-108, 5-150, 5-154, 5-158, 5-186, 6-26, 6-49, 6-50, 7-13 thru 7-15, 8-33
Bedform: 4-4, 4-6, 4-9, 4-10, 4-24, 5-44
Bedload sediment: 4-11, 5-44, 5-48, 5-56, 5-104, 5-112, 5-116, 5-120, 5-162, 5-186, 7-3, 7-6, 7-10, 7-11, 8-5 thru 8-8
—Ratio of bedload sediment to total load: 7-6, 7-7
—Bedload rating curves: 8-6
—Shift in bedload sediment rating curve: 7-10
—Bedload Size Distribution: 7-7
—Bed material load: 2-4
Bed material size distribution: 7-13, 7-15
—*see also pebble count*
Bedrock: 1-5, 4-5, 4-6, 4-10, 4-12, 5-2, 5-25, 5-36, 5-37, 5-41, 5-60, 5-61, 5-84, 5-85, 5-142, 5-143, 5-147, 5-155, 5-166, 5-167, 6-27, 6-41, 8-26 thru 8-31
Belt width: 2-5, 4-6, 4-9, 4-22, 5-2, 5-7, 5-11, 5-13, 5-122, 6-8, 6-13, 6-30, 6-49, 6-50, 8-34, 8-35
Blanco River: 3-7, 5-17, 8-37 thru 8-40
Boulder Placement: 8-16 thru 8-18, 8-26
Boulders: 2-4, 5-36, 5-37, 5-40, 5-41, 5-44, 5-45, 5-48, 5-49, 5-60, 5-61, 5-64, 5-65, 5-68, 5-72, 5-73, 5-89, 5-142, 5-143, 5-146, 5-167, 5-170, 6-29, 6-41, 8-16, 8-20, 8-26, 8-30, 8-34
Braided channels: 8-30
Broad life zones: 4-3, 8-42
Browsing: 8-10, 8-11

C

"C" stream types: 4-6, 4-8, 4-11, 4-29, 8-10, 8-11
—C1: 5-84 thru 5-88, 5-92, 5-96, 6-30, 8-9, 8-24 thru 8-30
—C2: 5-88 thru 5-92, 5-96, 6-30, 8-9, 8-24 thru 8-30
—C3: 5-7, 5-92, 5-93, 5-95, 5-96, 5-100, 6-26, 6-30, 6-36, 6-37, 6-42, 6-44, 6-45, 7-6, 8-2, 8-6 thru 8-10, 8-24 thru 8-30, 8-34
—C4: 5-7, 5-12, 5-14, 5-15, 5-96, 5-97, 5-99, 5-100, 6-3 thru 6-12, 6-15, 6-17, 6-27, 6-28, 6-30, 6-34 thru 6-38, 6-45, 6-46, 7-8, 8-6, 8-9 thru 8-13, 8-22 thru 8-30, 8-34 thru 8-40
—C5: 5-17, 5-100, 5-101, 5-103, 5-182, 6-11, 6-12, 6-15, 6-30, 6-35, 6-37, 6-38, 6-45, 6-46, 7-8 thru 8-10, 8-13, 8-24 thru 8-30

SUBJECT INDEX

—C6: 5-104, 5-106, 5-107, 6-26, 6-30, 8-9, 8-10, 8-24 thru 8-30
Canyons: 4-10, 4-12, 4-13, 4-14, 5-40
Channel:
  —Channel bed: 4-9, 4-10, 5-25, 5-26, 5-40, 5-52, 5-64, 5-68, 5-72, 5-76, 5-84, 5-88, 5-93, 5-108, 5-112, 5-116, 5-120, 5-122, 5-126, 5-130, 5-134, 5-166, 5-170, 6-14, 7-14
  —Channel classification: 4-23, 4-24, 5-13, 5-16, 5-25
  —Channel confinement: 4-22, 5-25
  —Channel constrictor: 8-17, 8-18, 8-27
  —Channel containment, vertical: 2-5, 5-19 thru 5-23
  —Channel containment, horizontal: 4-9
  —Channel entrenchment: 3-4, 5-7, 5-15 thru 5-23
    —see also river entrenchment
  —Channel evolution: 6-1 thru 6-12, 8-33
  —Channel geometry: 1-4
  —Channel length: 2-8, 4-23
  —Channel morphology: 1-4, 2-5, 3-3, 3-6, 5-22, 5-27, 6-7, 6-14, 6-15, 8-2, 8-15, 8-37
  —Channel pattern: 2-5, 2-7, 3-2, 4-8, 4-20, 4-24, 5-7, 5-30, 5-122, 5-124, 6-49, 8-15, 8-34
  —Channel profile: 2-8, 4-23, 8-16
  —Channel roughness: 8-3, 8-4
  —Channel restoration: 6-25
  —Channel shape: 4-20, 5-22
  —Channel stability: 1-3, 1-6, 3-1, 3-2, 3-4, 4-6 thru 4-9, 4-24, 5-22, 5-80, 5-126, 5-130, 5-134, 5-138, 6-3, 6-4, 6-13, 6-14, 6-19, 6-25 thru 6-30, 6-50, 7-1, 7-3, 7-10 thru 7-12, 8-6, 8-7, 8-20, 8-32, 8-42, 8-43
  —Channel stability evaluation: 6-29, 6-30
  —Channel stability rating: 1-6, 6-13, 6-26 thru 6-29, 6-50, 8-6, 8-7, 8-43
  —Channel stabilization: 8-32, 8-20, 8-21, 8-25
  —Channel types: 3-2, 4-10, 8-11, 8-26 thru 8-28
    —see also stream classification
Channel slope: 1-4, 2-8, 2-9, 4-6, 4-8, 4-10, 4-12, 4-20, 4-22, 4-23, 4-24, 4-29, 5-2, 5-11, 5-16, 5-21, 5-23, 5-29, 5-60, 5-80, 5-104, 5-109, 5-113, 5-117, 5-121, 5-122, 5-126, 5-130, 5-134, 5-138, 5-182, 5-186, 6-8, 6-13, 6-14, 6-26, 8-4, 8-37
Check dams: 6-25, 8-16, 8-20, 8-23, 8-26
Chutes: 4-4, 4-25, 5-36, 5-40
Circular arc: 2-6
Clay: 2-8, 4-20, 5-2, 5-53, 5-57, 5-76, 5-77, 5-80, 5-81, 5-97, 5-101, 5-104, 5-116, 5-117, 5-120 thru 5-123, 5-135, 5-138, 5-158, 5-162, 5-163 thru 5-187, 6-41, 7-6

Climate: 3-2, 4-1, 4-3, 4-29, 5-8, 5-142, 5-146, 6-7, 6-10, 6-11, 7-11
Coastal barrier: 4-19, 4-21, 4-22
Cobble: 2-4, 3-6, 4-20, 5-7, 5-36, 5-37, 5-40, 5-41, 5-44 thru 5-49, 5-60, 5-61, 5-64, 5-65, 5-68, 5-73, 5-85, 5-89, 5-92, 5-93, 5-97, 5-108, 5-109, 5-112, 5-113, 5-126, 5-127, 5-131, 5-142, 5-143, 5-146, 5-147, 5-150, 5-151, 5-154, 5-155, 5-167, 5-170, 5-171, 5-174 thru 5-179, 6-29, 6-41, 7-6, 7-7, 8-2, 8-4, 8-5, 8-7
Cohesive bank material: 4-8, 5-122, 5-162
Cohesive soils: 5-56, 5-163
Colluvial fan: 4-12
Colluvial slopes: 4-10, 4-12, 4-22, 4-24, 5-8, 6-21
Colluvial valley: 4-13, 4-29, 5-73, 5-76, 5-77, 5-175, 5-179, 5-183, 5-187
Colluvium: 4-12, 4-15, 5-37, 5-52, 5-65, 5-77, 5-175, 5-179, 5-183, 5-187
Companion inventories: 3-8, 4-3, 6-3, 6-4, 7-1, 8-41
Competence: 2-1, 6-4, 7-8, 8-33
Concentration: 7-3, 7-6, 8-6, 8-10, 8-11
  —see also suspended sediment
Confined meander scrolls: 6-21, 6-22
Confinement: 4-6, 4-9, 4-20, 4-24, 5-60, 6-21, 6-25, 8-26, 8-29, 8-31
  —see also channels, stream channel, rivers, lateral containment
Continuum: 3-3, 3-4, 4-10, 5-7, 5-22, 5-170, 5-182
Convergence: 4-5, 4-8, 4-23, 4-25, 5-2, 5-108, 5-112, 5-116, 5-120
Critical shear stress: 8-4
Cross-vane: 8-20, 8-21, 8-30
Cryoplanated: 4-10, 4-12
Cumulative effects: 1-5, 8-6, 8-38, 8-43

D

"D" stream types: 4-8, 4-11, 4-16, 4-20, 4-23, 4-29, 8-38
  —D3: 5-16, 5-17, 5-108 thru 5-111, 6-25 thru 6-27, 6-30, 6-46, 7-7, 8-2, 8-9, 8-24 thru 8-30
  —D4: 5-30, 5-34, 5-112 thru 5-115, 6-7, 6-11, 6-25, 6-27, 6-30, 6-35, 7-7, 8-5, 8-9, 8-24 thru 8-40
  —D5: 2-2, 5-2, 5-7, 5-16, 5-116 thru 5-119, 6-11, 6-25, 6-30, 7-7, 8-9, 8-24, 8-25, 8-29
  —D6: 5-120, 5-121, 6-25 thru 6-27, 6-30, 8-9, 8-25 thru 8-30
"DA" stream types: 4-8, 4-9, 4-23
  —DA4: 5-122, 5-123, 6-30, 8-9, 8-25
  —DA5: 5-122, 5-123, 6-30, 8-9
  —DA6: 5-122, 5-123, 6-30, 8-9, 8-25
Dams: 5-44, 5-48, 6-25, 8-16, 8-17, 8-24, 8-26, 8-34
Debris avalanche: 4-4, 5-44, 5-48, 5-52

Debris blockages: 6-13, 6-25 thru 6-28
Debris cones: 4-10, 4-14
Debris dams: 5-44, 5-48
Debris flows: 4-4, 5-48, 8-30
Debris torrents: 4-12, 5-52, 5-56, 8-30
Degradation: 1-3, 4-6 thru 4-8, 4-10, 5-182, 5-186, 6-4, 6-8, 6-13, 6-14, 6-25, 6-28, 6-34, 7-12, 8-10, 8-15
Delta features: 4-19 thru 4-22
  —Coastal barrier: 4-19 thru 4-22
  —Distributary channels: 4-12, 4-19 thru 4-22, 5-108, 5-112, 5-116, 5-120
  —Elongated, highly constructive: 4-19 thru 4-22
  —Delta front: 4-19 thru 4-22
  —Highly constructive, lobate configuration: 4-19 thru 4-22
  —Highly destructive, tide dominated: 4-19 thru 4-22
  —Highly destructive, wave dominated: 4-19 thru 4-22
  —Plain: 4-19 thru 4-22
  —Prodella: 4-19 thru 4-22
  —Shelf: 4-19 thru 4-22
  —Dunes on river bed: 4-19 thru 4-22
Departure point: 2-6
Deposition: 1-3, 2-1, 2-5, 2-9, 3-1, 3-2, 3-4, 4-1 thru 4-12, 4-15, 4-16, 4-20, 4-23, 4-24, 4-29, 5-1, 5-8, 5-22, 5-23, 5-30, 5-40, 5-41, 5-44, 5-45, 5-48, 5-49, 5-52, 5-53, 5-56, 5-64, 5-68, 5-72, 5-80, 5-84, 5-88, 5-92, 5-93, 5-96, 5-100, 5-101, 5-104, 5-108, 5-112, 5-116, 5-117, 5-120, 5-121, 5-123, 5-142, 5-146, 5-150, 5-151, 5-154, 5-155, 5-158, 5-159, 5-162, 5-163, 5-174, 5-178, 5-183, 5-186, 6-3, 6-4, 6-7, 6-11, 6-13, 6-14, 6-18, 6-19, 6-29, 6-30, 6-35, 6-50, 7-7, 7-8, 7-11, 7-14, 7-15, 8-22, 8-26, 8-30, 8-31, 8-33, 8-35, 8-36
Depositional features: 3-1, 3-4, 4-5, 4-20, 5-8, 5-30, 5-96, 5-100, 5-150, 5-154, 5-158, 6-18, 6-19, 7-7, 7-8
  —see also bars
Desired Future Conditions: 1-5, 6-2
Discharge: 2-2 thru 2-5, 3-1, 3-4, 3-6, 4-23, 5-7, 5-8, 5-11 thru 5-13, 5-20 thru 5-22, 5-138, 6-9, 6-13, 6-15, 6-30, 7-3, 7-6, 7-7, 8-1 thru 8-6, 8-15, 8-33
  —see also flow regime, bankfull
Dissected fluvial slopes: 4-10, 4-12, 4-15, 4-17
Distorted Meander Loops: 6-21
Distributary channels: 4-19 thru 4-22
Divergence: 4-5, 4-8, 4-23, 4-25, 5-2, 5-108, 5-112, 5-116, 5-120
Diversions: 2-5, 6-15, 6-17, 6-25, 8-31, 8-32, 8-38
Dominant discharge: 2-3
  —see also bankfull and effective discharge

SUBJECT INDEX

Double Wing deflector: 8-17, 8-27
Drainage area: 2-5, 2-9, 3-6, 5-13, 5-14, 5-16, 5-18, 6-18, 6-30, 6-47, 7-3, 7-4, 8-32, 8-37
Drainage basin: 4-1, 4-3
—*see also watershed*
Drainage density: 4-15

E

"E" stream types: 4-9, 4-10, 4-20, 4-24, 5-22, 8-10
—E3: 5-17, 5-126, 5-127, 5-129, 6-26, 6-28, 6-30, 8-2, 8-9, 8-24 thru 8-30
—E4: 5-30, 5-33, 5-130, 5-131, 5-133, 6-7 thru 6-9, 6-12, 6-30, 6-44, 8-9 thru 8-12, 8-14, 8-22, 8-24, 8-25, 8-29, 8-39
—E5: 5-134, 5-135, 5-137, 6-12, 6-30, 6-34, 8-9, 8-10, 8-13, 8-24, 8-25, 8-29
—E6: 5-138, 5-140, 5-141, 6-26, 6-30, 8-9, 8-10, 8-14, 8-24 thru 8-30
East Fork San Juan River: 5-14, 5-30, 7-3, 8-37, 8-38
Effective discharge: 2-3
—*see also bankfull, dominant discharge*
Energy dissipation: 1-4, 6-26
Energy distribution: 6-7
Entrainment: 5-44, 5-48, 5-56, 5-154, 5-158, 5-182, 7-7, 7-8, 7-14, 7-16, 8-3 thru 8-7
Entrenchment: 2-5, 2-7, 3-5, 3-6, 4-5, 4-11, 4-20, 4-22, 5-2, 5-7, 5-11, 5-15, 5-16, 5-19 thru 5-21, 5-37, 5-41, 5-45, 5-49, 5-53, 5-57, 5-61, 5-65, 5-69, 5-73, 5-77, 5-80, 5-81, 5-85, 5-89, 5-93, 5-97, 5-101, 5-105, 5-109, 5-113, 5-117, 5-121, 5-123, 5-127, 5-131, 5-135, 5-139, 5-142, 5-143, 5-146, 5-147, 5-150, 5-151, 5-154, 5-155, 5-158, 5-159, 5-162, 5-163, 5-167, 5-171, 5-175, 5-179, 5-183, 5-187, 6-8, 6-10, 6-30, 6-34, 8-4
Eolian deposition: 4-16, 5-100
Ephemeral channels: 8-26
Ephemeral flow: 6-17
Equilibrium: 2-7, 4-10 thru 4-12, 4-15, 5-122, 6-13, 6-25, 8-15, 8-31, 8-32
Erodibility: 2-5, 3-4, 5-27, 5-92, 5-96, 5-100, 5-104, 5-108, 5-112, 5-116, 5-120, 5-150, 5-154, 5-158, 6-3, 6-4, 6-40, 6-43 thru 6-46, 6-50, 7-14
Erosion: 2-1, 2-3, 2-5, 2-7, 3-2, 4-1 thru 4-5, 4-10 thru 4-12, 4-20, 4-29, 5-22, 5-44, 5-48, 5-52, 5-53, 5-166, 5-170, 5-182, 6-10, 6-28, 6-50, 7-14, 7-15, 8-9, 8-20, 8-22
—Bed scour: 4-8, 6-27, 8-26, 8-29
—Dry ravel: 5-44, 5-48, 5-52
—Fluvial entrainment: 5-44, 5-48, 5-56, 5-154, 5-158, 5-182
—Freeze-thaw: 6-35
—Liquifaction: 6-35
—Mass wasting: 5-44, 5-162, 6-29, 6-30, 8-32
—Streambank erosion rates: 4-6, 6-43
—Surface erosion: 5-52
Estuarine: 4-16
Estuarine depositional process: 4-16
Evolutionary stages: 6-1 thru 6-12, 8-33

F

"F" stream types: 4-9, 4-10, 4-16, 4-20, 4-23, 5-9, 5-22
—F1: 5-142 thru 5-146, 6-30, 8-9, 8-24 thru 8-30
—F2: 5-146 thru 5-149, 6-30, 8-9, 8-24 thru 8-30
—F3: 5-150, 5-151, 5-153, 5-162, 6-12, 6-30, 8-2, 8-9, 8-24 thru 8-30
—F4: 5-15, 5-154, 5-155, 5-157, 5-162, 6-7 thru 6-12, 6-30, 6-33, 6-34, 6-39, 6-46, 7-6 thru 7-10, 8-6 thru 8-14, 8-23 thru 8-30, 8-35
—F5: 5-158 thru 5-162, 6-12, 6-30, 6-34, 7-7, 8-9, 8-13, 8-24 thru 8-26, 8-29
—F6: 5-162 thru 5-165, 6-30, 8-9, 8-14, 8-24 thru 8-30
Falls: 4-4, 4-5, 4-23, 4-25, 5-7, 5-36, 8-17, 8-29
Fault controlled valley: 4-16
Field determination of bankfull stage: 5-8
Fish habitat: 1-4, 1-5, 3-4, 3-8, 5-1, 5-27, 6-3, 6-14, 6-25 thru 6-28, 6-39, 7-8, 7-10, 7-11, 8-1, 8-15, 8-16, 8-20 thru 8-42
Fish habitat improvement suitability guide: 8-20, 8-24 thru 8-30
Floating log cover: 8-17, 8-19, 8-28
Flood frequency: 2-3 thru 2-5, 5-11, 5-12
Flood-prone areas: 5-19, 5-21
Floodplain: 1-2, 2-2 thru 2-9, 3-1, 3-6, 4-5 thru 4-24, 4-29, 5-8, 5-11 thru 5-19, 5-21, 5-84, 5-92, 5-96, 5-100, 5-104, 5-122, 5-126, 5-130, 5-134, 5-138, 5-142, 5-146, 5-150, 5-154, 5-158, 5-182, 6-4, 6-8, 6-14, 7-13, 7-14, 8-11, 8-26, 8-35 thru 8-39
Floodplain encroachment: 6-47
Floods: 2-4, 5-21, 5-92, 5-142, 5-146, 6-18, 6-29, 8-31, 8-34, 8-37
Flow regime: 1-4, 1-5, 2-3 thru 2-5, 2-9, 3-4, 4-8, 5-1, 5-52, 5-162, 5-174, 5-178, 5-182, 5-186, 6-3, 6-4, 6-8, 6-13, 6-15 thru 6-18, 6-25, 6-50, 7-3
Fluvial processes: 2-1

G

"G" stream types: 4-10, 4-12, 4-15, 4-16, 4-24, 4-29, 5-9, 5-21, 6-10
—G1: 3-6, 5-166 thru 5-169, 6-30, 8-9, 8-24 thru 8-30
—G2: 3-6, 5-170 thru 5-173, 6-30, 8-9, 8-24 thru 8-30
—G3: 3-6, 5-30, 5-34, 5-174, 5-175, 5-177, 5-185, 6-12, 6-13, 6-30, 6-45, 7-7, 7-9, 8-9, 8-24 thru 8-30
—G4: 3-6, 5-15, 5-178, 5-179, 5-181, 6-7 thru 6-12, 6-30, 6-33, 6-34, 7-7, 8-9, 8-22 thru 8-25, 8-35
—G5: 3-6, 5-182 thru 5-186, 6-12, 6-30, 6-34, 6-46, 7-7, 8-9, 8-24, 8-25, 8-30
—G6: 5-186 thru 5-188, 5-189, 6-13, 6-30, 8-9, 8-24 thru 8-30
Gaging stations: 3-7, 5-7, 5-12, 6-19
Geologic control: 4-5
Geologic sediment supply: 8-35
Geology: 3-2, 4-29, 5-2, 7-11, 8-7, 8-42
Geomorphic characterization: 3-4, 4-1, 4-2, 4-3 thru 4-13, 4-29
Geomorphic threshold: 6-10, 7-10, 8-35, 8-43
Geomorphology: 8-15
Glacial: 4-5, 4-8 thru 4-12, 4-15, 4-16, 4-18, 4-20, 4-29, 5-37, 5-44, 5-48, 5-49, 5-88, 5-92, 5-93, 5-96, 5-100, 5-101, 5-104, 5-108, 5-109, 5-112, 5-113, 5-116, 5-117, 6-15, 6-17, 8-37
Glacial fed: 6-15
Glacial moraine: 5-40, 5-44, 5-45, 5-48
Glacial outwash terrace: 4-20
Glacial till: 4-4, 4-12, 4-15, 5-30, 5-48
Glacial trough valley: 4-8, 8-12, 4-15
Gorge: 4-10, 4-12, 4-14, 5-143, 5-147, 5-151, 5-155, 5-159, 5-163, 5-166, 5-170
Gradient: 1-4, 2-8, 2-9, 3-2, 3-3, 3-6, 4-5, 4-6, 4-8, 4-11, 4-12, 4-14, 4-15, 4-23, 4-24, 4-29, 5-2, 5-25, 5-29, 5-36, 5-60, 5-64, 5-68, 5-72, 5-76, 5-84, 5-85, 5-89, 5-92, 5-96, 5-97, 5-100, 5-101, 5-104, 5-108, 5-112, 5-116, 5-120, 5-126, 5-130, 5-134, 5-138, 5-142, 5-143, 5-146, 5-147, 5-151, 5-155, 5-159, 5-162, 5-163, 5-166, 5-170, 5-174, 5-178, 5-182, 5-186, 6-8, 6-26, 6-29, 6-30, 7-6, 8-2, 8-4, 8-6, 8-20, 8-26, 8-28, 8-33,
Gravel: 3-6, 4-20, 5-7, 5-36, 5-37, 5-40, 5-41, 5-44, 5-45, 5-48, 5-49, 5-52, 5-53, 5-61, 5-64, 5-65, 5-68, 5-72, 5-73, 5-76, 5-77, 5-84, 5-85, 5-89, 5-93, 5-96, 5-97, 5-101, 5-108, 5-109, 5-112, 5-113, 5-117, 5-122, 5-123, 5-127, 5-130, 5-131, 5-135, 5-142, 5-143, 5-146, 5-147, 5-150, 5-151, 5-154, 5-155, 5-158, 5-167, 5-170, 5-171, 5-174, 5-175, 5-178, 5-179, 5-183, 5-187, 6-29, 6-30, 6-41, 7-7, 7-8, 7-11, 7-16, 8-5, 8-7, 8-17, 8-19, 8-20, 8-25, 8-29, 8-38

SUBJECT INDEX

Gravel traps: 8-17, 8-19, 8-25, 8-29
Gravel bed stream: 8-29
Gravel traps: 7-8, 8-17, 8-25, 8-29
Gravitational acceleration: 6-13
Grazing: 1-1, 1-5, 2-5, 6-3 thru 6-9, 6-26, 8-1, 8-8, 8-10 thru 8-14, 8-35, 8-41
Gully: 4-5, 4-10, 5-182, 5-186, 6-8, 8-26, 8-35

H

Half Log cover: 8-17, 8-28
Hanging valley: 4-20
Headcut: 5-174, 5-178, 8-26
Hierarchical river inventory: 5-3
Hierarchy of river inventory: 3-5
High width/depth ratio: 4-5, 5-84, 5-88, 5-92, 5-96, 5-122, 5-142, 5-146, 5-150, 5-154, 5-158, 5-162, 6-9, 6-11, 6-12, 6-37 thru 6-39, 7-6, 8-12, 8-14, 8-22, 8-23, 8-26, 8-30, 8-36
Holocene river terrace: 4-16, 5-48
Hydraulic geometry: 3-4, 5-11 thru 5-13, 6-15, 6-17, 7-1, 7-3, 7-5, 8-1, 8-2 thru 8-3, 8-31, 8-34
Hydraulic Radius: 5-13, 8-3, 8-4
Hydro-physiographic province: 1-3, 3-3, 4-12, 5-9, 5-14, 5-18, 5-84, 5-162, 6-18, 7-3, 8-31, 8-43
Hydrograph: 2-3, 6-15, 7-10, 8-6, 8-32
Hydrology: 6-15, 7-3, 8-15, 8-33

I

Incision, degree of: 4-22
—see also bank height ratio
Increased sediment supply: 6-7, 8-8, 8-32, 8-37
Induced channel instability: 8-8
Instability: 1-3, 1-4, 2-7, 2-8, 4-10, 5-22, 5-174, 5-178, 6-7, 6-13, 6-14, 6-28, 7-11, 8-9, 8-27 thru 8-32
Intermittent flow: 6-17
Irregular Meanders: 6-21, 6-22

J

J-hook vane: 8-20, 8-21, 8-25, 8-30

L

Lacustrine: 4-16
Lacustrine deposition: 4-16, 5-116
Lacustrine soils: 4-5
Lacustrine valleys: 4-20, 5-131, 5-134, 5-135, 5-138
Lag deposit: 2-4, 5-40, 5-44, 5-48, 5-64, 5-65, 5-68, 5-88, 5-89, 5-92, 5-109
Landform: 2-2, 3-2, 3-4, 4-1 thru 4-16, 4-20, 4-24, 4-29, 5-1, 5-7, 5-15, 5-30, 5-37, 5-41, 5-44, 5-45, 5-48, 5-49, 5-52, 5-53, 5-56, 5-57, 5-61, 5-65, 5-72 thru 5-77, 5-81, 5-85, 5-89, 5-93, 5-97, 5-101, 5-109, 5-113, 5-117, 5-120 thru 5-123, 5-127, 5-131, 5-135, 5-142, 5-143, 5-146, 5-147, 5-151, 5-155, 5-159, 5-162, 5-163, 5-167, 5-171, 5-174 thru 5-179, 5-183, 5-186, 5-187, 6-29, 6-30, 8-42
Landslides: 4-29, 5-170
Landtype: 4-10 thru 4-12, 4-24, 8-38
Large woody debris: 1-4, 5-68, 6-15, 6-26
Lateral adjustment: 4-5, 5-92, 5-96, 5-100, 5-104, 6-19
Lateral bank erosion rate: 6-39
Lateral containment: 3-6, 4-4, 4-8, 4-20, 4-22, 5-25, 5-135, 5-139, 5-142, 5-146, 6-19, 6-21
Lateral migration rate: 4-8, 6-4, 6-35
Lateral stability: 7-13, 7-14, 8-20
Levees: 2-9, 4-16
Limitations, fish habitat: 8-15, 8-26 thru 8-30
Lithology: 3-2, 4-1, 5-186, 7-11, 8-6, 8-7
Log Vane: 8-20, 8-30
—see also native material revetment
Longitudinal profile: 2-8, 2-9, 4-10, 4-23, 5-2, 5-9, 5-11, 5-25, 5-27, 5-29, 6-7, 6-49, 8-33
Low sinuosity: 2-9, 4-20, 4-24, 5-44, 5-48, 5-52, 5-56
Low Stage Check Dam: 8-16, 8-18, 8-26
Low width/depth ratio: 4-4, 4-5, 4-23, 5-52, 5-126, 5-130, 5-134, 5-138, 5-166, 5-170, 5-174, 5-178, 6-4, 6-8, 6-13, 8-27, 8-30

M

Manning's equation: 8-3, 8-4
Manning's n: 8-3, 8-4
Meander geometry: 1-4, 2-5 thru 2-7, 2-9, 4-8, 5-11, 5-13, 5-23, 6-25, 8-2, 8-31 thru 8-34, 8-37
Meander pattern: 6-4, 6-13, 6-19, 6-21, 6-25, 6-50
Meander radius of curvature: 6-25, 8-34, 8-35
Meander scrolls: 3-1
Meander wavelength: 2-5, 2-7, 4-8, 5-23, 5-25, 5-27, 6-26, 8-33, 8-35
Meander width ratio: 4-5 thru 4-12, 4-22, 4-24, 4-29, 5-2, 5-7, 5-11, 5-13, 5-84, 5-92, 5-96, 5-126, 5-130, 5-134, 5-138, 5-142, 5-146, 5-162, 6-7, 6-8, 6-10, 6-13, 6-14, 6-19, 6-25, 6-35, 8-32, 8-34, 8-35
Meanders: 2-3, 6-21, 6-22, 8-16, 8-17, 8-28
Medium stage check dams: 8-26
Migration barrier: 6-25, 8-17, 8-19, 8-22, 8-29
Mining: 3-3, 3-7, 4-6, 4-23, 5-9, 5-19, 5-20, 5-23, 5-26, 6-14, 7-11, 8-3, 8-10, 8-27, 8-28, 8-38
Moderate sinuosity: 4-10, 5-126, 5-142, 5-146

Moderate width/depth ratio: 4-10, 5-64, 5-68, 5-72, 5-76
Momentary maximum flow: 2-3
Multiple thread channels: 5-22
—see also braided, anastomosed
Monitoring: 5-7, 6-3, 6-5, 7-3, 7-10 thru 7-12, 7-15, 7-16, 8-11, 8-20, 8-38, 8-42, 8-43
Moraine: 4-15
—see also glacial
Morphological description: 3-4, 4-2, 4-10, 5-1 thru 5-191
Morphology, rivers: 1-2, 1-4, 2-1, 2-5, 2-7, 2-9, 3-2 thru 3-6, 4-3 thru 4-12, 4-20, 4-22 thru 4-24, 4-29, 5-1, 5-20, 5-23, 5-36, 5-60, 5-64, 5-68, 5-72, 5-76, 5-112, 5-116, 5-120, 5-170, 5-174, 5-178, 5-186, 6-1, 6-7 thru 6-9, 6-14, 7-2, 7-3, 7-10, 7-11, 8-1, 8-3, 8-20, 8-23, 8-37, 8-38, 8-41 thru 8-43
—see also river morphology

N

Native material revetment: 8-20
Natural channel design: 4-1, 7-3
Natural stability concepts: 1-3, 8-31
Near-bank region: 6-13, 6-41, 6-43, 7-14, 8-20
Near-bank shear stress: 6-35, 6-36, 6-39, 6-41, 6-43
Non-point source pollution: 8-32

O

Over-steepened: 4-6, 8-35
—see also rejuvenated
Over-width channels: 8-33
—see also high width/depth ratio
Oxbow cutoffs: 6-21, 6-22

P

Parent material: 4-12
Particle detachment: 5-182
Particle entrainment: 7-16
Particle size distribution: 4-10, 5-8, 5-12, 5-25, 6-7, 7-7, 7-8, 7-9, 7-15
Pavement: 5-25, 5-26, 5-100, 5-104, 7-8, 8-5
Pebble count: 5-11, 5-14, 5-16, 5-25, 5-26, 5-27, 5-28, 6-50, 7-8, 7-15
Perennial flow: 6-15, 6-17
Pleistocene: 4-20, 5-92
Plunge pool: 5-166, 8-16
Point bars: —see bars, dispositional features
Pool spacing: 1-4, 2-5, 4-6, 5-36, 5-60, 5-68, 5-72, 5-76, 6-26, 8-15, 8-33
Pools: 1-4, 2-5, 2-7, 2-9, 4-5, 4-6, 4-9, 4-23 thru 4-25, 5-2, 5-25 thru 5-27, 5-36, 5-44, 5-48, 5-60, 5-64, 5-68, 5-72, 5-76, 5-84, 5-88, 5-112, 5-116, 5-120, 5-143, 5-147, 5-178, 6-26, 6-29, 7-15,

SUBJECT INDEX

7-16, 8-15 thru 8-17, 8-20, 8-26, 8-27, 8-30, 8-33, 8-34, 8-42
Precipitation: 5-150, 5-154, 5-158, 6-17, 7-11
Probable stable form: 1-2, 6-7, 8-33
Profile: —see longitudinal and bank
Processes, erosional: 4-12, 5-44, 5-48, 5-52
Proper functioning condition: 1-5, 6-2, 6-3, 8-41

R

Radius of curvature: 2-4 thru 2-7, 5-7, 5-13, 5-23, 6-49, 8-32, 8-34, 8-35
—see also meander geometry
Random boulder placement: 8-16, 8-18, 8-23, 8-26
Rapids: 1-4, 4-5, 4-6, 4-12, 4-23 thru 4-25, 5-2, 5-60, 5-64, 5-68, 5-72, 5-76, 5-112, 5-116, 5-120, 5-166, 5-170, 6-14, 8-33
Rating curve: 5-11, 5-20, 5-21, 6-27, 7-6 thru 7-10, 8-6 thru 8-8, 8-42, 8-43
Ratio of Bedload to Total Load: 7-6
Rearing Habitat: 8-16
Recurrence interval: 2-3, 2-4, 5-8, 5-11 thru 5-14, 5-20
Redds: 7-15, 8-20
Reference reach: 3-4, 5-1, 5-2, 5-7, 5-9, 5-12 thru 5-16, 5-23, 5-27, 5-30, 7-11, 7-12, 8-33
Regime: 2-2, 2-5, 3-2, 4-8, 4-20, 5-7, 5-8, 5-36, 5-96, 5-100, 5-104, 6-3, 6-7, 6-15, 7-11, 8-31 thru 8-33
Regional curves: 5-18, 7-3, 8-34, 8-42
Regular Meanders: 6-22
Regulated flow: 6-17
Regulated rivers: 6-15, 6-17
Rejuvenated: 4-13, 5-44, 5-48, 8-30
—see also over-steepened
Relative roughness: 5-88, 6-17, 8-4
Residual soils: 4-6, 4-12, 4-15, 5-53, 5-57, 5-80, 5-183, 5-187
Resistance: 2-5, 2-7, 5-25, 5-84, 5-88, 5-126, 5-130, 5-134, 5-138, 6-9, 6-13, 7-3, 8-3, 8-4, 8-31
—see also Manning's n
Riffle/pool channels: 5-9, 5-27
Riffles: 1-4, 2-7, 2-9, 4-6, 4-23 thru 5-27, 7-15 thru 8-17, 8-33, 8-42
—see also bed features
Rip-rap: 2-9
Riparian vegetation: 3-4, 3-8, 4-11, 4-12, 4-29, 5-8, 5-9, 5-76, 5-80, 5-92, 5-96, 5-100, 5-104, 5-122, 5-150, 5-154, 5-155, 5-158, 5-162, 5-186, 6-3 thru 6-18, 6-50, 8-8, 8-10, 8-11, 8-20, 8-35, 8-36
River:
—River classification: 3-1
—River entrenchment: 2-7
—see also channel entrenchment
—River inventory: 3-5, 4-2, 5-3, 6-2, 7-2
—River management: 2-7
—River morphology: 2-5, 3-2 thru 3-4, 4-3, 4-10, 6-7, 8-1, 8-37, 8-43
—River pattern: 2-7, 4-20, 4-24
—River profile: 2-2
—River restoration: 1-3, 1-4, 5-27, 7-3, 8-1, 8-31, 8-34, 8-35, 8-37
—River stabilization: 8-34
—see also channels, streams
Riverine: 1-5, 4-16, 5-100, 5-104, 5-105, 5-117, 5-131, 5-135, 5-139
Rock vanes: 8-30
Root wads: 8-16, 8-20, 8-25, 8-30, 8-34
Rooting depth: 5-150, 5-154, 5-158, 6-14, 6-15, 6-35, 6-40, 8-8, 8-10
Roughness coefficients: 5-88
Runoff: 4-8, 4-11, 5-64, 5-68, 5-108, 5-112, 5-116, 5-120, 6-15, 6-17, 7-11, 7-13, 7-14, 8-6, 8-37
—see also flow regime

S

Sand: 1-6, 3-2, 4-15, 4-18, 4-20, 5-25, 5-41, 5-44, 5-45, 5-48, 5-49, 5-52, 5-53, 5-57, 5-60, 5-64, 5-65, 5-68, 5-72, 5-73, 5-76, 5-77, 5-81, 5-84, 5-85, 5-89, 5-93, 5-97, 5-100, 5-101, 5-108, 5-109, 5-112, 5-113, 5-116, 5-117, 5-122, 5-123, 5-127, 5-131, 5-135, 5-143, 5-147, 5-150, 5-151, 5-154, 5-155, 5-158, 5-159, 5-170, 5-174, 5-175, 5-178, 5-179, 5-182, 5-183, 6-29, 6-41, 7-8, 8-12, 8-30, 8-31
Saturation: 5-56, 5-162, 8-32 thru 8-33
Schlimmbesserung: 8-31
Sediment:
—Sediment budgets: 6-4, 6-14
—Sediment concentration: 7-6, 8-6
—Sediment entrainment: 7-7, 7-8, 8-4, 8-5
—Sediment load: 1-3, 2-7 thru 2-9, 3-2, 4-9, 5-174, 5-178, 7-6, 7-10, 7-16, 8-11, 8-33
—Sediment particle size distribution: 7-7 thru 7-9
—Sediment rating curves: 6-27, 7-6, 7-7, 8-6, 8-7, 8-8, 8-42, 8-43
—Sediment standards: 7-11
—Sediment supply: 1-1, 1-5, 3-2 thru 3-4, 3-6, 4-5 thru 4-16, 5-1, 5-22, 5-30, 5-36, 5-40, 5-44, 5-48, 5-52, 5-72, 5-80, 5-84, 5-88, 5-92, 5-96, 5-100, 5-104, 5-108, 5-112, 5-116, 5-120, 5-122, 5-126, 5-130, 5-134, 5-138, 5-142, 5-146, 5-150, 5-154, 5-158, 5-162, 5-166, 5-174, 5-178, 5-182, 5-186, 6-3, 6-5, 6-7, 6-8, 6-13, 6-14, 6-18, 6-25 thru 6-30, 6-33, 6-50, 7-3, 7-8, 7-10, 7-12, 7-16, 8-6 thru 8-8, 8-15, 8-26 thru 8-29, 8-34, 8-35, 8-42
—Sediment, suspended: 2-4, 4-10, 5-122, 5-186, 6-15, 7-3, 7-6, 7-10, 8-6
—Sediment thresholds: 8-42
—Sediment transport capacity: 2-9, 5-126, 5-130, 5-134, 6-7, 6-9, 6-13, 8-8, 8-20, 8-35
—see also bedload sediment, suspended sediment
Sensitivity to disturbance: 5-1, 8-8, 8-9
Shear stress: 2-9, 5-21, 5-22, 5-44, 6-13, 6-26, 6-41, 7-8, 8-2, 8-4, 8-5, 8-20
Single-thread channels: 5-6
Single Wing Deflector: 8-16, 8-17, 8-27
Sinuosity: 2-7, 2-8, 3-1, 3-2, 4-5, 4-6, 4-23, 4-24, 5-2, 5-7, 5-11, 5-12, 5-16, 5-23, 5-25, 5-29, 5-30, 5-36, 5-37, 5-40, 5-41, 5-45, 5-49, 5-53, 5-57, 5-60, 5-61, 5-64, 5-65, 5-68, 5-72, 5-73, 5-76, 5-77, 5-81, 5-84, 5-85, 5-89, 5-93, 5-97, 5-101, 5-109, 5-113, 5-117, 5-121, 5-123, 5-127, 5-130, 5-131, 5-134, 5-135, 5-138, 5-143, 5-147, 5-151, 5-155, 5-159, 5-163, 5-166, 5-167, 5-170, 5-171, 5-175, 5-179, 5-183, 5-187, 6-7, 6-8, 6-10, 6-13, 6-14, 6-19, 6-25, 6-30, 6-49, 6-50, 8-2, 8-34 thru 8-36
Slope: 1-4, 1-5, 2-2, 2-5 thru 2-9, 3-1, 3-2, 4-5 thru 4-29, 5-2, 5-5 thru 5-187, 6-3, 6-7 thru 6-14, 6-19, 6-21, 6-25, 6-26, 6-29, 6-30, 6-41, 6-48, 6-49, 6-50, 7-8, 7-13, 8-2 thru 8-10, 8-15, 8-20, 8-26, 8-30, 8-31 thru 8-37
Slope measurements: 5-27, 5-29
—see also gradient, profile
Spawning gravel placement: 8-17, 8-19, 8-25, 8-29
Spawning habitat: 8-17, 8-20, 8-29
Specific weight of water: 5-21, 6-41, 8-4
Spring-fed flow: 6-17
Snowmelt dominated flow: 6-17
Step/pool channels: 4-12, 4-23, 5-9
Stratification, streambanks: 1-3, 4-20, 4-22, 6-18, 6-41, 6-47, 7-16, 8-2, 8-4
Stormflow dominated flow: 6-17
Stream:
—Adjustment: 1-4
—Classification: 1-4, 2-10, 3-1 thru 3-8, 4-2 thru 4-4, 4-11, 4-22, 4-29, 5-3, 5-7, 5-12, 5-16, 5-25 thru 5-30, 6-4, 6-15, 6-26, 6-29, 6-30, 6-50, 7-2, 7-14, 8-1 thru 8-6, 8-31, 8-33, 8-37, 8-41
—Stream competence: 6-4, 7-7, 8-4, 8-5
—Stream condition: 3-4, 3-7, 5-2, 6-1 thru 6-50, 7-1, 8-10, 8-33, 8-42
—Stream confinement: 4-9
—Stream entrenchment: 5-15 thru 5-23
—Stream order: 2-9, 4-6, 4-11, 6-13, 6-15, 6-17, 6-18, 6-30, 6-50
—Stream power: 5-22, 5-44, 6-9, 6-13, 7-8, 8-28

SUBJECT INDEX

—Stream recovery potential: 8-8, 8-9, 8-11
—Stream restoration: 6-7, 8-31, 8-34
—Stream Size: 6-3, 6-4, 6-13, 6-15, 6-17, 6-18, 6-50, 7-3, 8-4, 8-7
__Stream state: 3-4, 6-2, 6-3, 6-49, 8-6, 8-42
—Stream type: 1-2, 1-4, 1-5, 2-2 thru 2-4, 3-2 thru 3-7, 4-3 thru 4-28, 4-29, 4-30, 5-1 thru 5-187, 6-3 thru 6-39, 6-44, 6-49, 6-50, 7-1 thru 7-16, 8-1 thru 8-43
—Stream width: 1-5, 2-4, 2-5, 3-6, 6-13, 6-26, 6-30, 7-14, 8-33
—see also channels, river
Streambank angle: 6-35
Streambank erosion: 2-4, 2-5, 4-6, 5-186, 6-4, 6-13, 6-43, 6-50, 8-8
Streambank erosion hazard index: 6-35
Streambank stability: 1-3, 5-150, 5-154, 5-158, 6-9, 8-11
Streambank stability assessment: 6-35 thru 6-43
Streambank stabilization: 8-20, 8-21, 8-25, 8-32
Streamflow: 1-5, 2-1, 2-3 thru 2-9, 3-2, 4-2, 4-6, 5-7 thru 5-9, 5-11, 5-12, 5-14, 5-20 thru 5-22, 5-25, 5-48, 5-126, 5-130, 5-134, 5-138, 5-174, 5-178, 6-3, 6-7, 6-10, 6-15, 6-17, 6-25, 6-26, 6-28, 7-1, 7-3, 7-10, 7-16, 8-2, 8-3, 8-6, 8-7, 8-9, 8-16, 8-17, 8-31 thru 8-33, 8-37, 8-41
Streamgage: 2-4, 8-37
Structural control: 4-3, 4-22, 5-1, 5-41, 5-61, 5-65, 5-85, 5-89, 5-167, 5-171
Submerged shelters: 8-17, 8-19, 8-25, 8-28
Subterranean flow: 8-14, 6-17
Suspended sediment: —see sediment

T

Tectonically active forelands: 4-8
Temperature: 6-14, 6-15, 8-10, 8-42
—see also stream
Terraces: 4-5, 4-6, 4-15 thru 4-17, 4-20, 4-22, 4-24, 4-29, 5-8, 5-12, 5-44, 5-92, 5-93, 5-96, 5-174, 5-178, 6-21, 7-13, 8-10, 8-37
Thalweg: 5-19, 7-7, 7-14, 8-17, 8-26 thru 8-28
Threshold: 4-8, 6-7, 6-47
Tidal flats: 4-16
Timber harvest: 3-7, 6-3, 7-10, 8-41, 8-43
Tortuous meanders: 6-22
Total load: 2-4, 5-182, 7-6, 7-7
—see also sediment
Tributary streams: 5-44, 5-48
Trout: 8-17, 8-38
Truncated meanders: 6-21

U

Unconfined Meander Scrolls: 6-21, 6-22
Unconsolidated: 3-6, 4-11, 5-44, 5-45, 5-48, 5-49, 5-60, 5-92, 5-96, 5-174, 5-175, 5-178, 5-179, 5-183
Uplift: 4-8, 4-9, 4-12, 5-143, 5-147, 5-151, 5-155, 5-159, 5-163

V

Validation: 4-24, 5-7, 6-3, 6-30, 7-2, 7-3, 7-6, 7-11, 7-14, 7-16, 8-41, 8-42
Valley types: 3-2, 4-3, 4-4, 4-6, 4-8 thru 4-11, 4-20, 4-24, 4-29, 5-11, 5-30, 5-40, 5-44, 5-48, 5-52, 5-56, 5-60, 5-64, 5-68, 5-72, 5-76, 5-80, 5-84, 5-88, 5-92, 5-96, 5-100, 5-104, 5-108, 5-112, 5-116, 5-120, 5-122, 5-126, 5-130, 5-134, 5-138, 5-150, 5-154, 5-158, 5-162
Variables: 1-1 thru 1-5, 2-2, 2-9, 2-10, 3-2, 3-3, 3-6, 3-7, 4-2, 4-3, 4-24, 4-29, 5-2, 6-3, 6-5, 6-7, 6-10, 6-13 thru 6-15, 6-49, 7-1, 7-3, 7-11, 8-1, 8-2, 8-4, 8-6, 8-15, 8-32, 8-34, 8-41
Vegetation: 2-5, 3-1, 3-2, 3-4, 3-8, 4-1, 4-8, 4-9, 4-11, 4-12, 4-29, 5-8, 5-9, 5-25, 5-76, 5-80, 5-92, 5-96, 5-100, 5-104, 5-122, 5-126, 5-130, 5-134, 5-138, 5-150, 5-154, 5-155, 5-158, 5-162, 5-182, 5-186, 6-3 thru 6-18, 6-29, 6-30, 6-50, 8-8 thru 8-11, 8-20, 8-25, 8-30, 8-31 thru 8-33, 8-35, 8-36, 8-42
—Alnus spp. (Alder): 5-9
—Carex, spp. (Sedges): 8-10
—Vegetation alteration: 8-33
—Vegetation composition: 6-14
—Vegetation controlling influence: 8-8
—Vegetation density: 6-14, 6-29, 6-30, 6-35, 6-40, 6-41
—Vegitation effects on streambank erosion: 6-14, 6-35 thru 6-46
—Juncus, spp. (rushes): 8-10
—Populus spp. (cottonwood): 8-10
—Salix, spp. (Willow): 8-10
—Vegetation overstory: 6-14
—Vegetation transplants: 8-20, 8-30
—Vegetation understory: 6-14
Velocity distribution: 2-9, 6-42, 8-20
Velocity estimation: 8-1 thru 8-4
Velocity gradient: 2-9, 5-21, 5-22, 6-41
Velocity isovels: 6-42
Vertical stability: 5-96, 5-100, 5-104, 6-26, 7-11, 7-14
Vortex: 8-22

W

Wash load: 4-5, 5-56, 5-122, 7-6
—see also sediment, load
Water density: 6-13
Water quality: 1-5, 6-14, 6-15, 7-3, 7-10, 7-16
Water surface slope: 5-2, 5-11, 5-27, 5-29, 6-49, 6-50
Watershed: 1-1, 1-2, 1-4 thru 1-6, 2-1, 2-4, 2-5, 2-7 thru 2-10, 3-7, 3-8, 4-1, 4-3, 4-4, 4-8, 4-10, 4-11, 4-20, 4-23, 4-24, 5-11, 5-13, 5-22, 5-27, 5-30, 5-34, 5-96, 5-100, 5-104, 5-162, 5-174, 5-178, 5-182, 5-186, 6-2 thru 6-4, 6-7, 6-9, 6-10, 6-14, 6-15, 6-17, 6-26, 6-27, 7-10, 7-11, 7-16, 8-1, 8-16, 8-31 thru 8-38, 8-41 thru 8-43
Watershed management: 1-2, 1-5, 1-6, 6-7, 7-16
Wavelength, meander: 2-5, 2-7, 4-8, 5-23, 5-25, 5-27, 6-26, 8-33, 8-35
Weminuche River: 8-35, 8-36
Wetlands: 4-5, 4-16, 4-20, 4-22, 5-122
Width/depth ratio: 1-3, 1-5, 2-5, 2-7 thru 2-9, 3-2, 3-6, 4-4 thru 4-10, 4-20, 4-23, 4-29, 5-2, 5-7, 5-16, 5-21 thru 5-24, 5-36, 5-37, 5-40, 5-41, 5-45, 5-49, 5-52, 5-53, 5-56, 5-57, 5-60, 5-61, 5-64, 5-65, 5-68, 5-72, 5-73, 5-76, 5-77, 5-80, 5-81, 5-84, 5-85, 5-88, 5-89, 5-92, 5-93, 5-96, 5-97, 5-100, 5-101, 5-104, 5-108, 5-109, 5-112, 5-113, 5-116, 5-117, 5-120 thru 5-123, 5-126, 5-127, 5-130, 5-131, 5-134, 5-135, 5-138, 5-142, 5-143, 5-146, 5-147, 5-150, 5-151, 5-154, 5-155, 5-158, 5-159, 5-162, 5-163, 5-166, 5-167, 5-170, 5-171, 5-174, 5-175, 5-178, 5-179, 5-182, 5-183, 5-187, 6-4, 6-7 thru 6-14, 6-19, 6-25, 6-28, 6-30, 6-35, 6-37, 6-38, 6-39, 6-49, 6-50, 7-6, 8-2, 8-4, 8-8, 8-9, 8-12, 8-14, 8-20, 8-22, 8-23, 8-26, 8-27, 8-30, 8-33 thru 8-36
W-weir: 8-20, 8-22, 8-25, 8-30

NOTES

NOTES

NOTES

NOTES

NOTES

NOTES

NOTES

To Order Additional Books, Call (970) 731-6100, Fax (970) 731-6105, or contact:
Wildland Hydrology Books • 1481 Stevens Lake Road • Pagosa Springs, CO 81147

❑ Send Information on the Wildland Hydrology Short Courses: Applied Fluvial Geomorphology • River Morphology and Applications • River Assessment and Monitoring • River Restoration • Grazing in Riparian Ecosystems

Your Billing Information
Company _____
Name _____
Street _____
City _____ State _____ Zip _____
Phone _____
Fax _____
PO# _____
Payment: ❑ MC ❑ Visa ❑ AMEX ❑ Check
Card# _____ Exp. Date: ___ / ___
Issuing Bank: _____
Signature _____

Your Shipping Information
Company _____
Name _____
Street _____
City _____ State _____ Zip _____

Call for quantity discount pricing.
Colorado residents add 6.9% sales tax per book.
Shipping and Handling:
Your order will be shipped via UPS ground.
1 book...$7.50
Please send _____ copies @ $89.95 each = _____
Shipping and Handling + CO sales tax = _____
TOTAL = _____

Other publications are available. Call *Wildland Hydrology* for more information.

To Order Additional Books, Call (970) 731-6100, Fax (970) 731-6105, or contact:
Wildland Hydrology Books • 1481 Stevens Lake Road • Pagosa Springs, CO 81147

❑ Send Information on the Wildland Hydrology Short Courses: Applied Fluvial Geomorphology • River Morphology and Applications • River Assessment and Monitoring • River Restoration • Grazing in Riparian Ecosystems

Your Billing Information
Company _____
Name _____
Street _____
City _____ State _____ Zip _____
Phone _____
Fax _____
PO# _____
Payment: ❑ MC ❑ Visa ❑ AMEX ❑ Check
Card# _____ Exp. Date: ___ / ___
Issuing Bank: _____
Signature _____

Your Shipping Information
Company _____
Name _____
Street _____
City _____ State _____ Zip _____

Call for quantity discount pricing.
Colorado residents add 6.9% sales tax per book.
Shipping and Handling:
Your order will be shipped via UPS ground.
1 book...$7.50
Please send _____ copies @ $89.95 each = _____
Shipping and Handling + CO sales tax = _____
TOTAL = _____

Other publications are available. Call *Wildland Hydrology* for more information.

To Order Additional Books, Call (970) 731-6100, Fax (970) 731-6105, or contact:
Wildland Hydrology Books • 1481 Stevens Lake Road • Pagosa Springs, CO 81147

❑ Send Information on the Wildland Hydrology Short Courses: Applied Fluvial Geomorphology • River Morphology and Applications • River Assessment and Monitoring • River Restoration • Grazing in Riparian Ecosystems

Your Billing Information
Company _____
Name _____
Street _____
City _____ State _____ Zip _____
Phone _____
Fax _____
PO# _____
Payment: ❑ MC ❑ Visa ❑ AMEX ❑ Check
Card# _____ Exp. Date: ___ / ___
Issuing Bank: _____
Signature _____

Your Shipping Information
Company _____
Name _____
Street _____
City _____ State _____ Zip _____

Call for quantity discount pricing.
Colorado residents add 6.9% sales tax per book.
Shipping and Handling:
Your order will be shipped via UPS ground.
1 book...$7.50
Please send _____ copies @ $89.95 each = _____
Shipping and Handling + CO sales tax = _____
TOTAL = _____

Other publications are available. Call *Wildland Hydrology* for more information.